PEST
CONTROL
F · O · R
HOME
and
GARDEN

PEST
CONTROL
F · O · R
HOME
and
GARDEN

The Safest
and Most Effective Methods
for You and the Environment

M I C H A E L   H A N S E N ,   Ph . D .
AND THE EDITORS OF CONSUMER REPORTS BOOKS

CONSUMER REPORTS BOOKS
A Division of Consumers Union
YONKERS, NEW YORK

Copyright © 1993 by Consumers Union of United States, Inc.,
Yonkers, New York 10703.

Published by Consumers Union of United States, Inc.,
Yonkers, New York 10703.

Library of Congress Cataloging-in-Publication Data

Hansen, Michael.
Pest control for home and garden :   the safest and most effective
methods for you and the environment   /   by Michael Hansen and the
editors of Consumer Reports Books.
p.   cm.
Includes bibliographical references and index.
ISBN 0-89043-423-9
1. Household pests—Control.   2.Garden pests—Integrated control.
I. Consumer Reports Books.
TX325.H25   1992
648′.7—dc20   92-2263
                                                               CIP

Design by Ruth Kolbert

First printing, March 1993

Manufactured in the United States of America

This book is printed on recycled paper ♲

Pest Control for Home and Garden is a Consumer Reports Book published by Consumers Union, the nonprofit organization that publishes *Consumer Reports*, the monthly magazine of test reports, product Ratings, and buying guidance. Established in 1936, Consumers Union is chartered under the Not-for-Profit Corporation Law of the State of New York.

The purposes of Consumers Union, as stated in its charter, are to provide consumers with information and counsel on consumer goods and services, to give information on all matters relating to the expenditure of the family income, and to initiate and to cooperate with individual and group efforts seeking to create and maintain decent living standards.

Consumers Union derives its income solely from the sale of *Consumer Reports* and other publications. In addition, expenses of occasional public service efforts may be met, in part, by nonrestrictive, noncommercial contributions, grants, and fees. Consumers Union accepts no advertising or product samples and is not beholden in any way to any commercial interest. Its Ratings and reports are solely for the use of the readers of its publications. Neither the Ratings, nor the reports, nor any Consumers Union publications, including this book, may be used in advertising or for any commercial purpose. Consumers Union will take all steps open to it to prevent such uses of its material, its name, or the name of *Consumer Reports*.

ACKNOWLEDGMENTS

I would like to thank Edward Groth III for his intellectual input and guidance, as well as his comments on the manuscript. His suggestions and insights made this a significantly better book. I also thank David Pimentel for his comments on the manuscript.

Special thanks are due Roslyn Siegel and Sally Smith for their mercilessly effective editing of the text and tables, respectively, and for their patience.

CONTENTS

FOREWORD

None of us likes to find creeping, crawling, or flying creatures in our homes or on our pets. We hate to see snails and insects destroy our flowers and vegetables, or weeds choke the lawn. However, the pesticides we use to fight these pests are not just lethal to the pests, they may also pose a danger to us and our environment.

Consumers currently use more than 73 million pounds of pesticide active ingredients in and around their homes each year. And while most of us assume that the easy availability of these chemicals means they are safe, that's not necessarily so. In fact, many ingredients in pesticides can be toxic to people. They may cause cancer, nerve damage, birth defects, mutations, or other adverse effects in laboratory animals. Many also pollute the environment and kill nontarget organisms (plants and animals).

The truth is, for most active ingredients in pesticides, not enough information is available to know what short- and long-term hazards these chemicals really pose. Health risks aside, chemical pesticides often simply aren't very effective at controlling pest problems. Consumers may find themselves on the "pesticide treadmill," using more and more pesticides to control more resistant and more numerous pests.

In recent years, scientists have devised a safer, more effective approach to pest control, called Integrated Pest Management (IPM), which uses a combination of short- and long-term controls, including employing natural enemies, planting insect-resistant varieties, changing local environmental conditions, and, as a last resort, using environmentally safer pesticides.

To aid consumers in choosing environmentally safer pest control techniques, Consumers Union carried out a four-year project, directed by Michael Hansen, Ph.D., in the Public Service Division of our Technical Department. We sent shoppers out to hardware and garden stores in the Northeast, Southeast, Midwest, and on the West Coast to buy hundreds of pesticide products. We compiled extensive data on the ingredients used in more than 500 different products. We then collected and reviewed scientific data available to the U.S. Environmental Protection Agency on 89 active ingredients and assessed each compound's toxicity and its potential health and ecological effects.

We used information on comparative hazards to choose the pesticide products listed in each chapter. If you need a chemical pesticide, those in our listings (and others with the same ingredients) should be the safest choices.

More important, though, our study compiled information on the biology and ecology of common household, lawn, and garden pests. We describe techniques for stopping pest invasions or nipping them in the bud by safe, nonchemical means. In short, we explain how to apply IPM to home pest problems.

It is not only possible but essential, we think, to control garden pests without creating needless hazards to health and environment. This book is designed to help you become an active, supportive practitioner of Integrated Pest Management.

PART ONE

A

BETTER

SOLUTION

T·O

PEST

CONTROL

1

THE CASE AGAINST PESTICIDES

Everybody has had a problem with pests in the home or garden at one time or another. No one likes finding roaches in the kitchen, mice in the pantry, moths in the closet, fleas on the dog, termites in the walls, slugs in the garden, or crabgrass in the lawn. Indeed, pests of all varieties, both indoors and outdoors, can cause major headaches for the average person. Pests can damage property, carry disease, ruin food and clothes, destroy gardens and lawns, and make life generally miserable for people and their pets. No wonder we want to get rid of pests as quickly and easily as possible.

Americans are increasingly interested in gardens and lawn care. Indeed, lawns are becoming a national obsession. We have over 5 million acres of home lawns, and we collectively spend some $6 billion annually to keep them looking good. This concern with having a well-manicured lawn is reflected in the dramatic increase in the lawn-care industry, which, between the mid-1960s and mid-1980s, grew at a rate of 25 to 30 percent *per year*.

With the increased interest in lawns and gardens comes greater attention to combating outdoor pests.

3

For most people, synthetic chemical pesticides are the dominant approach. More than 73 million pounds of pesticides are used by consumers in and around their homes, and this figure does not include the pesticides applied by professional exterminators or lawn-care companies. Furthermore, in pounds per acre, suburban homeowners use pesticides more intensively on their lawns than farmers do on their crops. And use of pesticides indoors is even higher. Over 90 percent of all homes use pesticides, applying them indoors two to four times more frequently than outdoors. These figures reflect the fact that most homeowners overuse pesticides—using them when they are not really needed and in amounts greater than are recommended.

Most people assume that the government requires extensive safety testing before permitting pesticides to be sold. They are wrong.

Why the overuse of pesticides? The simple attitude that "if some is good, more is better" probably explains a lot of it. Consumers also feel that the products must be safe, since they are freely available in grocery and hardware stores, and no special precautions are required for their use. Most people assume that the government requires extensive safety testing before permitting such products to be sold.

HUMAN HAZARDS

Acute Effects

Unfortunately, such faith in the relative safety of pesticides for the home and garden is not justified. The Environmental Protection Agency's (EPA) national household pesticide survey, performed in 1976–77, estimated that 2.5 million poisonings occur each year. Yet, poison control centers reported only 67,000 cases of poisoning for 1991. Why the discrepancy in these two numbers? Many cases of pesticide poisoning involve symptoms such as dizziness, vomiting, headaches, heart palpitations, sweating, shortness of breath, fatigue, and muscle tremors. Such poisoning cases are vastly underreported because people often do not seek medical treatment or, if they do, the doctor attributes the symptoms to a cold, flu, or some other illness. Nevertheless, internal organs such as the liver and heart can be damaged by pesticide toxicity.

Indeed, pesticides are the second most common source of childhood poisonings. A 1981 study found more than 20,000 cases of pesticide poisoning were treated in hospital emergency rooms that year, with more than half the victims being children. Eighty-five percent of the accidents occurred in and around the home, and about 2,000 required hospitalization. And these data are only for the immediate acute effects.

Chronic Effects

In addition to acute, readily noticed effects, many pesticides can also cause subtler, more distant in time, chronic effects. Chronic effects occur as a result of long-term exposure to pesticides at levels that have no immediately obvious effects. Many serious long-term effects—reproductive problems, birth defects, cancer, mutations, nerve damage, suppressed immune systems—have been demonstrated in laboratory animals. But controversy exists over the degree to which these effects might occur in humans and how to apply data obtained from experimental animals to people. Neither animal tests nor human studies are capable of *proving* whether pesticides cause cancer, for instance. We are left with the possibility that they *may* have a role in causing certain kinds of human cancer, and a great deal of scientific uncertainty about how real and how large the risk really is.

Neurotoxicity
A number of studies have demonstrated that acute exposure to pesticides can cause neurobehavioral effects in people, including memory loss, personality and mood changes, brain damage, and impairment of brain function. The neurotoxic hazards of chronic low-level exposure are essentially unknown, but the evidence for neural damage is clear enough that the National Academy of Sciences has recommended more than once that the EPA gather such data before permitting a pesticide to go on the market. Yet at present the EPA requires pesticides to be tested only minimally for neurotoxicity.

Potential chronic effects of pesticide poisoning can also be a result of synergism, breakdown products and contaminants, and inert ingredients.

Synergism
Synergism occurs when the effect of a mixture of chemicals is greater than the sums of the effects of the individual chemicals. Although we know that synergism occurs with other toxic substances—the greater risk of cancer from the combined effects of smoking and exposure to asbestos, the greater sedative effect of alcohol combined with barbiturates—it has not been actively studied in pesticides. Nevertheless, manufacturers use synergism to enhance the effectiveness of certain insecticides.

We know that chemical synergists used in formulated pyrethroid-based products dramatically increase the killing power of the main pesticide ingredient. Virtually nothing is known about the effects of mixtures of chemicals on health, and it is a very difficult problem to study. Yet many home pesticide products contain more than one active pesticide ingredient, and most people are exposed to several different pesticide products over time.

Breakdown Products and Contaminants

Breakdown products are the compounds formed when a pesticide is digested or degraded in the environment; contaminants are generally unwanted by-products created in the manufacture of a given active ingredient. Either may pose risks of chronic or acute toxicity as great as or greater than those of the original chemicals. Take, for example, the furor over the chemical daminozide (Alar) found largely in apple products. Daminozide itself is unlikely to be a carcinogen, but the primary breakdown product, unsymmetical dimethylhydrazine (UDMH), has been shown to be carcinogenic in rodents. Examples of contamination are the nitrosamines, which are established animal carcinogens, produced as by-products of the synthesis of the herbicide 2,4-D.

Inert Ingredients

These are substances added to the pesticide formulation for purposes other than killing pests—a wetting agent, a stabilizer, a solvent, etc. The inert ingredients are a particular concern in household pesticides because they commonly make up well over 90 percent of the product, but they don't have to be identified on the label. Sometimes the inert ingredients can cause greater health problems than the active ingredients. Take the herbicide glyphosate, for example. While glyphosate itself has a fairly low acute toxicity, an "inert" wetting agent has been shown to cause severe skin irritation in agricultural workers who regularly used a particular formulation of Roundup. Many of the solvents used in aerosol pesticides—substances such as xylene, benzene, toluene—are suspected or known carcinogens and neurotoxins.

Long-term Effects

Most of the commonly used pesticide ingredients were developed and registered before the law required testing for chronic effects. The National Academy of Sciences estimates that only 10 percent of all pesticide active ingredients have been adequately tested for possible long-term health effects. Furthermore, the EPA has virtually no reliable data on exposure to pesticides in and around the home. This makes it impossible to estimate accurately the potential health risks to humans from exposure to these products.

There is some evidence that the hazards are not trivial. Although good studies are few and far between, some results are ominous. For instance, the National Cancer Institute (NCI) found that children whose parents sprayed pesticides in the home were 3.8 times more likely to develop leukemia than children whose parents did not use sprays. Use of garden pesticides appears to increase the risk of leukemia 6.5 times. In both cases the risk increased the more often the sprays were used. Furthermore, childhood leukemia was more closely associated with the mother's use of pesticides than with the father's use. Another NCI study found that dogs whose owners used the herbicide 2,4-D had twice the cancer rate of dogs whose owners did not use lawn pesticides.

ENVIRONMENTAL HAZARDS

In addition to affecting health adversely, pesticides can contaminate the environment and lead to pest resistance and resurgence. These problems have been well documented for agricultural pesticide use. Drifting spray from aerial applications of pesticides contaminates land and water adjacent to target fields. Unexpectedly high concentrations of a number of pesticides have been found in rainwater and fog in agricultural areas. The more persistent pesticides leave residues in the soil that continue to contaminate crops for many years.

Herbicides and insecticides—primarily in the soil, but also in rainwater and fog—may eventually contaminate both surface water and groundwater. A study in Iowa found that 91 percent of samples of public drinking water supplies from surface water sources contained one or more pesticides. Pesticide residues have been found in the groundwater in 26 states. In addition, the U.S. Department of Agriculture (USDA) estimates that 46 percent of all U.S. counties contain groundwater susceptible to contamination from fertilizers and pesticides. Indeed, contamination of surface water and groundwater is such a problem that some states, such as Iowa and Massachusetts, are trying to reduce their use of pesticides and fertilizers to protect drinking water.

A consequence of environmental contamination is the widespread killing or sickening of wildlife, cattle, birds, fish, bees, and pests' natural enemies. Affecting nature's food chain can result in unforeseen consequences. For example, honeybees and wild bees pollinate some $20 billion worth of fruit, vegetables, and forage crops each year. A recent study estimates that $320 million is lost annually because of honeybees killed by pesticides; this figure reflects both loss of honey and reduced crop yields owing to lack of pollination.

Although comparable research has not been done on the environmental consequences of home and garden pesticide usage, ecologically speaking it makes sense to expect similar problems might occur from applying pesticides to lawns and gardens.

 # PESTICIDE FAILURE

The search for safer alternatives has ironically been more directly stimulated by the failure of pesticides to control pests than by the environmentalists' concerns for people's health. Pesticide failure is attributable to the dual phenomena of pest resistance and pest resurgence. After farmers had used some pesticides for a number of years, they began to observe that pest outbreaks were recurring with renewed vigor. Pest populations were developing genetic resistance to pesticides.

The number of pest species resistant to pesticides is increasing. Five hundred insect and mite species, almost 270 species of weeds, at least 150 species of plant pathogens, 5 species of rodents, and 2 species of nematodes are now resistant to one or more pesticides. In fact, some insect populations are resistant to most pesticides. For instance, populations of the Colorado potato beetle on Long Island have rapidly evolved resistance to every insecticide that has been used to control them.

Resistance

Resistance evolves in a process much like natural biological evolution, but with pesticides in the role of Darwin's natural selection. When a pest population is sprayed with a pesticide, most individuals are killed, but a few survive—generally those that, for one reason or another, are best able to withstand the chemical attack. These survivors pass on their resistance to the next generation. As this cycle is repeated, each new pest population grows more resistant to the pesticide.

Resurgence

Pesticides, particularly insecticides, have also caused resurgence of primary as well as secondary pests. Primary pest resurgence occurs when pest populations rebound to previous or even higher levels after insecticide use. Secondary pest resurgence occurs where insects that were not initially problems suddenly increase dramatically and attain pest status after insecticide use to control another pest.

Ironically, pest resurgence generally occurs because pesticides are usually much more toxic to a pest's natural enemies than to the pest itself. Plant-feeding insects—those most likely to be crop or garden pests—are better able to tolerate poisons than predacious or parasitic insects—those most likely to be the pest's natural enemies—because many plants contain toxic chemicals to protect them from herbivorous insects. Predators and parasitoids that feed on other insects do not routinely encounter poisons in their food, so they have developed less resistance to such poisons than plant feeders have.

Pesticide Treadmill

The twin problems of pest resistance and resurgence have led to a situation called the "pesticide treadmill." Increased pesticide use in agriculture results in decreased effectiveness, which in turn leads to even

Use of pesticides leads to resistance and resurgence of pests, which results in increased use of pesticides, until ultimately the pesticide fails.

more intensive use. Fifty years ago, approximately 50 million pounds of pesticides were used and an estimated 7 percent of the preharvest crop was destroyed by insects. Today, pesticide use has grown to 600 million pounds and insect damage has risen to 13 percent. Furthermore, a study of the 25 most serious insect pests in California found that 17 were resistant to one or more classes of insecticides and that 24 were cases of resurgence, primarily of secondary pests. Clearly, the chemical approach to pest control has not worked.

2

INTEGRATED
PEST
MANAGEMENT

Given the drawbacks of most pesticides, safer, more effective approaches to pest control are sorely needed. Such an approach is Integrated Pest Management, or IPM.

IPM is largely scientists' response to the growing recognition of the problems caused by pesticide use. IPM is an ecological approach to pest control. It looks at the components of a specific area (ecosystem)—both living (crops, pests, their natural enemies, soil flora and fauna, surrounding vegetation) and nonliving (soil characteristics, weather)—and how these components interact as a whole.

Although developed for agricultural systems, IPM can readily be applied to home and garden pest problems. By understanding the biology and ecology of pests and the environments in which they live, IPM tries to increase natural pest control factors and decrease the need for artificial measures such as synthetic pesticides. It is based on the premise that the vast majority of insects, for example, are not pests and that their numbers are usually kept low by natural mortality factors. It also aims to use control measures that pose the least possible risk to humans, pests' natural enemies, nontarget organisms, and the environment. Finally, it assumes that a variety of methods will usually be needed to control pests in a given ecosystem. In essence, IPM works with nature rather than against it.

There are six basic steps in an IPM program:

1. Identify the pests.
2. Decide when to take action.
3. Monitor and forecast the key pests or signs of their damage, and identify their natural enemies.
4. Devise long-term nonchemical methods to keep key pests under basic control.
5. Establish remedial measures for emergency situations that cause minimum ecological disruption.
6. Develop training programs to ensure that the IPM program is actually carried out.

 # IDENTIFY THE PESTS

The first step in an IPM program is to identify the organisms, or pests, that are causing a problem. For example, are the little holes in your flowers caused by a beetle, a caterpillar, a bee, or something else? Pay particular attention to key pests, those that appear regularly, usually every season, and cause significant damage if they are not controlled. If you don't recognize the pest, collect specimens and have them identified. After identifying the source of the problem, read about the pest's biology and ecology either in your gardening books or in some of the books listed in References and Resources. This information can be used against the pests.

 # DECIDE WHEN TO TAKE ACTION

Central to IPM is the notion that not all pest damage causes significant losses. Some pests, at certain levels, cause no appreciable damage and even stimulate yields. Plants are far more resilient to pest damage than most people think. For example, you can remove over half of the leaf area on growing squash plants without affecting the yield at all. A certain amount of weeds in a highly manicured lawn is not a threat to the health of the lawn. Rather than automatically using pesticides because a pest is present, IPM sets thresholds to help determine when to take action. Particularly for home and garden pest control, you must decide when the level of damage or nuisance becomes intolerable.

For example, consider the question of flea control on pets. How many fleas will you accept on your dog or cat before you take some kind of action? Should your pet be treated when you find a single flea, when the animal shows discomfort, or when the fleas start infesting the house and jumping on people? In fact, most people and pets are not aware of low-level flea infestations, which means they cause little damage. Cockroaches are an even better example. How many cockroaches are too many? Most people would say one. But many homes have a small population of roaches that are never seen and never bother anyone. It may be preferable to see an occasional roach than to fumigate your living quarters. You need to consider both the cost of pest control and the potential health risk to you and your pets.

 ## MONITOR AND FORECAST

After you have discovered the identity of the pest or pests causing a given problem and decided on an action threshold, carefully monitor the situation and the pest population. In order to monitor effectively, you need to find an appropriate way to determine pest or damage level. For gardeners, monitoring may simply mean keeping an eye out for changes in pest numbers, damage levels, condition of the plants, and so forth. You can use traps containing a baited sticky substance to monitor cockroach populations, or flea combs to monitor flea levels on pets. You should also keep track of weather conditions, which affect pest numbers, particularly fungal diseases and weeds.

 ## DEVISE LONG-TERM CONTROLS

Once you've established your threshold, devise techniques to keep the key pests below that level. The aim of these techniques is to alter the system intrinsically and naturally, using pesticides only as a last resort.

Numbers of key pests and their damage normally fluctuate throughout the season, usually exceeding the action threshold at some point. The average level of pest density is known as the *equilibrium level*. The goal of an IPM program is to reduce a key pest's equilibrium level to keep it permanently below the action threshold. Control techniques that achieve such a reduction are called *primary management components*. Think of these as long-term control strategies. The three primary management components are *genetic control*, *cultural control*, and *biological control*.

Genetic control involves breeding for resistance to pest attack. For the home gardener, this means buying resistant varieties when possible. For example, have a problem with nematodes or bacterial wilt on tomato plants? Choose varieties of tomatoes that can withstand both.

Cultural controls involve removing the pest or modifying the environment to reduce a pest's numbers, to make its natural enemies more effective, or to decrease a pest's breeding, feeding, or shelter habitats. Cover holes to keep out rodents, move the woodpile that contains a colony of carpenter ants away from the house, or cover the garden with various types of mulch to keep weed populations down.

Finally, biological controls use natural enemies specifically introduced for the purpose or attracted by cultural techniques to reduce pest populations. For example, use milky spore disease to control Japanese beetles and white grubs in various areas of the country.

ESTABLISH SHORT-TERM EMERGENCY MEASURES

Occasional fluctuations may push pest populations above the action threshold, despite long-term controls. When that happens, try short-term measures, choosing steps that will disrupt the ecosystem as little as possible. Examples of short-term control measures, in order of preference, include physical interventions, releasing natural enemies, using insect-specific chemicals such as pheromones or juvenile hormones, or using microbial or chemical pesticides.

Physical Controls

If the situation is appropriate, and the area is small enough, physical intervention of some sort may be all that is needed. You can deal with a sudden increase in weeds in the garden by hand weeding. A sudden influx of flies, mosquitoes, or wasps into the house may indicate a torn screen or other hole in a window that simply needs to be patched.

Selective Pesticides

Synthetic pesticides are the most widely used short-term measures. When used in IPM programs, the types of pesticides and their dosages and application times aim to minimize the negative impact on natural enemies, other nontarget organisms, the surrounding environment, and nearby ecosystems. Some pesticides are made from natural substances that do little harm to the surrounding environment. For example, fatty acid soaps are particularly valuable against soft-bodied insects as well as some weeds and fungi. If you must use a synthetic pesticide, IPM recommends selective pesticides—chemicals that are as specific as possible to the targeted pest—rather than pesticides that kill organisms indiscriminately and can harm useful natural enemies. For a complete list of pesticides evaluated, their toxicity ratings, and cautions concerning their use, see chapters 9 and 10.

Remedial Measures

A number of remedial measures are available for use in the home and garden, especially microbial pesticides such as bacteria, fungi, and viruses. The most prominent of these, the bacterium *Bacillus thuringiensis,* called *Bt,* attacks a wide range of caterpillar pests. New strains of *Bt* have been developed to attack mosquitoes, black flies, and a number of beetles.

Milky spore disease is a bacterial disease that attacks a variety of white grubs, but particularly Japanese beetle grubs. (Be sure to wear a dust mask when applying powder or using spray that contains these substances.) Numerous species of nematodes (tiny, soil-dwelling roundworms) are effective against Japanese beetle grubs, white grubs, grasshoppers, termites, and other soil-dwelling insects.

There are also pheromone/insecticide traps to attract and kill gypsy moths. (A pheromone is a chemical substance released by an organism and used for communication purposes.) Many products use pheromones, often sex lures and other chemical odors, to attract roaches or flies to a trap, where they are killed by a pesticide.

Finally, some products, particularly for roaches and fleas, contain an insect growth regulator that prevents the insects from becoming adults, thereby effectively preventing reproduction. These products are usually advertised as providing birth control for the pests.

DEVELOP TRAINING PROGRAMS

This book is designed to teach you the IPM approach to pest problems as well as specific IPM techniques for a number of common home and garden pests.

To develop your own IPM system requires four steps. First, monitor the problem area (yard, garden, kitchen) on a regular basis. In the beginning, monitor once every two weeks or once a week. When you discover pest problems, increase the frequency of your monitoring. Second, study the ecology of the pests you most frequently encounter. Third, keep records (the more detailed the better) of your monitoring. Try drawing a map of the area you are monitoring as well as recording any observations you make on pest damage and/or numbers. By keeping written records of your observations, you will be able to tell rather quickly when pest numbers are increasing. Finally, be creative in the pest control techniques you use. Experiment on your own; try different pest control measures and record the results of your experiments. Train other family members to follow the system you develop.

PART TWO

OUTDOOR PESTS

3

LAWN
PEST
CONTROL

The obsession over lawns is an American phenomenon. But does our vision of an ideal lawn make ecological sense? We seem to think that a lawn must be a dense, green carpet, cut very short, fertilized and watered frequently, and *completely* free of weeds. We think that the lawn requires a high level of maintenance to keep it looking well groomed. But lawns don't have to look like carpets. A few dandelions add color. Slightly longer grass is healthier for the lawn and looks more natural, smothers weeds, and tolerates insects.

 ## THE LAWN AS AN ECOSYSTEM

The first step in developing an IPM system for your lawn is to look at it as an ecosystem. This ecosystem is composed of

- the grass plants themselves

- the organisms associated with grass plants, either pests—such as weeds, insects, and diseases—or their enemies
- the soil and the organisms that live in it

- the climate, both in the immediate vicinity and in the general locality.

An IPM approach is based on understanding the components of the lawn ecosystem and on using this knowledge to keep the ecological balance tipped in favor of the grass and beneficial organisms and against the pest species.

CREATING A HEALTHY LAWN

A healthy lawn is better able to resist diseases, weeds, and insects than an unhealthy lawn. Stress caused by too much or too little heat, water, nutrients, soil compaction, wear, and mowing makes plants more susceptible to damage.

The basic components of a healthy lawn are the right genetics and the right environment. Pick the grass varieties most suited to your climate, your specific ecological requirements, and your intended use. In the North, for example, grow a mix of cool-season grasses (those that grow best in spring and fall) such as one of the Kentucky bluegrass varieties, a perennial ryegrass, and a fescue. If you have strong shade, plant shade-resistant varieties. If your lawn gets a lot of wear, pick a grass variety with high wear tolerance (see References and Resources).

Types of Grass

The many grass varieties on the market tend to fall into two broad groups: cool season and warm season. Cool-season grasses grow best at temperatures of 60 to 75 degrees, do most of their growing in the spring and fall, grow little or not at all during the summer, and stay green in the winter. Warm-season grasses grow best at temperatures of 80 to 95 degrees, do most of their growing in the summer, and turn brown when the weather turns cold. As their names imply, the cool-season varieties do best in the North and the warm-season varieties do best in the southern third of the country. Imagine a line running roughly from South Carolina westward through southern California. Above this line, Kentucky bluegrass is the most common species; below it, Bermuda grass is most widespread. A belt 100 to 300 miles wide running along this line is a transitional zone where either turfgrass group can survive.

The two major zones are further divided according to precipitation into a humid and dry-to-semidry zone, with the same dividing line, to make four turfgrass adaptation zones.

There is a strong coastal effect, with the warm, humid zone extending as far north as Delaware in the East, and a cool, humid zone extending in the West from the Canadian border to southern California.

The appropriate grasses for these four zones often differ. In general, though, lawns in the cool, humid zone are polycultures of two or more grass species—usually some combination of Kentucky bluegrass, various fescues, and a ryegrass, either annual or perennial—although lawns composed primarily of bentgrass are common, particularly in the Pacific Northwest, and of zoysia grass and Bermuda grass in the transition area.

The cool, dry-semidry zone has two subareas: the central Plains area to the east, which has moderate rainfall, and the mountains to the west, where there is little rain. In the drier areas, where irrigation is common, Kentucky bluegrass and creeping and colonial bentgrass are the dominant turfgrasses. In areas without irrigation, drought-tolerant grasses such as buffalograss and wheatgrass predominate. Pure stands of Burmuda grass are grown throughout the warm, humid zone; zoysia grass is planted mostly in the North; centipedegrass, Bahia grass, and St. Augustine grass play an important role in parts of the Southeast.

Further complicating this picture is the fact that local ecological conditions, such as soil type, soil acidity, drainage, rainfall, or shade, also significantly affect the kind of grasses grown. In addition, the use pattern of your lawn—do you use it mainly for appearance, or do you play football or volleyball on it?—influences the appropriate grass.

Advances in lawn grass genetics, as well as the increased interest in lawns, have led to an explosion in the number of varieties available. Many of them have been bred for increased tolerance to different kinds of stress. Therefore, there are differences in such traits as nitrogen retention, tolerance to heat, drought, water, compacted soil, and wear, as well as inclination to thatching. If you have one of these problems and want to know which varieties can remedy it, contact the Lawn Institute, a seedsman's trade association.

The Soil

By the right environment, we mean the appropriate soil, water, nutrients, and weather conditions. But the soil is the key. It provides not only passive support and rooting for plants, but it is also alive and has a reciprocal relationship with the plants, with the soil affecting the plants, and vice versa. There is a lot of truth to the adage that a healthy soil creates a healthy plant. Soils contain living and dead organic material, minerals, water, and air; a lack of any of these can cause problems. The most important characteristics of soil are its structure and texture, nutrient level (both quantity and relative amount), relative acidity (pH), and soil organisms. You should know something about each.

The basic question is what kind of soil you have, since this determines the problems you may encounter and what will grow most effectively. There are four basic soil types: sand, loam, clay, and gravel. We will deal with only the first three; no one would plant a lawn in gravelly soil.

For a rough test, pick up some soil and feel it. When dry, sandy soil is gritty, loam soil is powdery, and clay soil is hard. Wet a lump of the soil, then try to roll it into a cylinder with your palms. If the soil doesn't retain any shape, it's mostly sand. If it forms a nice thin tube, it's mostly clay. Anything in between is mostly loam. For a more precise measurement, put 5 inches of dry soil in a large jar, add water, shake it up, and let it settle overnight. The various particle types settle out at different rates, so that there will be a layer of sand on the bottom, silt in the middle, and clay on top. Measure the height of each layer in inches and multiply by 20 to give you the percentage of sand, silt, and clay in your soil. Of the various types, loamy soils are the most fertile and best for most plants, because they contain the highest amount of organic matter.

Soil Problems

After you've determined your soil type, check for general problems with the soil, as indicated by compaction, grass root length, and nutrient level.

Compaction

Compacted soil prevents water or air from infiltrating into the soil and to a plant's roots, slowing down root growth and reducing microbial activity. To check for compaction, try to push a screwdriver into the dry soil. If you can't, or if you encounter a lot of resistance, your soil is too compacted and needs to be aerated.

You can aerate your soil in a number of ways. Any type of cultivation decreases compaction. Or you can use an aerator, which pokes holes in the soil. Aerators are gas- or foot-powered. You can rent or buy gas-powered aerators at better garden centers; foot-powered ones, which are simpler in design (and cheaper), can be bought at garden centers or by mail from garden supply companies.

Root Length

Grass root length gives you some information about your lawn's health. Dig a small hole with sharp sides so that you can easily see the grass roots. Although optimum root length differs among various types of grasses, we can make some generalizations. If the roots are longer than 6 inches, you have fairly healthy soil. Roots between 4 and 6 inches indicate that your grass has been overwatered or overfertilized and perhaps mowed too short. Roots shorter than 4 inches mean that your lawn needs aerating and, perhaps, less fertilizer.

Nutrient Level and pH

The nutrient level and relative acidity (also called pH) can be determined by a chemical analysis of your soil, which can be done at your state's agricultural college or the local Co-operative Extension Service. This will tell you the relative amounts of needed nutrients—nitrogen, phosphorous, and potassium are the most important—as well as the pH. If it is very detailed, the analysis will also give you information on a range of *macronutrients* (such as sulfur, calcium, and magnesium) and *micronutrients* or trace elements (such as iron, boron, and zinc) needed in small quantities by plants. Keep in mind that the amounts of nutrients your lawn needs to be healthy can be lower than the recommendations included with the soil analysis.

The soil acidity should be relatively neutral, with a pH between 6.5 and 7.5. If the soil is too acid, nitrogen and potassium do not degrade readily into forms most plants can use.

Flora and Fauna

Flora and fauna are the living parts of the soil, consisting of an astronomical number of macro- and microorganisms that perform a variety of functions, chiefly decomposition. Decomposers interact with dead organic matter, such as grass clippings and other plant parts, as well as dead microbes and minerals from the inorganic part of the soil, and convert both into nutrients that plants can use. The result is *humus*, the key to a fertile, healthy soil.

Water

The proper amount of water is vital to a healthy environment. Too much water, particularly early in the season, creates problems later on. Plentiful water discourages roots from growing deep in search of moisture, which means the grass will be more vulnerable to drought during the dry summer months. Therefore you need to water in a way that encourages root growth. Soak the soil thoroughly in the spring—from 6 to 18 inches, or the full depth of the roots—and then let it dry out almost completely before watering again.

Fertilizers

For proper nutrient management, organic fertilizers (i.e., materials that are biological in origin) have a number of benefits over synthetic chemical fertilizers.

- Organic fertilizers improve soil structure as they turn into humus with the help of soil decomposers. Synthetic chemical fertilizers, particularly the kinds that come in small particles, tend to deplete the soil's organic matter, leading to poor soil structure and soil compaction. Synthetic fertilizers also tend to release nutrients more rapidly, which can cause more groundwater pollution than use of organic fertilizers does.

- Since most synthetic fertilizers release nutrients all at once, any quantity of mobile nutrients, such as nitrogen, not taken up by the plant within a few days leaches through the soil and into the groundwater. This pulse of nutrients can lead to population outbreaks of many plant-feeding pests, as well as some diseases. This is not as true for organic fertilizers.
- Organic fertilizers harbor beneficial decomposer microorganisms. Synthetic fertilizers supply no microorganisms and tend to upset the soil ecology with their quick nutrient pulses, often favoring harmful microorganisms and inhibiting or killing beneficial ones. In addition, synthetic fertilizers usually change the soil chemistry, acidifying it and slowing down or killing some important biological processes.
- Organic fertilizers often supply

the macro- and micronutrients, or trace elements, that growing plants need. While synthetic fertilizers do supply nitrogen, phosphorous, and potassium, they do not supply these other nutrients unless they are specifically added.

- Organic fertilizers fit in with a recycling ethic. Indeed, you should compost all your plant wastes and garden scraps and use the compost as a high-quality soil conditioner. For information on how to construct a compost pile, contact your local organic gardening club or association.

There are a number of sources for and types of organic fertilizer. For all nutrients, compost or manure is the best. Organic sources high in nitrogen include fish meal, blood meal, and canola seed meal. Sources high in phosphorous include bone meal and single super phosphate. Sources high in potassium include kelp meal and liquid seaweed. Seaweed and kelp also provide many minor nutrients.

 # LAWN STRESSES

By having both the right genetics and the right environment, you can minimize lawn stress. Two prominent sources of lawn stress are thatch and improper mowing.

Thatch

Thatch appears right at the soil surface and looks like a layer of straw. It is composed mainly of parts of the

Lawn stress can be caused by thatch and improper mowing.

grass plant that live underground or creep along the soil's surface, although it also contains lawn clippings and other undecayed material. In healthy soil, these plant parts decompose relatively quickly into humus. But in soil that is compacted, overwatered, overfertilized with chemicals, heavily treated with pesticides, or underpopulated with earthworms and soil insects, you get thatch.

Synthetic fertilizers can cause or exacerbate thatch. Earthworms—which pull thatch underground, eat it, and turn it into humus—are frequently killed by pesticides. Earthworms avoid acidic soils with high nitrogen levels, often the result of overuse of synthetic fertilizers. Synthetic fertilizers also stimulate a shallow root system, thereby building up the thatch layer; the nutrients applied on the top of the soil make roots proliferate there. Too thick a thatch layer contributes to drought stress, because water has a harder time reaching the soil and runs off instead of sinking in. A small thatch layer is beneficial because it slows water runoff and can lessen the probability of drought.

Numerous insect pests, including sod webworm and chinch bug, live and breed in the thatch layer. The humid thatch climate harbors fungal diseases, especially brown patch, dollar spot, and Fusarium. Light can't easily penetrate thatch, making grass spindly; bare patches may open up, where weeds can invade. In sum, a thick thatch layer is the sign of a sick lawn.

The presence of some thatch is not inherently a problem, however. Thatch can act as a mulch. A thatch layer less than ¼ inch thick poses no problem. Consider treatment when the mat measures more than ½ inch.

Combating Thatch

If you do not have a severe thatch problem, simply add a top dressing of organic material that contains decomposer microorganisms—mulch or an equal mix of sand and manure. Make sure the material has not been sterilized. (Ringer Company's Lawn Rx and Lawn Keeper are effective at decomposing thatch.) Stop using chemicals that kill earthworms. If these remedies don't work, you can physically remove the thatch. A stiff rake or a thatch rake should do a thorough job on a moderate thatch layer. You can also buy a dethatching blade for your rotary lawn mower. For a very thick thatch layer, you may need a verti-cutter or verti-mower, which looks like a lawn mower with vertical blades.

Mowing

Improper mowing is another source of lawn stress. Mowing affects grass vigor, disease resistance, weediness, and water and nutrient needs. Proper mowing can kill weeds or retard their growth, prevent or cure diseases, save water, and provide nutrients to your lawn.

Most people mow too short and too often. Close, frequent mowings, combined with drought, lack of nutrients, or excessive heat or cold, make the grass grow smaller and thinner, creating open patches ideal for weed invasion. In addition, the cut end of the grass blade provides an entry point for disease.

Cutting places a stress on plants by removing growing, food-producing plant tissue that the grass has to replace. Initially, when grass blades are cut, root growth temporarily stops, food production and storage slow, water loss from the cut ends of the grass increases, and water uptake decreases. Then, in three or four days, the cut grass blades begin to grow again. After that, the grass begins to initiate new leaves. In general, the longer the blades, the longer the roots. Each time you cut grass, you stress the root system. If you cut it too short, the roots stop growing and may never grow again.

Mowing Dos and Don'ts

Mowing also has positive effects. Proper mowing induces a form of asexual reproduction called *tillering,* which makes the lawn thicker and denser, thereby making it harder for weeds to invade. The key is to adjust your mowing to the biology of your lawn. Do not mow on a schedule; adjust mowing intervals to suit lawn conditions. To minimize pests, keep the grass at least as tall as the recommended mowing height, which differs among grass types and over the growing season.

Appropriate grass heights change with the seasons. In summer heat, grass grows slowly. Let it grow higher to encourage the roots to grow deeper, thereby helping your lawn to withstand drought. Mow infrequently, but *never* remove more than one-third of the height of the grass at one time—mow whenever your grass grows 50 percent higher than the recommended height. Since grass grows more slowly in the shade, let grass in shaded areas grow about 1 inch higher. Rotary lawn mowers are best for cutting grass at higher heights; reel-type mowers are best at cutting low heights. Keep the blades sharp; dull blades tear grass, leaving a larger area of cut tissue for disease organisms to invade.

INSECTS

Insects proliferate in warmth and humidity. Therefore, turfgrasses in the cool, dry-semidry zone have the least problems with insects. We deal with the most common insect pests among and between zones: chinch bugs, sod webworm, white grubs, Japanese beetles, and billbugs.

TABLE 3-1

▪ RECOMMENDED MOWING HEIGHTS, IN INCHES ▪			
Grass	Cool Months	Warm Months	Final Mowing
Cool season			
Canada bluegrass	3	4	3
Kentucky bluegrass	2½	3	2
Annual bluegrass	½	1	½
Tall fescue	2½	4	2
Fine fescue	1½	2½	1
Annual ryegrass	2	2½	2
Perennial ryegrass	1½	2½	1
Creeping bentgrass	1⅓	⅔	⅓
Velvet bentgrass	¼	½	¼
Warm season			
St. Augustine grass	2	3	1½
Bahia grass	2	3	1½
Centipedegrass	1	2	1
Buffalograss	1½	2½	1
Bermuda grass	½	1	½
Zoysia grass	½	1	½

Remember that the IPM approach consists of understanding the biology and ecology of a pest species and using that knowledge against the pest. As a rule, survey your lawn every week or two throughout the season to look for early signs of pest damage, so that you can nip any problems in the bud.

Use your knowledge of the biology and ecology of a pest species to control the pest.

Chinch Bugs

Natural History

Chinch bugs damage turfgrass in all parts of the country except for the Northwest and the Plains states and they are found throughout the year. The hairy chinch bug is found in all the northern states in and east of Minnesota and throughout New England and the mid-Atlantic states south into Virginia. It is a major pest in Connecticut, New Jersey, New York, and Ohio. It feeds on most cool-season turfgrasses, particularly Kentucky bluegrass, red fescues, bentgrass, and perennial ryegrass, as well as zoysia grass. The southern chinch bug is found in Alabama, Florida, Georgia, Louisiana, Mississippi, North and South Carolina, and Texas. Although it feeds on a number of warm-season grasses, it is particularly damaging to St. Augustine grass. Both species like hot, dry weather, becoming particularly active when the temperature reaches the high seventies.

The adults, eggs, and young of both species look basically the same. The young, called *nymphs*, are at first red with a white stripe, the size of a pinhead; later they turn black with a white stripe. The adults are about ⅙ inch long, orange-brown to black with wings. They have piercing mouthparts and feed by sucking nutrients from grass stems and blades; they also inject saliva into the plant while feeding, causing further problems.

The two species differ in terms of their strategies for surviving the winter. In late summer to early fall the adult hairy chinch bugs begin to seek hibernation sites. Only adults can survive the winter; nymphs that have not become adults die. Favored sites are thatch or tall grass near the edge of lawns, plant debris, and space around the foundations of houses and under shingles and clapboards. The adults become active and leave their sites when the temperature exceeds 45 degrees. The adults mostly crawl (but may fly) from these sites to the lawn, where they immediately begin feeding.

The southern chinch bug does not really hibernate; it just becomes dormant in cool weather. On warm winter days the adults are active and may even lay eggs. All stages can potentially be found in the lawn during the winter.

Warning Signs

Look for signs of damage in the spring, every week or two. The farther south you live, the earlier you should begin monitoring. When the weather is hot and dry, monitor more frequently. Hairy chinch bug nymphs appear in May to early June. Southern chinch bug nymphs can occur as early as February in southern Florida to

---

Chinch bugs give off an offensive odor when crushed, so if you have a high infestation, you will notice a stench as you walk over your lawn.

---

early April in Louisiana. Check around lawn edges, especially near patios, sidewalks, driveways, and streets, where initial damage is likely to occur because of the warming effect of reflected heat from the concrete or asphalt (this is particularly true of the hairy chinch bug). In the South, look in sunny, open areas of the lawn.

General monitoring for either species is simple. Chinch bugs give off an offensive odor when crushed, so if you have a high infestation, you will notice a stench as you walk over your lawn. If you notice such a smell, take a closer look at the grass and thatch for adults or nymphs. Also look for irregular patches of yellow that eventually turn brown. If you find damage or suspect that you have a chinch bug problem, check the level of infestation systematically. Cut off both ends of a coffee can, push it into the soil, and fill it with soapy water. Do this either in areas with visible damage or in apparently unaffected areas nearby. Within five minutes, any chinch bugs will float to the surface.

When to Take Action

The level at which you take action depends on the lawn condition. If your lawn is in good condition, it can tolerate 15 to 20 bugs per coffee can. If the number is higher, or if you have fewer bugs but they are in a damaged area (remember that other pests may cause similar symptoms), you need to take action.

Long-term Controls

Genetic Controls. There are grass varieties, particularly of perennial ryegrass and tall fescue, that resist hairy chinch bugs. These grasses' blades and stems contain a naturally occurring fungus, called an *endophytic fungus,* which doesn't hurt the plant but confers some level of resistance against insect pests and diseases. In addition, the Kentucky bluegrass varieties Baron and Newport, and the perennial ryegrass varieties Score, Pennfine, and Manhattan are espe-

cially tolerant of hairy chinch bug damage. The St. Augustine grass cultivar Floratam is also resistant to southern chinch bug. If you have severe problems, consider reseeding your lawn with one of these varieties. Ask your local nursery, county Cooperative Extension Service, or the Lawn Institute (see References and Resources) to recommend resistant grasses appropriate for your area.

Cultural Controls. Use your knowledge of the chinch bug's biology to outwit it. Chinch bugs like hot, sunny lawns, dry roots, lots of thatch, and either low soil nitrogen or high levels of synthetic, or soluble, nitrogen. To decrease their prime habitat, dethatch your lawn when the thatch layer exceeds ½ inch. Use a thatch rake or a verti-cutter. Dethatch in the fall, as it creates a temporary stress on the lawn. Apply some compost after dethatching to minimize this stress.

Since the bugs like dry roots, increase irrigation and aerate the lawn to allow oxygen and water to penetrate and restore proper drainage.

Minimize your use of nitrogen fertilizer, which can stimulate chinch

---

To protect against chinch bugs, increase irrigation and aerate the lawn, allowing oxygen and water to penetrate and restore proper drainage.

---

bug populations. If you must fertilize, use an organic fertilizer or a slow-release version of a synthetic nitrogen fertilizer (such as some form of urea).

For preventive measures, destroy the chinch bugs' overwintering sites or kill the adults just after they have emerged in the spring. In the fall, remove plant debris around the house foundation, cut tall grass around the edge of the lawn, and thoroughly dethatch the lawn. As soon as the temperature regularly exceeds 45 degrees, look for the adults and try one of the controls suggested below to kill them.

Short-term Controls

Physical Controls. For light infestation, mix one capful of liquid dishwashing detergent per gallon of water and drench the problem area. Cover it with a large white or light-colored flannel sheet. The bugs will cling to the sheet as they attempt to get away from the detergent. In 15 to 20 minutes, take off the sheet and either scrape or rinse off the bugs into a stronger sudsy solution, or use a wet/dry vacuum to remove them.

Biological Controls. Big-eyed bugs are important predators of chinch bugs. Minimize your use of synthetic chemicals that kill these beneficial bugs. Plant alfalfa, carrot, goldenrod, or oleander—their flowers attract big-eyed bugs.

The fungus *Beauvaria bassiana* naturally infects chinch bugs, particularly during warm, humid weather, which is conducive to the spread of the fungus. Use *B. bassiana* for particularly large infestations, although the real impact of this fungus is not seen until the next generation (see References and Resources).

Last Resorts

Chemical Controls. Stay with the least toxic compounds. Insecticidal soaps sold by Safer, Ringer, and Ortho are effective against chinch bugs. You can also use a product containing pyrethrum, pyrethrins, or synthetic pyrethroids. Limit your use of the chemical to infested areas. Do not use pyrethroids near water; they are highly toxic to fish and some aquatic invertebrates, such as the water flea (*Daphnia*) and dragonfly nymph. *Caution:* Some people are allergic to the pulverized flowers that are the source of pyrethrum. Wear a dust mask or respirator if you are allergic to pollen and are spraying pyrethrum, or choose a different chemical.

Sod Webworm

Natural History

Sod webworms are major lawn pests in all parts of the country except for the Southwest. There are 60 to 80 webworm species in the United States, but only 10 to 15 are lawn pests. Sod webworms particularly attack Kentucky bluegrass, perennial ryegrass, bentgrass, tall and fine fescues, and zoysia grass. The adults are

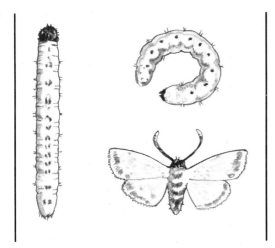

CUTWORMS AND ARMYWORMS

Cutworms and armyworms hide under plant debris or soil during the day and feed at night. When disturbed, cutworms curl up into a C shape.

smallish (¾ to 1 inch long) dingy brown to whitish moths with spotted wings that fly in a zigzag fashion. They have two common names indicative of their appearance—closed-wing moth and snout moth (so-called because of the two fingerlike horns on the head). The moths themselves do not cause any damage; the damage is done by the larval, or caterpillar, stage of the moth. They overwinter as mature caterpillars, feed a little in the spring, and then pupate. The adults emerge around June in the Northeast and Midwest. The females fly over lawns at dusk and drop their eggs preferentially in humid areas of succulent grass.

The caterpillars do all their feeding at night, so damage often goes unnoticed for a while. Young caterpillars build a silk-lined tube in the thatch and pull grass blades in to eat, often eating up to twice their weight each night.

Warning Signs

Most of the damage is done in either spring or fall, when mature caterpillars are most abundant, although in the South and Southwest the damage can also occur during the summer. Since the caterpillars like hot, dry weather, you can expect large infestations during the summer.

The caterpillars stay close to their burrows, so to catch early damage, look for quarter- to half-dollar-size brown patches that look like pockmarks in late spring. Since webworm caterpillars prefer hot, dry weather during the day, check particularly around patios, driveways, sidewalks, and streets because these areas warm up fastest. Also, you may notice that the damage is restricted to sunny areas of the lawn. If you do not monitor early, the population can increase dramatically, with the small pockmarks expanding and spreading.

Since other insects and diseases can cause similar damage, you need to make sure that webworm caterpillars

In late spring, quarter- to half-dollar-size brown patches that look like pockmarks may indicate the presence of webworm caterpillars.

are present. First, investigate the thatch layer in these patches and look for missing grass blades, caterpillar feces (small dark-colored pellets), and small caterpillars in silk-lined tubes. If they are present, you can monitor more systematically by using a soap drench. Add 2 tablespoons of dishwashing liquid to 1 gallon of water and drench the area. Soap irritates the caterpillars and they quickly crawl to the surface to escape. After 5 or 10 minutes, count the caterpillars. Since there is a possibility of other caterpillars being present, make sure you count only the webworm caterpillars. They are grayish brown, greenish, or dirty white, with four parallel rows of dark brown spots, and are ¾ to 1 inch long.

When to Take Action

A lawn in good condition can have two to three caterpillars per square foot and not suffer significant damage. A stressed lawn shows damage with just one caterpillar per square foot. Take action in areas with visible damage. If you do have damage, also randomly check undamaged areas of your lawn with the soap drench to determine webworm population levels. Check weekly during the spring and fall, if you find any evidence of webworms. When populations reach two per square foot, take action.

Long-term Controls

Genetic Controls. Some varieties of perennial ryegrass (such as All Star, Citation II, Commander, Pennant,

Repell, and Sunrise) and tall fescue contain endophytic fungi, which confer some resistance to webworms. If you have severe webworm problems, consider reseeding your lawn with a resistant variety.

Cultural Controls. Caterpillars hide in thatch. If your thatch layer is higher than ½ to ¾ inch, dethatch it in the fall.

While hiding during the day, the caterpillars like hot, dry areas. Soil dryness can result from a lack of rain or irrigation, or from water not getting into the soil because of compaction or a thick thatch layer—both indications of a stressed lawn. Aerate the soil with an aerating tool. If necessary, irrigate (or water) the area with a hose, sprinkler, or irrigation system until the top inch or two of the ground is wet.

Short-term Controls

Physical Controls. In highly damaged areas, drench the entire area with sudsy water (2 to 3 tablespoons dishwashing liquid per gallon of water). When the caterpillars crawl to the surface, rake them up into a pile, scoop them up with a shovel, and dump them into a bucket of very soapy water.

Biological Controls. A number of native natural enemies feed on webworm caterpillars: many bird species (particularly robins and starlings), parasitic wasps and flies, ants, earwigs, numerous predatory beetles, paper wasps, and beneficial nematodes. Try to conserve these natural enemies by minimizing your use of synthetic chemicals. A wide range of plants, from wildflowers and weeds to shrubs and trees, attract one or more of the webworm's natural enemies. Choose a range of plants to ensure that some are always flowering (this will help attract parasitic insects). (See References and Resources).

A wide range of plants, from wildflowers and weeds to shrubs and trees, attract webworm's natural enemies.

If you have chickens, or can borrow some, let them roam over the lawn. They will quickly clear up a heavy infestation (and add some organic fertilizer). But make sure they are fenced in.

You can buy the fungus *Beauvaria bassiana* for heavy infestations. By the time the caterpillars die, however, they have done most of their damage. So the impact of the fungus is not apparent until the following year, when fewer caterpillars are present at the beginning of the season.

The microbial product *Bacillus thuringiensis* (or *Bt*) specifically attacks caterpillars. *Bt* attacks only the caterpillars, not the moths, so don't buy it to combat moths during the summer. You can use *Bt* at almost any population level, but it takes a couple of days or even a week to work. Apply it liberally to the infested areas.

Predatory nematodes, particularly the genera *Heterorhabditis* and *Steinernema*, are promising control agents against webworm caterpillars, and can be purchased commercially. They need a film of water to move around in and find hosts. So use a watering can to apply them to areas with significant pest populations, and then irrigate. The nematodes take about a week to work and are especially good for moderate infestations.

Last Resorts

Chemical Controls. For large infestations, use an insecticidal soap. Treat the infected area every two or three days for a couple of weeks, since the soap is rapidly broken down by soil microorganisms. Saturate the thatch layer with the solution; it must penetrate to where the webworms are in their silken tubes. You can also use pyrethrum, pyrethrins, or pyrethroids, but be careful—they also kill beneficial insects. Apply them only to infested areas. Do not use pyrethroids near water; they are highly toxic to fish and some aquatic invertebrates. *Caution:* Some people are allergic to the pulverized flowers that are the source of pyrethrum. Wear a dust mask or respirator if you are allergic to pollen and are spraying pyrethrum, or choose a different chemical.

White Grubs

Natural History

White grub is a catchall term for the larvae of many different species of beetles, including scarab beetles, june bugs, Japanese beetles (discussed separately below), and various chafers (rose, oriental, masked). They live in the soil and feed on plant roots. White grubs primarily attack the roots of bluegrasses, bentgrasses, and tall and fine fescues, although other grasses may also be attacked. White grubs present large problems in all areas of the United States except the Gulf and Plains states.

The larvae all have a similar appearance: plump, C-shaped bodies, ¼ to ¾ inch long, that are blunt-ended and creamy white with hard yellow or brown heads. They overwinter in the larval stage. The most severe damage usually occurs during late summer and early fall, when water stress and grub populations are highest.

Warning Signs

Monitor soil grub populations in early spring before the adults emerge. Walk the lawn and look for signs of feeding damage—irregular patches of slow-growing, yellowing grass that

Irregular patches of slow-growing, yellowing grass and spongy, loose turf may indicate grubs.

eventually turn into brown dead patches. Other pests cause similar damage. You can distinguish grub damage by the spongy or loose feel of the turf, which, owing to the cut roots, you can often roll back or pick up. Systematically sample both damaged and undamaged areas by marking out square pieces of turf 1 foot on a side. Cut the turf on three sides to a depth of 3 to 5 inches and turn it over to expose the roots. Count the grubs.

When to Take Action

A lawn in excellent condition can withstand up to 15 grubs per square foot in the spring and up to 10 per square foot in the fall. If your lawn is stressed, damage can occur at lower levels. To be on the safe side, take action when the grub levels exceed 10 per square foot in the spring and 6 per square foot in the fall.

Long-term Controls

Cultural Controls. Check your lawn for excessive thatch and compacted soil, both of which inhibit water and nutrients from reaching plant roots. If necessary, dethatch the lawn, aerate the soil, and top-dress with a ¼- to ½-inch layer of 50 percent organic mulch (such as steer or chicken manure) and 50 percent sand. If your lawn shows severe damage, water the damaged areas frequently to lessen drought stress (remember that the grass in this area has most of its roots cut off). It also pays to add some dilute fertilizer to the water to help the plants get some nutrients.

Short-term Controls

Physical Controls. One simple control consists of turning over the damaged sod, picking up the grubs, and dropping them into a pail of sudsy water. Afterward, tamp the sod back down and give it some water and a little fertilizer. You can reduce Japanese beetle populations in late spring, when grubs are feeding close to the surface, by walking over the area two to three times with spiked shoes (spikes are 2 to 4 inches long). This ingenious control for grubs has been developed by researchers at the University of Colorado, who found it kills almost as many grubs as synthetic pesticide does. The shoes have the added benefit of aerating the soil.

A grub problem can be partially solved by turning over the damaged sod and removing the grubs by hand.

A machine used to aerate the soil would have the same effect. Make sure through your monitoring that the grubs are close to the surface (i.e., during spring and fall) before using this method.

Another simple control for the adults, which frequently attack bushes and flowers, is to handpick them off bushes and trees early in the morning, when it is cooler and they are sluggish. (You can also use a handheld vacuum.) This has the greatest chance for success with light infestations early in the season. Handpicking will reduce the number of females laying eggs, thereby reducing grub populations and, consequently, adult populations the following year.

Biological Controls. Preserve the grubs' natural enemies, particularly birds, soil microorganisms, and parasitic wasps. Encourage birds to visit your area by planting bushes and trees. Some white grubs are attacked by the milky spore disease, which primarily attacks Japanese beetle grubs.

Predatory nematodes, particularly *Heterorhabditis* and *Steinernema*, can control white grubs, especially the masked chafers. These nematodes need a film of water to move around in and find hosts. With a watering can, apply them to areas with significant populations, and then lightly irrigate the area. They take about a week to work and are especially good for moderate infestations.

Last Resorts

Chemical Controls. If you must use a chemical, use the least toxic one. Neem (BioNeem), an extract from the neem tree, repels adult beetles as well as kills some white grubs. A pyrethrum, pyrethrins, or pyrethroid product also kills adults. Do not use pyrethroids near water; they are highly toxic to fish and some aquatic invertebrates. *Caution:* Some people are allergic to the pulverized flowers that are the source of pyrethrum. Wear a dust mask or respirator if you

are allergic to pollen and are spraying pyrethrum, or choose a different
chemical.

Japanese Beetles

Natural History

Japanese beetles are a type of white
grub. As the name suggests, these
beetles were accidentally introduced
into the United States from Japan.
They were initially discovered in Ri-
vington, New Jersey, in 1916. Since
then, this pest has spread west to the
Mississippi and south to the Gulf
Coast. The adults are ¼ to ½ inch in
length, with a metallic green or blue
body and coppery wing covers. They
feed voraciously on a large number of
plant species. The grubs are 1 to 1½
inches in length, C-shaped, plump,
and white with brown heads. They
eat decaying vegetation and roots and
are often the most serious pest in the
East.

Since other grubs, particularly the
chafers, look very similar, you need
to be sure that the grubs you are deal-
ing with are Japanese beetle grubs, as
some biological controls (e.g., milky
spore disease) are specific to them.
Look closely at the underside of the
grubs. There is a V-shaped arrange-
ment of stiff hairs on the last abdom-
inal segment of a Japanese beetle
grub, but not on other white grubs.
You can also collect some large grubs,
put them into a jar, wait until an adult
emerges, then take it to a Cooperative
Extension office for identification.

Japanese beetle grubs overwinter

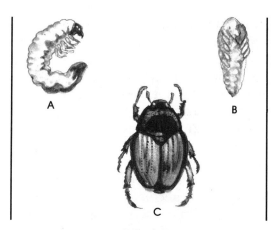

JAPANESE BEETLE LIFE CYCLE

*Eggs are laid in the ground during late
summer. They hatch in about two weeks,
and the larvae (A) burrow into the
ground, where they feed on roots. When
the weather turns cold, they hibernate.
The following spring, they feed actively
on roots and then pupate (B). The adult
(C) emerges from the larval burrow in
early summer and feeds on leaves and
fruits.*

deep in the soil, feed on roots early in
the spring, pupate in May or June,
and emerge as adults in mid-May to
mid-July (emergence is later as you
move north). Adults live four to six
weeks, with populations usually
peaking in July. The adults feed,
often voraciously, on the leaves of
various shrubs and plants. Females
search for grassy areas, especially
closely cropped grass, with moist soil
on which to lay their eggs. The eggs
need the water to hatch successfully.

Warning Signs

Begin monitoring for grubs just
prior to adult emergence in mid-May
to June. Ideally, you should begin

monitoring before you see a lot of damage. Randomly choose five or six areas on your lawn and mark off square sections of turf 1 foot on each side, cut on three sides to a depth of 3 to 5 inches, turn over the sod, and expose the roots. Count the grubs.

Also look for feeding damage, which first shows up as irregular patches of slow-growing, yellowing grass that eventually turn into brown dead patches that feel spongy or loose and can often be rolled back or picked up because of the cut roots.

When to Take Action

A lawn in excellent condition can withstand 15 grubs per square foot in the spring and up to 6 to 10 per square foot in the fall. Stressed lawns show damage at lower levels. Monitor your lawn and take action when you see damage. To be on the safe side, take action if grub population exceeds 10 per square foot in the spring or 6 per square foot in the late summer or early fall.

Long-term Controls

Cultural Controls. Check your lawn for excessive thatch and compacted soil, which prevent water and nutrients from reaching plant roots. Dethatch your lawn if the thatch layer is over ¾ inch thick. Also aerate your soil and top-dress with a ¼- to ½-inch layer of 50 percent organic mulch (steer or chicken manure) and 50 percent sand (see References and Resources).

Since the adults lay eggs in moist soils and the eggs need moisture to hatch, irrigate infrequently and deeply in July. This minimizes the amount of time that the soil is moist. In the spring and fall, irrigate regularly to prevent drought, help offset the stress of any damage, and encourage the spread of milky spore disease, a natural enemy of Japanese beetle grubs. (See page 38.)

Short-term Controls

Physical Controls. You can use physical controls against both adults and grubs. It is important to attack the adults, particularly the females, both to reduce the damage they do as well as to reduce the eggs laid, and thus the future grub population. For adults, handpick them off plants and bushes and drop them in a bucket of sudsy water (use dishwashing liquid), which kills them. Do this on a cool morning when the adults are sluggish. If you have a lot of area to cover, or a very high infestation, take advantage of the fact that the adults, when sluggish, drop to the ground and feign death. Put a large sheet underneath a plant or bush and then shake it. All the beetles will fall on the sheet, and you can dump them into a pail of sudsy water. For the more technically inclined, the BioQuip Company sells a modified hand-held vacuum to re-

---

Walking repeatedly over infested areas of sod with spiked shoes can kill many Japanese beetle grubs.

---

move adults from plants. A small home vacuum also works. This approach works primarily with light infestations early in the season.

SEX PHEROMONE TRAPS. A number of companies sell traps that contain sex pheromone and floral (or food) scents designed to attract adults. A recent study suggests that they do not work well. One or two traps in a yard caught only slightly more than half the number of beetles attracted to them, leading to a net increase in the number of beetles around the trap. Further, even though some traps caught up to 20,000 adults in a day, there was no decrease in the number of grubs in the soil around the trap. Therefore, one or two traps do not affect grub populations and may dramatically increase adult populations. Other work has shown that the traps can be effective if used in larger numbers throughout an entire community. So either do not use these traps (particularly if you don't have much land), or be prepared to buy a lot of them and put them all over your property. You'll need to get your neighbors to put them out too. One study suggests that one trap be placed every 150 feet on the perimeter of the area to be protected so that adults are caught before flying into the area.

INSECTS. There are also a number of native insect predators and parasitoids that feed on Japanese beetles. A fly that attacks the adults and a parasitic wasp that attacks the grubs may become important natural control agents in areas with high beetle densities. The adult flies and wasps feed on nectar, while the larvae feed on Japanese beetles.

The wasp, particularly, likes flowers in the wild carrot family. Plant or encourage wildflowers such as wild carrot, fennel, or dill to attract and maintain parasitic insects. Also, refrain from using conventional synthetic pesticides because you may kill these natural enemies.

BACTERIA. The bacteria that cause milky spore disease, particularly *Bacillus popillae*, constitute the major source of mortality for Japanese beetle grubs. Spores from the bacteria, which occur in the soil, germinate when ingested by grubs feeding on plant roots. The bacteria multiply rapidly inside the grub and kill it, turning its insides a milky white in the process (hence its name). When the grub dies and disintegrates, billions of spores are released into the soil. They live in the soil for a long time, but eventually need to be replenished. Low populations of grubs are actually beneficial, as they serve as a reservoir for the milky spore disease.

You can buy the disease spores in powder form at garden stores. Be sure to wear a dust mask when applying the powder. After application, water the lawn lightly to help wash the spores into the thatch and soil. In warm (70 degrees and above), moist soils with a large number of actively feeding grubs, milky spore disease can sweep through a population of Japanese beetle grubs within a week or two. Under less ideal conditions, the disease takes longer to work. View this as a longer-term control; it can take a season or two before it has a significant impact on grub populations. It is less effective and slower acting in areas where soil temperatures do not reach 70 degrees when grubs are actively feeding.

NEMATODES. Predatory nematodes, particularly *Heterorhabditis* and *Steinernema*, are most effective against low to moderate Japanese beetle grub populations (10 to 15 per square foot). (See References and Resources.) With a watering can, apply nematodes to areas with significant populations, then irrigate the area so that it gets ¼ to ½ inch of water, enough to wash the nematodes into the soil. Keep the soil moist so that the nematodes can move around and do their work. This is particularly important in the late summer and early fall, when soil drought is more common, and there are larger numbers of grubs.

Last Resorts

Chemical Controls. Neem, an extract from the neem tree and the active ingredient in BioNeem, repels adult beetles. Indeed, they will starve to death rather than eat leaves sprayed with neem extract. Thus, neem preparations sprayed on the

shrubs and trees that adults feed on may help repel them from the area. You can also spray pyrethrum, pyrethrins, or pyrethroid-based products on adults. Use these products sparingly or you may kill some of the beetles' natural enemies along with the beetles. Do not use pyrethroids near water because they are highly toxic to fish and some aquatic invertebrates. *Caution:* Some people are allergic to the pulverized flowers that are the source of pyrethrum. Wear a dust mask or respirator if you are allergic to pollen and are spraying pyrethrums, or choose a different chemical.

Billbug

Natural History

Billbugs are weevils that attack turfgrasses. The bluegrass billbug and the hunting billbug cause the bulk of the problems. The bluegrass billbug attacks cool-season turfgrasses. It thrives wherever its primary host, Kentucky bluegrass, is grown, but is primarily a problem in Kansas, Massachusetts, Nebraska, New York, Ohio, Pennsylvania, South Carolina, Washington, and Wisconsin. The hunting billbug is a major turfgrass pest in the Southeast, occurring in the Atlantic coastal states from Virginia to Florida and along the Gulf of Mexico from southern Texas north to Kansas and Missouri. It most severely damages zoysia grass and the improved hybrid Bermuda grasses, but also feeds on St. Augustine grass and centipedegrass, as well as on a number of grass-type plants, including crabgrass, nutsedge, barnyard grass, wheat, and corn.

Bluegrass billbugs and hunting billbugs are nearly identical in appearance and have similar biologies. Adults range in color from gray through brown to black and are small with a long snout. Hunting billbug adults are slightly larger (⅓ to ½ inch) than bluegrass billbug adults (¼ to ⅓ inch). The eggs of both species are oblong and clear to creamy white. The larvae, also called grubs, are ¼ to ⅜ inch in length and mostly white, with a black blotch on their backs and an orange-brown head capsule. They look like puffed wheat.

Both billbugs overwinter as young adults produced during the summer and early fall. During the fall the adults seek hibernating quarters in weeds and leaf litter, hedgerows, surface litter around the foundations of buildings, and the junction of turf and sidewalk. The adults emerge during warm, sunny periods in spring, crawl to the nearest turf areas, and start feeding on grass stems. Eggs are laid in the leaf sheaths just slightly above the crown in May to June. The larvae hatch within two weeks and initially feed on stem tissue. As the larvae grow, they hollow out the stem, burrow downward, and feed in and on the crowns. The larger larvae eventually move through the thatch layer to feed on the roots. The oldest larvae burrow down 3 to 4 inches into the soil, where they pupate. A light tan to whitish sawdustlike excrement accumulates in areas of heaviest larval feeding.

Warning Signs

Look for the adults, particularly on sidewalks and driveways, when they are leaving their overwintering sites and migrating back to turf areas during May and June. If you see any adults in these areas, systematically look through the lawn for adults, larvae, or damage. Young larvae occur higher up in the thatch layer, while older larvae move down into the soil and are hard to spot.

Damage appears as yellow or brown patches of dying grass. Patches damaged by bluegrass billbugs are spotty and often first appear along sidewalks and driveways or near trees. In heavy infestations, these patches appear throughout the lawn and enlarge to cover virtually the entire area. Damaged patches from hunting billbugs are elongated or rounded and occur irregularly throughout the lawn. Since billbug larvae attack roots, the affected grass can often be pulled up in mats. Unlike damage caused by white grubs, where the soil is soft and spongy, the soil under billbug-damaged areas is firm. Look for a brownish or whitish sawdustlike material around the roots or stems; this is a definite indication of billbug larval damage.

To sample larvae, mark off a number of square-foot plots, both in damaged and undamaged areas, and

Brownish or whitish sawdustlike material around the roots or stems of the grass is an indication of larval billbug damage.

carefully look through the grass and thatch layer. Look at the soil layer by either cutting three sides of the square foot and peeling back the sod, or, if there is enough damage, simply pulling up the mat of grass.

When to Take Action

A healthy lawn can withstand 1 adult or up to 10 billbug grubs per square foot. If the number of larvae exceed this, take action. In spring, count the number of adults crawling over paved areas next to the lawn for 5 to 10 minutes. Take action if you count 2 or more adults per minute.

For the bluegrass billbug, July to early August is the most critical period for damage because the larvae are at peak feeding stage and the grass is often under stress from summer drought. For the hunting billbug, extended dry periods throughout summer can make the damage more noticeable.

Long-term Controls

Genetic Controls. Some varieties of perennial ryegrass (Citation II, Commander, Pennant, Regal, Repell, and Sunrise) and tall fescue contain endophytic fungi, which confer some resistance to billbugs. For Kentucky bluegrass, Kenblue, Nebraska Common, and the Park are tolerant of billbug damage. If you have severe billbug problems, consider reseeding your lawn with a resistant or tolerant variety. Ask your local nursery, county Cooperative Extension Service, or the Lawn Institute for the most suitable varieties.

Cultural Controls. Billbugs prefer dry, stressed lawns. First, dethatch the lawn if the thatch layer exceeds ¾ inch. Then, if the soil is compacted, aerate it and add a top dressing of ¼ inch fine compost or other organic manure, such as 50 percent chicken or steer manure and 50 percent sand. Add mulch to the soil to help improve water retention. You may also need to irrigate your lawn.

As a preventive control, remove leaf litter, weedy debris, and other surface debris around the foundations of buildings and the roots of trees, and at the junction between lawn and sidewalk. Particularly for bluegrass billbugs, use one of the chemical controls listed below on turf adjacent to paved areas in spring if you see more than two adults per minute on the paved areas.

Short-term Controls

Biological Controls. The major biological controls are predatory nematodes, particularly the genera *Heterorhabditis* and *Steinernema*. They need a film of water to move around in and to find hosts, so apply them to areas containing significant

Predatory nematodes can be used to control billbugs.

populations with a watering can, and then lightly irrigate with ¼ to ½ inch of water.

Last Resorts

Chemical Controls. Pyrethrum, pyrethrins, or pyrethroid-based products suppress adult populations. Do not use pyrethroids near water; they are highly toxic to fish and some aquatic invertebrates. *Caution:* Some people are allergic to the pulverized flowers that are the source of pyrethrum. Wear a dust mask or respirator if you are allergic to pollen and are spraying pyrethrum, or choose a different chemical.

Chart 3-1 lists some commercial pesticide products for controlling in-

sect pests in lawns. The listed products are a small fraction of those pesticides marketed for this purpose. Products were selected for listing based on CU's judgment that they can be effective when used in the context of an Integrated Pest Management strategy, and that they pose the least risk of harm to humans, pets, or the environment, based on the active ingredients they contain.

Some unlisted products contain exactly the same active ingredients as listed products and may be substituted for them. But many other widely available products are not listed because they contain active ingredients that, in CU's judgment, pose greater potential risks to health or the environment than the ingredients of listed products. In our view, effective pest control does not require taking the risk of using the more toxic pesticides, and we have chosen not to list products that contain them.

 # DISEASES

Diseases can be difficult to control once they have infected a lawn, so you need to focus on prevention, particularly if you want to avoid using synthetic fungicides. To focus on prevention, you must be able to diagnose damage caused by diseases and identify the organism or factor causing the disease. Use the IPM method to control disease in your lawn by fol-

lowing the same steps used for insect control.

Four different agents cause disease: environmental factors, fungi, bacteria, and viruses. Environmental factors—too much or too little sun, excess water, a soil deficiency (inappropriate pH, inadequate or unbalanced nutrients)—are the most frequent causes. In addition, environmental factors can stress a plant enough to make it susceptible to a disease-causing organism. The other causal agents of disease are different types of organisms. Fungi are the most prevalent disease-causing organisms. Environmental (especially weather) conditions often stimulate fungal diseases, which can usually be controlled or eliminated by proper lawn care. Unlike fungal diseases, whose symptoms may take weeks to develop and spread, bacterial and viral diseases' symptoms develop in a couple of days.

Monitor Your Lawn. The first step in developing an IPM program for diseases involves monitoring your lawn for symptoms at least once a week or every two weeks, depending on its susceptibility to disease. To do this, you must recognize the general as well as specific symptoms of lawn diseases. There are six symptoms of lawn diseases: yellowing of leaves, dead or brownish areas on leaves, rotten spots, a water-soaked or greasy appearance, a wilting plant, or a plant that dies mysteriously. Although a number of these symptoms (particularly the dead, brown patches) are similar to ones caused by certain insect pests, diseases do not usually sever the roots of grass completely. Therefore the affected patch cannot usually be lifted up, as it can if the damage is caused by insect pests.

Diagnosis. After diagnosing disease as the cause of your problem, determine the specific disease and its cause. Get a color key to local lawn diseases from a county Cooperative Extension agent, a garden store, or one of the books mentioned in the References and Resources section. For lawns, the possibilities are narrowed by the fact that environmental problems and fungi cause the vast majority of diseases. These two causes are closely related because environmental factors are a major contributor to fungal diseases. Plant pathologists often talk of the "three legs" of plant disease—the *causal organism* (usually a fungus for lawns), the *susceptible host* (a grass plant), and the *environmental conditions* conducive to the disease. The disease needs all three legs to develop.

In the case of fungal lawn diseases, the fungi are usually present in the environment. However, a healthy soil can keep the disease-causing fungi at low enough levels so that they do not cause a problem. Susceptibility of the grass plant to the disease is important. Plant breeding has made a number of inroads in this area. At present, more than 200 varieties of grasses are resistant to at least one disease, and many varieties are resistant to more than one disease. Given these recent advances in turfgrass breeding, the key factor determining the existence of disease becomes whether or not the conditions conducive to the disease are present.

Stress. Conducive conditions are either human-induced or environmental stress, with the former often

CHART 3-1

Ammunition Against Common Insects   *As published in* Consumer Reports, *June 1990.*

TYPE	WHERE AND WHEN	CHECK FOR
Japanese beetles 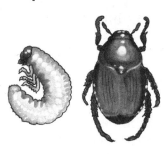	Attack lawns in a larval stage, as white grubs. They eat the roots of bluegrass, bentgrass, and fescues anywhere in the United States except the Gulf and Plains states, doing most of their damage in the spring and fall.	Irregular patches of slow-growing, yellowing grass and turf that is loose because its roots have been eaten. Cut a foot-square piece of sod 5 inches deep and turn it over. If you see 10 or more grubs, take action.
Billbugs 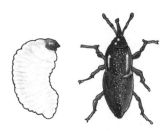	A problem in almost all of the United States, especially with Kentucky bluegrass, Bermuda grass, and zoysia grass. The larvae, which look like puffed wheat, do the damage. Preferred meal: dry grass.	Adults on sidewalks in May and June; larvae in thatch and soil. Damaged areas are small yellow or brown circles. Larvae attack roots, so the affected grass can often be pulled up in mats. Peel back a piece of sod. If you see more than 10 larvae per square foot, take action.
Chinch bugs  	Pests everywhere except the Northwest and Plains states. They suck the nutrients from bluegrass, fine fescues, bentgrass, St. Augustine grass, and zoysia grass, especially in hot, dry weather.	Yellow patches that turn brown. The bugs emit an odor when crushed; a heavily infested lawn will stink as you walk on it. To sample, cut off both ends of a coffee can, push into the soil, and fill with soapy water. If more than 15 bugs float, take action.
Sod webworm 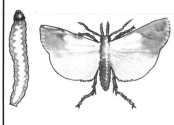	Pests everywhere except the Southwest, they attack several grasses. Caterpillars build a silk-lined tube in thatch and pull grass blades in, eating up to twice their weight each night. They proliferate in dry, hot weather.	Small brown pockmarks in late spring. Damage often starts around pavement. If not checked, webworms can multiply quickly until the pockmarks coalesce. If damage is visible, take action.

NONCHEMICAL OPTIONS	CHEMICAL OPTIONS
Attract birds, natural enemies, with berry bushes, trees, and feeders. ■ Plant wildflowers to lure parasitic insects, also enemies. ■ Apply milky spore disease, a biological control for grubs. ■ Use predatory nematodes to control grubs. Nematodes attack while swimming, so pour them on damaged areas with a watering can. ■ Don't use beetle traps, which can attract more than they kill.	Neem tree extract repels adult beetles and kills grubs. ■ Pyrethrum/pyrethrins kill adults but may kill their enemies too. Apply only where needed.
Use predatory nematodes, pouring them on the lawn along with water. ■ Add organic matter to the soil to improve water retention. ■ Irrigate. ■ If the problem just won't go away, consider reseeding with perennial ryegrass or tall fescue. Certain strains tend to resist billbugs.	Pyrethrum/pyrethrins should suppress adult populations.
For a quick (and dirty) fix, mix one capful of dishwashing soap per gallon of water and drench the problem area. Cover it with a large flannel sheet. The bugs will cling to the sheet; wait 15 minutes, remove the sheet, and scrape the bugs into the trash. ■ Preserve natural enemies such as big-eyed bugs by limiting synthetic pesticides. ■ Consider reseeding with resistant varieties of perennial ryegrass or tall fescue.	Use insecticidal soap, or, as a last resort, pyrethrum/pyrethrins.
Another quick and dirty fix: Drench damaged areas with soapy water; caterpillars will crawl to the surface; rake them up; throw them out. ■ Avoid chemical insecticides, which can kill natural enemies. ■ For a heavy infestation, use the bacteria *Bacillus thuringiensis (Bt)* or predatory nematodes (poured on along with water). ■ Consider reseeding with varieties of perennial ryegrass and tall fescue that resist webworms.	Treat damaged areas with insecticidal soap. Use pyrethrum/pyrethrins as a last resort, only on infested areas.

creating or interacting with the latter. Examples of environmental stress are inappropriate weather or microclimate. Examples of human-induced stress are over- or underwatering; over- or underfertilizing; changes in soil produced by the use of chemical fertilizers and pesticides, which may make the soil more receptive to the disease organism or disrupt soil ecology and decrease the disease's antagonists; and, often most important, mowing too short and too often.

Prevention. To treat a disease, correct the underlying source of the problem; do not treat just the symptoms. Remember that the appearance of the disease in your lawn is the result of more than just the presence of a fungus, virus, or bacterium; the environmental conditions are key. If you simply spray fungicide and do not treat the underlying environmental conditions that are the source, the disease will recur, making you even more dependent on lawn-care chemicals.

Keeping a lawn healthy and using cultural controls minimizes its disease problems. Use the following strategies to maximize lawn health:

- Use grass varieties resistant to the most prevalent diseases.
- Minimize shade.
- Check for problems with soil fertility, soil chemistry, thatch layer, soil compaction, or water drainage, and correct them.
- Mow regularly and properly.

- Stop using synthetic pesticides, which kill off beneficial soil organisms.

Brown Patch

Natural History

Brown patch disease, caused by the fungus *Rhizoctonia solani*, is the most prevalent disease of turfgrasses. It occurs everywhere in the United States except the Pacific Northwest and is most severe east of the Mississippi River. Brown patch can attack all grasses, although it hits bentgrasses and St. Augustine grass particularly hard. Kentucky bluegrasses are often hit, but not severely. It attacks cool-season grasses in the early and late summer and warm-season grasses in the spring and fall. Brown patch invariably attacks highly maintained turf, particularly turf on waterlogged soils with excessive thatch. It thrives in humidity and temperatures above 80 degrees and in areas with either too much or too little nitrogen.

Warning Signs

During the appropriate season, look for "frog's eyes," circular spots up to 2 feet in diameter surrounded by a discolored ring of grass, with a thinning of grass in both areas.

When to Take Action

Take action when you see the "frog's-eye" damage.

🎋 *"Frog's eyes," circular spots up to 2 feet in diameter surrounded by a discolored ring of grass, with a thinning of grass in both areas, is an indication of brown patch disease.*

Long-term Controls

Genetic Controls. If you experience severe problems with the disease year after year, and poor drainage is not your problem, plant a resistant or tolerant grass, if available. The perennial ryegrasses All Star, Barry, Manhattan II, Prelude, Premier, and Yorktown II and the tall fescues Brookston, Jaguar, and Mustang are resistant to brown patch; check with your local nursery, Cooperative Extension office, or local state university for varieties appropriate for your area.

Cultural Controls. Check your lawn for excessive thatch and compacted soil. Dethatch your lawn if the thatch layer exceeds ½ inch, using either a dethatching tool or a product such as Ringer's Lawn Rx, which hastens thatch breakdown. Aerate compacted soil with a machine or foot tool and top-dress with an organic compost or manure. For severely waterlogged soil, you may need to install drainage tiles. (Contact a local landscape architect or local garden supply store.) Make sure you are not adding too much nitrogen fertilizer.

Short-term Controls

Physical Controls. If an extended period of high humidity strikes during the summer, knock the dew off the grass blades by dragging a rope or garden hose over your lawn in the morning.

Last Resorts

Chemical Controls. If other methods haven't brought brown patch under control, use a flowable sulfur fungicide every three to five days until the symptoms begin to disappear. (A *flowable* is a solid powder mixed with, not dissolved in, water.) Use ground sulfur rock or a liquid form of sulfur. Do not use lime sulfur; it causes problems when temperatures exceed 80 degrees. *Caution:* Wear gloves and protective clothing and a dust mask when applying sulfur; do not spray into the wind.

Leaf Spot

Natural History

Leaf spot, or melting out, caused by the fungus *Drechslera poae*, is the second most prevalent turfgrass disease in the United States, causing problems in the Southeast, Midwest, Southwest, and Plains states. Leaf spot can attack most grass varieties but particularly likes Kentucky bluegrass and the fine fescues. Leaf spot usually hits Kentucky bluegrasses during cool, moist weather in the spring and fall, although it can cause

🦋 *Reddish brown to purplish black spots on leaf blades is an indication of leaf spot disease.*

problems in New Jersey throughout the entire growing season.

Warning Signs

During wet, cool weather, first look for irregular patches of thin grass. Then look more closely for reddish brown to purplish black spots on the leaf blades. As the disease spreads, the grass blades shrivel, the crown and roots rot, and irregular patches of thin grass develop.

When to Take Action

Take action as soon as you see signs of the disease.

Long-term Controls

Genetic Controls. For severe or persistent problems, plant a resistant or tolerant variety, if available. A number of Kentucky bluegrasses (Bonnieblue, Challenger, Majestic, Midnight, and Nassau), perennial ryegrasses (Belle, Blazer, Cowboy, and Ranger), tall fescues (Adventure, Brookston, Houndog, Jaguar, and Mustang) and a fine fescue (Reliant) are resistant; check with your local nursery, Cooperative Extension office, or local state university for the appropriate varieties.

Cultural Controls. Since an excessive thatch layer increases the humidity at soil level, dethatch your lawn if there is more than ¾ inch. Avoid heavy application of nitrogen fertilizer, especially in hot weather.

Short-term Controls

Last Resorts

Chemical Controls. Apply flowable sulfur, either in a powdered rock form or as a liquid, every three to five days until the symptoms disappear. You can also use lime sulfur. Use both of these just after it has rained or the lawn has been watered so there is a layer of water on the leaf blades. A fixed-copper fungicide (copper sulfate, copper oxychloride, or cuprous oxide) or Bordeaux mixture also works against leaf spot. Use these compounds only during warm, dry weather. Wear a dust mask and do not spray into the wind or during windy weather.

Dollar Spot

Natural History

Dollar spot, caused by the fungus *Sclerotina homeocarpa*, occurs east of the Mississippi, and attacks particularly annual ryegrass, bluegrass, bentgrass, Bermuda grass, zoysia grass, Saint Augustine grass, and red fescues. It usually appears in early and late summer. Dry, acidic soils low in nitrogen are conducive to the development of dollar spot. The ideal conditions for its growth are warm days and cool nights, with temperatures between 60 and 80 degrees.

Warning Signs

The early damage appears as tan or straw-colored spots, with a bleached center and brown margin, about the size of a half-dollar. As the disease develops, the spots grow until they form irregular patches. If you look closely at infected grass, you can usually see white, fluffy strands of fungi on the turf and light tan, red-bordered spots on the tips of the grass blades.

When to Take Action

Take action as soon as you see damage.

Long-term Controls

Genetic Controls. For severe or persistent problems, plant a resistant or tolerant variety. A number of Kentucky bluegrasses (Adelphi, America, Bonnieblue, Eclipse, the Park, Primo, and Windsor), perennial ryegrasses (Barry, Dasher, Exponent, Linn, Manhattan II, Regal, and Sprinter), and fine fescues (Agram, Checker, Reliant, Shadow, and Tournament) are resistant; check with your local nursery, the Lawn Institute, Cooperative Extension office, or local state university for the appropriate varieties.

Cultural Control. Since dollar spot likes acidic soils, don't use synthetic fertilizers, most of which acidify the soil. Test the soil pH and add lime to the soil if the pH is 6.0 or below. Since dollar spot likes dry soil, water your lawn deeply during early and late summer.

To combat dollar spot disease, regularly spray a seaweed extract on the lawn.

Short-term Controls

Cultural Controls. Since low nitrogen levels make the plant more susceptible, improve plant nutrition with a light watering of organic nitrogen fertilizer; better yet, regularly spray the lawn with a seaweed extract, which provides nitrogen and other nutrients and a beneficial plant hormone. Mow the lawn higher than you normally would once the disease has struck. Mowing slows spread of the disease by removing the infected areas (the tips of the blades).

Last Resorts

Chemical Controls. Apply flowable sulfur, in a powdered rock form or as a liquid, every three to five days until symptoms go away. Wear a dust mask and do not spray into the wind or during windy weather.

Red Thread/Pink Patch

Natural History

Red thread and pink patch are caused by two different but related fungi: *Laetisaria fuciforme* and *Limonomyces roseipellis.* These two diseases are a problem in the Pacific Northwest and the Northeast, although they also occur in the Southeast. They attack a wide range of grasses, particularly annual and Kentucky bluegrasses, Bermuda grass, perennial ryegrass, red fescue, and some bentgrasses. Outbreaks usually occur in the spring and fall in the Pacific Northwest and the Northeast, and in the winter as far south as Mississippi. Both fungi demand high humidity, dew, drizzle, or fog, temperatures in the fifties or sixties, and low soil nitrogen.

Warning Signs

These diseases have a characteristic appearance, as their names suggest. Red thread causes rusty or red threads to come out of the leaf blade tip in infested patches 2 inches to 2 feet in diameter. Pink patch shows up as pink gelatinous masses on the leaves.

When to Take Action

Take action at the first sign of the disease.

Long-term Controls

Genetic Controls. For severe or persistent problems with red thread, plant a resistant or tolerant variety of grass. A number of Kentucky bluegrasses (Adelphi, Bonnieblue, Eclipse, Majestic, Nassau, and Primo), perennial ryegrasses (Acclaim, Belle, Dasher, Exponent, Linn, Regal, and Sprinter), and fine fescues (Argenta, Barfalla, Highlight, Jade, Ruby, or Waldorf) are resistant; check with your local nursery, the Lawn Institute, Cooperative Extension office, or local state university for appropriate varieties.

Cultural Controls. For either red thread or pink patch, water regularly and deeply to minimize high humidity levels near the soil. Test the soil for nitrogen content and add nitrogen, in an organic form or a slow-release formulation, if needed.

Short-term Controls

Physical Controls. Mow the lawn regularly and remove the clippings to remove the infected leaf tips. Chemical controls are rarely if ever needed.

Chart 3-2 lists some commercial pesticide products for controlling lawn diseases. The listed products are a small fraction of those pesticides marketed for this purpose. Products were selected for listing based on CU's judgment that they can be effective when used in the context of an Integrated Pest Management strategy, and that they pose the least risk of harm to humans, pets, or the environment, based on the active ingredients they contain.

Some unlisted products contain exactly the same active ingredients as listed products and may be substituted for them. But many other widely available products are not listed because they contain active ingredients that, in CU's judgment, pose greater potential risks to health or the environment than the ingredients of listed products. In our view, effective pest control does not require taking the risk of using the more toxic pesticides, and we have chosen not to list products that contain them.

WEEDS

In common parlance, a weed is any unwanted plant. However, the definition of what constitutes a weed is variable and has changed over time. Indeed, many of the plants we now consider weeds were once deliberately introduced.

If you think your lawn should look like a golf course, then any plant other than certain bentgrasses is a weed (golf courses are stands of pure bentgrasses). Sometimes such ideas are engendered by companies that have something to gain. Until about 40 years ago, for example, clover was considered a useful and highly prized lawn plant and was as abundant as bluegrasses in many lawn mixes. Indeed, having a clover lawn was a sign of prestige. Clover has many beneficial characteristics: it is soft, it mows well, it smothers many other weeds; and it fixes nitrogen, which it eventually adds to the soil.

In the 1950s, a major lawn seed company started a campaign to persuade the public that clover was a noxious weed that did not belong in a well-maintained "good" lawn. The company was successful, and people now expend a lot of energy and money to get rid of clover. Is it coincidence that the lawn seed company introduced a chemical to kill the clover plant? We think not. Similarly, lawn-care companies promote the notion that a healthy and beautiful lawn consists of one or a few grass species and nothing else.

You need to decide what you consider a weed. Some people welcome violets in their yards, while others consider them weeds that must be removed. As the first step in your weed IPM program, decide what plants you will tolerate, and at what levels. Next, identify those plants that have now become weeds. Then learn a little about those plants' biology and ecology (see References and Resources) and use that knowledge against the plants, to deny them the conditions they need to grow and prosper.

The best defense against weeds is a good offense; a healthy, vigorously growing lawn can fend off most weeds. In general, weeds are a symptom of stress. To make the environment as inhospitable to your selected weeds as possible, you need to know a weed's identity; the conditions that work against one weed can benefit another. Some weeds do well in areas that are shady, too wet, too dry, under- or overfertilized, mowed too closely, or have compacted soil. For example, buttercup and dock indicate too much moisture; lamb's-quarters, disturbed soil and not enough lawn seed; morning glory, too-close mowing and/or disturbed, dry soil of low fertility; purslane, too much fertilizer and not enough grass.

Avoid using herbicides to kill weeds. They do not deal with the underlying causes of stress, and they ultimately favor the weed over the grass.

CHART 3-2
Ammunition Against Common Fungi and Weeds

FUNGI

TYPE	WHERE AND WHEN	CHECK FOR
Brown patch 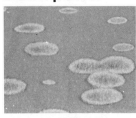	Occurs everywhere in the United States except the Pacific Northwest and is most severe east of the Mississippi. It hits bentgrass and St. Augustine grass hard. Waterlogged soil and thick thatch exacerbate the problem.	A "frog's eye"—a circular spot up to 2 feet wide surrounded by a discolored ring of grass.
Dollar spot 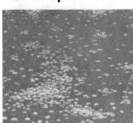	Dollar spot occurs east of the Mississippi and attacks a wide variety of grasses. It usually appears in early and late summer, in dry, acidic soil.	Tan spots with a bleached center and brown margin, about the size of a half-dollar. They can coalesce into patches.
Leaf spot 	Leaf spot, or melting out, is partial to Kentucky bluegrass and fine fescues in the Southeast, Midwest, Southwest, and Plains states. It often appears during cool, moist weather.	Irregular patches of thin grass, and red/brown to purple/black spots on the leaf blades.

WEEDS

TYPE	WHERE AND WHEN	CHECK FOR
Crabgrass	A familiar interloper in much of the United States, it is attacked with a barrage of chemical herbicides. A recent study showed that nonchemical controls work as well, though they take longer.	The grass you see here; decide how much of it you're willing to tolerate.
Dandelion 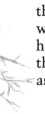	Dandelions bloom primarily in the spring. Lawn owners who have tried hacking off their heads know that a new plant can sprout from the roots.	Those cheery yellow blooms and, later, those fluffy white seedheads. If you have them, you'll know it.

Photographs courtesy of the Crop Science Society of America.

NONCHEMICAL OPTIONS	CHEMICAL OPTIONS
Make sure you're not adding too much nitrogen to the soil. ■ In high humidity, try removing dew by dragging a garden hose over the lawn in the morning. ■ If the soil is chronically waterlogged, consider installing drainage tiles. ■ Consider reseeding with perennial ryegrasses and tall fescues that resist brown patch.	Use flowable sulfur fungicide or a sulfur-based fungicidal soap every three to five days until the symptoms begin to disappear.
Water the lawn deeply and infrequently during early and late summer. ■ Fertilize lightly, especially with a seaweed extract, which will add nitrogen. ■ Don't use synthetic fertilizers, which tend to acidify the soil. ■ Test the pH, and add lime if necessary. ■ Consider reseeding with a Kentucky bluegrass, perennial ryegrass, or fine fescue that resists dollar spot. ■ Mow an infected lawn at a higher height.	Apply flowable sulfur or a sulfur-based fungicidal soap every three to five days until symptoms start to disappear.
Avoid heavy application of nitrogen fertilizer, especially in hot weather. ■ Consider reseeding with a Kentucky bluegrass, perennial ryegrass, tall fescue, or fine fescue that resists leaf spot.	Try flowable sulfur, lime sulfur (only after rain), a sulfur-based fungicidal soap, or a copper fungicide (only during warm, dry weather).
Mow 1 inch higher to decrease the amount of light that gets to young crabgrass. ■ Fertilize only in late fall and early spring. ■ Cover an especially heavy patch with black plastic for 10 days. When you uncover the area, water and fertilize it. ■ For small patches, pull by hand.	Shouldn't be needed, but if all else fails, try an herbicidal soap.
If you dig up a dandelion while it's flowering, removing 5 inches of the taproot, there is an 80-percent chance that you have killed the plant. If you cut it at ground level, you may have to cut three or four times before it dies.	Try an herbicidal soap, but apply it only to the dandelions themselves.

In an ecological sense, weeds are plants that are particularly adapted to disturbed habitats. In nature, places such as riverbanks (where the soil may periodically slough off) or turned-over soil when a tree falls are disturbed habitats. Weedy species, primarily annuals, invade patches of bare soil before other plants get there, and then quickly grow and reproduce. They produce huge numbers of seeds that can sit in the soil for a long time, waiting for the appropriate conditions (disturbed soil, sun, etc.) to germinate. Indeed, it is estimated that the top 6 inches of soil in an average lawn have about 5,000 weed seeds per square foot.

The general strategy for controlling weeds, in addition to knowing something about their biology and ecology, is to use a combination of physical controls (pulling, digging, or cutting), cultural controls (fertilizing, mowing high and watering properly, aerating, or top dressing), and genetic controls (some grass varieties are better at outcompeting weeds than others).

Newly Established Lawns

New lawns often have bad weed problems, in large part because there is a great deal of bare soil that has been disturbed, and herbicides may have been used both before and after emergence of the grass seeds. You can minimize your weed problems when starting a lawn and without using any herbicides.

 When starting a lawn, you can minimize weed problems without using herbicides.

Weed Types

Weeds are grouped according to three characteristics: growth cycle, life cycle, and plant type. Since the basic thrust of an IPM program for weeds is to use their biology and ecology as ammunition, knowledge of each of these characteristics is vital.

Growth Cycle

Growth cycle refers to whether the weed is a cool-season or warm-season plant. As the name implies, cool-season weeds do best in cool conditions. They germinate and/or grow fastest in the spring and fall in the North and during the winter in the South.

Life Cycle

Is the plant an annual, a biennial, or a perennial? Annual plants live for only one year or one season; they germinate, grow, flower, and set seed within one year. They also tend to produce huge numbers of seeds. Biennials live for two years; typically they germinate and grow in the first year and then reproduce in the second year. Perennials live for two or more years and reproduce sexually (they have flowers and set seeds) and/or asexually (they expand and grow vegetatively, like most grasses). Bienni-

TO MINIMIZE WEED PROBLEMS

1. Cultivate or rototill the soil and water the area until the top 2 inches of soil are wet (this is the depth to which most weed seeds germinate). By turning over the soil, you bring weed seeds to the surface.

2. Wait 7 to 10 days for the weeds to germinate, then destroy the weeds either by rototilling the soil again or using a flame weeder. If you rototill, set it to 1 inch, or as shallow as possible, so that you do not bring more weed seeds to the soil surface. If a huge number of weeds germinate the first time, repeat the whole process.

   For particularly weedy soil, use a flame weeder (usually called a flamer) rather than a rototiller for the second cycle. This eliminates the risk of bringing up new weed seeds to germinate. Flamers consist of a backpack or hand-held fuel tank and a long handle with a tiny flame at the end. The flamer heats weeds to 2,000 degrees, causing the sap in the cell to expand and burst the cell. Walk over the area and hold the flamer several inches above the shallow green carpet of newly germinated weeds. Move the flamer steadily and slowly over the weeds; do not sear them. Research has shown that seared weeds, particularly perennials, have a greater chance of resprouting than weeds where the flamer was just passed over them. After flaming the weeds the first time, irrigate your bare soil to encourage any ungerminated weeds to germinate. Wait 7 to 10 days and flame these weeds. *Caution:* Keep the flamer away from clothing and flammable substances, and store the fuel out of reach of children.

3. After one or two cycles, level the soil, add lawn seed, cover the seed with an organic mulch, and water so that the grass seeds germinate.

als and perennials, especially the latter, are often a problem because of their vegetative spread, which can be difficult to control using methods for annuals.

Type

Plant type refers to grass (monocot) or nongrass (dicot). The term *grass plant* includes grasses and related plants. A nongrass plant, also called a

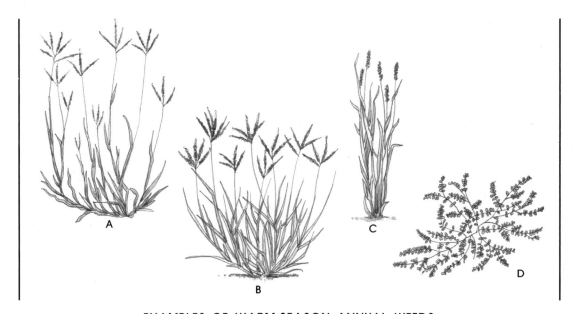

EXAMPLES OF WARM-SEASON ANNUAL WEEDS

(A) crabgrass, (B) goose grass, (C) foxtail, (D) prostrate spurge

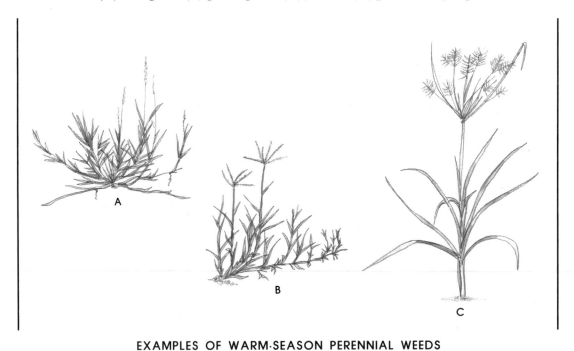

EXAMPLES OF WARM-SEASON PERENNIAL WEEDS

(A) zoysia grass, (B) Bermuda grass, (C) yellow nutsedge

broad-leaf weed, differs physiologically from a grass.

Since the above three categories are independent, all combinations are possible—there are cool-season annual grassy weeds, cool-season annual broad-leaf weeds, cool-season perennials, and so on.

Control

The general approach to weed control is to make environmental conditions inhospitable for the weed and hospitable to the lawn grass, and then concentrate on physical and cultural controls: Suppress germination of weed seeds, minimize their access to water and fertilizer when they are growing most vigorously, and prevent their reproduction.

Cool-Season Weeds

Cool-season weeds usually plague vigorously growing northern lawns and invade winter lawns in the South. Minimize water and fertilizer in spring and fall in the North, and during winter in the South.

To combat weeds, suppress germination of weed seeds, minimize their access to water and fertilizer when they are growing most vigorously, and prevent their reproduction.

Annuals

Cool-season annual grasses include annual bluegrass and downy bromegrass. Cool-season annual broad-leaf weeds include mallow, violets, and speedwell.

Since annual plants need lots of sunlight to germinate, they tend to invade thin lawns with open spaces. To make your lawn thick with no open spaces, reseed or overseed with a mixture of grass varieties suitable to your climate and use pattern. Also, shade the soil as much as possible during germination times. In the North, mow high, which increases shade at the soil level, during the spring and fall. In the South, mow high during the late summer and then overseed your lawn in the fall with a cool-season turfgrass, such as Kentucky bluegrass or perennial ryegrass, which usually outcompetes any cool-season weed.

Since annuals usually produce huge numbers of seeds, you must stop these weeds before they set seed. Remove them by pulling, cutting, or hoeing before flowers and seeds appear. If you don't have many of these weeds, you can probably remove them by hand. If an open space is left behind after removing a clump of grass, put some sod in the space to prevent another annual weed from invading.

If you have a lot of weeds, mow before flowering. You can use a flamer for cool-season annual broad-leaf weeds, but be sure not to sear the plant; just pass the flame lightly over the top of it. Keep the flame away from adjacent grass as much as possible.

Perennials

Cool-season perennial broad-leaf weeds include plantain, dandelion (see page 59), ground ivy, white clover, and Canada thistle.

A number of cool-season perennial broad-leaf weeds have a low, creeping growth habit so they stay close to the ground and root at various places along their runners. This makes them difficult to reach, either by pulling or by mowing. The best way to attack them is to rake them up vigorously in the spring and discard them (do not leave them on the lawn, as they can reroot themselves). Then mow at the low end of the recommended height for your grass and, again, discard the weed clippings. Fertilize your lawn after the first frost, so that it starts to grow early the next spring. You can use a flamer for cool-season perennial broad-leaf weeds, but be sure not to sear them; just pass the flame lightly over the top of each plant. (Resprout is more common when the plant is seared.) You may need to treat these weeds a couple of times with the flamer if they resprout.

Cool-season perennial grass weeds include the bentgrasses, tall fescue, timothy, wild garlic, and wild onion.

They can be very hard to control in northern lawns because they are so similar to the turfgrasses themselves. Indeed, turfgrass species are often weeds in lawns composed of other turfgrass species. Thus, tall fescue is a weed when it occurs in a lawn of creeping bentgrass, and vice versa. Ideally, scout your lawn in the early spring and fall. Pull or dig up weeds before they set seed. To prevent a cool-season annual weed from invading, put sod in any bare patches created by removing the weed. Mow high in the spring and fall to minimize germination. Do not water your lawn during these times unless absolutely necessary.

Warm-Season Weeds

Warm-season weeds can be severe in dormant heat- and water-stressed summer lawns in the North. In the South, warm-season annual grasses can be a problem; they are favored by close mowing, summer fertilization, and frequent light watering—all elements of a pampered lawn.

Annuals

Warm-season annual broad-leaf weeds include common purslane, prostrate spurge, and puncture vine. Warm-season annual grass weeds include crabgrass, barnyard grass, foxtails, and goose grass.

For warm-season annual weeds, mow at the recommended height and remove the clippings if any flowers or seed heads are present. Avoid summer fertilization, water only when necessary during the summer, and aerate the soil to reduce compaction. You can use a flamer, but be sure not to sear the plant; just pass the flame lightly over it.

Warm-season annual grasses thrive in lawns with thin turf caused by excess traffic or disease. In this case, you may need to reseed the lawn with a more wear-tolerant variety or cure the disease.

Perennials

Warm-season perennial grasses include Bermuda grass, zoysia grass, and yellow nutsedge.

Of the warm-season perennial weeds, only grasses present a problem. Popular southern turfgrasses such as Bermuda grass and zoysia grass are considered weeds in northern lawns. They occur in northern lawns stressed by heat and drought, and in pampered southern lawns. Since some of these grasses often grow in clumps, dig out the clumps and reseed the area or install sod. Mow high during the summer to inhibit germination, fertilize only in the spring and fall, and water deeply only when needed.

Dandelion

Natural History

The dandelion is a cool-season perennial broad-leaf weed that has bright yellow flowers and a white puffball seed head. It does best in disturbed habitats and slightly acidic soils as well as in lawns with low soil fertility, occasional drought, or thin grass. Dandelions bloom primarily in the spring, although some populations flower throughout the entire growing season. They reproduce sexually via seeds and asexually by a deep taproot, which can regenerate a new plant if you remove only the head.

Warning Signs

Look for the dandelion itself.

When to Take Action

Take action when there are more dandelions than you can tolerate.

Long-term Controls

Cultural Controls. Since dandelions prefer slightly acid soils, check your soil pH and add lime if the pH is 6.5 or lower. If the grass is thin in areas prone to dandelions, add grass seed. Water more frequently if the soil is perennially dry. Check soil fertility in dandelion-prone areas by sending a soil sample to a local lab (ask your county Cooperative Extension office or local gardening center for the location). Since dandelions prefer low-mown areas, mow your grass as high as possible.

Short-term Controls

Physical Controls. Dandelions can be effectively controlled without chemicals, but it takes elbow grease. If you dig up the plant with either a short- or long-handled dandelion knife when it is flowering, and remove 4 to 5 inches of the taproot, there is an 80-percent chance that the plant will not resprout. The second technique is to cut off all the aboveground portion of the plant. The plant will resprout from the roots; again cut it off at ground level. After doing this five to six times throughout the season, the plant will have exhausted all its reserves and will not resprout. Both techniques rid a lawn of dandelions within a year.

Last Resorts

Chemical Controls. The least toxic chemical is an herbicidal soap such as Safer's Sharpshooter. Spray it only on the dandelion itself. This destroys only the aboveground portion of the plant, so you have to reapply it a few times throughout the season. An herbicide such as glyphosate (Roundup) kills the dandelion with only one application. Use a small paintbrush to dab some onto the plant leaves or squirt it only onto the leaves of the dandelions themselves. *Caution:* Wear gloves and protective clothing to avoid skin contact.

Crabgrass

Natural History

Crabgrass is a warm-season annual grass that has short, pointed, light-green, hairy leaves and a creeping growth habit. Although it grows throughout the United States, it causes special problems in lawns seeded with cool-season grasses. Indeed, in some parts of the South, it is deliberately grown as a turfgrass. Crabgrass seeds begin germinating in the spring when soil temperature ex-

ceeds 55 degrees. It grows most vigorously during the summer and flowers from late summer throughout the fall. Flowers are borne on the main stem, with three to nine spikes per flower head. It does best in thin turf that is mowed low. Persistent crabgrass indicates compacted soil, low fertility, drought, thin grass, and hot spots.

Warning Signs

Look for the crabgrass itself, which usually stands out distinctly from the cool-season grasses.

When to Take Action

Take action as soon as you notice crabgrass, or when it exceeds your personal threshold of tolerance.

Long-term Controls

Cultural Controls. Nonchemical control works just as well as herbicides, but takes longer. Shade and fertilization schedules can dramatically reduce crabgrass infestations. Increasing shade is the key, because of the creeping growth habit of crabgrass, which makes it easy to shade. Mowing higher increases shade and thus decreases the amount of light available for photosynthesis. If this is all you do, within five years the crabgrass will be virtually gone. If you increase mowing height and use fertilizer, it takes two to three years. If the lawn is thin, reseed or overseed it to make the grass denser, thereby increasing shade at the ground level. Reseed any bare spots.

Higher mowing and a proper fertilizing schedule can virtually eliminate crabgrass from a lawn in two to three years.

Since crabgrass grows most vigorously during the summer, when the cool-season grasses are dormant, fertilizing during summer benefits the crabgrass at the expense of the lawn. If you fertilize only in the late fall and early spring, you'll reduce crabgrass by as much as 75 percent the next year. Water deeply and infrequently during the spring to encourage growth of turfgrass roots.

Finally, have your soil checked at a laboratory (ask your local nursery or Cooperative Extension Service for the location) and correct any deficiencies.

Short-term Controls

Physical Controls. If the crabgrass is not extensive, you can simply pull it up. For a moderate infestation, cover the infected area with black plastic for 10 days. The crabgrass will die. The bluegrass will yellow; give it some water and fertilizer after you take off the plastic.

CHART 3-3
Products to Control Lawn Diseases

The products listed below are a fraction of the pesticides marketed to control lawn diseases. These products were selected for listing based on CU's judgment that they can be effective when used in the context of an Integrated Pest Management strategy, and that they pose the least risk to humans, pets, or the environment, based on the active ingredients they contain.

Some products not listed here contain one or more of the same active ingredients as these products and may be substituted for them. But many widely available products are not listed because they contain active ingredients that, in CU's judgment, pose greater potential risks to health or the environment than the ingredients of products listed. In our view, effective pest control does not require use of more toxic pesticides, and we have chosen not to list products that contain them. Products are listed in alphabetical order.

Recommended products listed below contain the least hazardous active ingredients, including one or more of the following: Bordeaux mixture (BM), copper (CU), lime sulfur (LS), microorganisms (MO), or sulfur (SU).

Diseases listed below include brown patch (BP), leaf spot (LF), dollar spot (DS), and red thread (RT).

BRAND NAME	ACTIVE INGREDIENT(S)	DISEASE
Acme Bordeaux mixture	BM	LF
Acme lime sulfur spray	LS	LF
Acme Wettable Dusting Sulfur	SU	BP, DS, LF
Dexol Bordeaux Mixture	BM	LF
Earl May Lime Sulfur Spray	LS	LF
Gro-Well Bordeaux Mixture	BM	LF
Gro-Well Wettable Sulfur	SU	BP, DS, LF
Ortho Dormant Disease Control Lime Sulfur Spray	LS	LF
Ortho FLOTOX Garden Sulfur	SU	BP, DS, LF
Ringer Lawn Restore	MO	BP, DS, RT
SA 50 Liquid Copper Fungicide	CU	LF
That Liquid Sulfur	SU	BP, DS, LF
Top Cop Liquid Copper	CU	LF
Top Cop with Sulfur	CU, SU	LF

CHART 3-4
Products to Control Lawn Pests

The products listed below are a fraction of those pesticides marketed to control lawn pests. These products were selected for listing based on CU's judgment that they can be effective when used in the context of an Integrated Pest Management strategy, and that they pose the least risk to humans, pets, or the environment, based on the active ingredients they contain.

Some products not listed here contain one or more of the same active ingredients as these products and may be substituted for them. But many widely available products are not listed because they contain active ingredients that, in CU's judgment, pose greater potential risks to health or the environment than the ingredients of products listed. In our view, effective pest control does not require use of more toxic pesticides, and we have chosen not to list products that contain them. Products are listed below in alphabetical order.

Recommended products listed below contain the least hazardous active ingredients, including one or more of the following: *Bacillus thuringiensis kurstaki* (BTK), fatty acids (FA), neem (or azadirachtin [AZ]), nematodes (*Steinernema feltiae*, or SF), milky spore disease (MS), sulfur (SU), pyrethrins (PY), or pyrethroids, such as allethrin (AL), phenothrin (PH), resmethrin (RS), or tetramethrin (TM).

Other products listed below also contain synergists, such as MGK 264 (M2) or piperonyl butoxide (PB), the insecticide rotenone (RO), and/or petroleum distillates (PD), whose chronic health effects are not completely known.

Pests controlled listed below are the following: billbug adults (BA), larvae (BL); chinch bug (CB); Japanese beetle adults (JBA), larvae (JBL); sod webworm (SW); white grub adults (WGA), larvae (WG).

BRAND NAME	ACTIVE INGREDIENT(S)	PESTS
Recommended		
BioLogic SCANMASK	SF	BL, JBL, SW, WG
Bonide Dipel .86% W.P.	BTK	SW
Dexol Worms Away	BTK	SW
Dipel Worm Killer	BTK	SW
Fairfax Biological Laboratories Doom	MS	JBL, WG
Fairfax Biological Laboratories Japidemic	MS	JBL, WG
Gro-Well Thuricide-HPC Insect Control	BTK	SW
Hydro-Gardens Guardian	SC	BL, JBL, SW, WG
Hydro-Gardens Lawn Patrol	HH	BL, JBL, SW, WG
Ortho Biosafe Soil Insect Control	SF	BL, JBL, SW, WG
Ortho Insecticidal Soap	FA	CB, SW

Products to Control Lawn Pests (*continued*)

BRAND NAME	ACTIVE INGREDIENT(S)	PESTS
Perma-Guard Plant & Garden Insecticide	DE	SW
Raid Multi-Bug Killer D39	AL, RS	BA, CB, JBA, SW, WGA
Real-Kill House & Garden Bug Killer	PH, TM	BA, CB, JBA, SW, WGA
Safer BioNeem	AZ	BA, CB, JBA, SW, WGA
Safer Bt Caterpillar Attack	BTK	SW
Safer Caterpillar Attack	BTK	SW
Safer Fruit & Vegetable Insect Attack	FA	CB, SW
Safer Tree & Shrub Insect Attack	FA	CB, SW
Safer Vegetable Insect Attack	BTK	SW
Safer Yard & Garden Insect Attack	FA, PY	CB, SW
SA 50 Thuricide-HPC	BT	CB, SW
Spectracide Lawn & Garden Insect Control 1	PM	BA, CB, JBA, SW, WGA

Other

Ace Hardware Flower & Vegetable	PB, PY	BA, CB, JBA, SW, WGA
Ace Hardware House and Garden Bug Killer II	PB, PY	BA, CB, JBA, SW, WGA
Acme Garden Guard	RO	BA, CB, JBA, SW, WGA
Gro-Well 1% Rotenone Dust	RO	BA, CB, JBA, SW, WGA
Gro-Well Organic Insecticide	PB, PD, PY, RO	BA, CB, JBA, SW, WGA
Gro-Well Rose & Flower Insect Killer	PB, PD, PY	BA, CB, JBA, SW, WGA
Gro-Well Vegetable & Tomato Insect Killer	PB, PD, PY	BA, CB, JBA, SW, WGA

Brand Name	Active Ingredient(s)	Pests
Gro-Well Vegetable Tomato Dust	RO, CU	BA, CB, JBA, SW, WGA
Meijer's Rose & Flower Insect Killer	PB, PY	BA, CB, JBA, SW, WGA
Ortho Home & Garden Insect Killer Formula II	PB, PY, TM	BA, CB, JBA, SW, WGA
Ortho Rose & Flower Insect Killer	PB, PD, PY	BA, CB, JBA, WGA
Ortho Rose & Flower Insect Spray	PB, PY	JBA, WGA, BA, CB
Ortho Tomato & Vegetable Insect Killer	PB, PY	BA, CB, JBA, SW, WGA
Raid House & Garden Formula 8	AL, PD, RS	BA, CB, JBA, SW, WGA
Raid House & Garden Formula II	PB, PY, TM	BA, CB, JBA, SW, WGA
Spectracide Rose & Garden Insect Killer	PB, PY	BA, CB, JBA, SW, WGA

4

GARDEN
PEST
CONTROL

The basic IPM approach to garden pest control is the same as for lawns, except that in the garden you can give more attention to each plant. The garden, whether you are growing flowers or vegetables, is also a more diverse ecosystem, which helps both to control pests and to make their control a bit more complex. A healthy garden needs good soil, proper watering, enough light, appropriate levels of nutrients, and so on. Meeting these needs creates healthy plants and is essential for minimizing pest problems, since healthy plants can fend off pests more easily than unhealthy or stressed plants.

In the garden, IPM strategy has five components:

1. Monitor the garden.
2. Identify the problem.
3. Establish thresholds.
4. Choose long-term controls.
5. Consider short-term or emergency controls.

THE IPM STRATEGY

Warning Signs

The first vital element in IPM garden pest control is monitoring.

Monitoring includes checking your plants regularly for signs of damage or for the pests themselves. Make it a habit at least once a week to inspect your plants. Do you see any overt signs of damage, such as leaves or flowers with holes, chewed sections, spots, or a greasy or wilted appearance? Are there insects or fungi on them? Be inquisitive. Call your county Cooperative Extension agent, local botanical garden, or state agricultural college to ask what pests attack the kinds of plants you're growing and how to recognize the damage they do.

When monitoring your garden, try to quantify pest numbers and damage so you'll be able to detect changes from week to week. How many leaves per plant are damaged, or infested? What proportion of the plants are damaged or infested? How much damage, or how many pests, occur per leaf or per plant? If possible, also note the plant's growth stage and any natural enemies of the pest that are present.

Problem Identification

Once you see damage, you need to identify the source. Sometimes the cause is obvious—you see insects on the plant. If you don't see any pests, the damage could be caused by either an organism or an environmental factor such as soil pH or mineral levels; too much or not enough water, light, or fertilizer; or physical damage. For instance, yellowing of the youngest leaves could be caused by a lack of iron or manganese, while yellowing of all leaves could be caused by a lack of nutrients, too much light, or high temperatures.

If the problem is a pest, identify the organism and learn about the life cycles of the host plant and the pest, and their natural control factors. At what point in its life is the pest most susceptible to attack? Controls can be environmental (temperature, climate, and nutritional factors, including the nutritional quality of the host plant) and natural enemies (predators, parasitoids, and diseases).

For example, mites like hot, dry conditions. You can modify the area around the plant to make it cooler and wetter. Most soft-bodied sucking insects—particularly aphids, mealybugs, whiteflies, and scales—respond dramatically to surges in plant nitrogen levels and have many natural enemies. You can decrease the nitrogen content of your plants by reducing fertilizer use. Powdery mildew likes warm humid weather but a film of water inhibits it. Simply spraying the plant with water during warm humid weather goes a long way toward preventing the disease.

Biological controls consist of utilizing natural enemies. The habitat can be modified to increase the numbers and/or effectiveness of natural enemies. For example, you can plant certain wildflowers that act as a food source for many adult predators and parasitoids. Or you can simply spray an artificially prepared food on your plants that also serves as a food source for certain natural enemies. Finally, you can buy natural enemies and release them into your garden.

When to Take Action

For any pest problem, you need to decide when control action is required. Most garden plants can stand some pests, so the mere presence of a pest is no reason to panic. Many plants are most sensitive just after germination and during flowering, when the plant is putting the bulk of its energy into producing fruit and/or seeds. How much damage, or how many pests, your garden can tolerate depends on the kind of plant, the kind of pest, and the kind of damage.

For many insect pests, action thresholds have already been determined through research; for others they have not, so you'll need to make your own judgments. To find out if an action threshold exists, call your Cooperative Extension Service, botanical garden, or the nearest land-grant university and speak to someone knowledgeable about IPM. Otherwise you'll have to do some experimenting on your own. Sometimes the action thresholds are arbitrary or highly personal. This is especially true for weeds. Unless they are abundant, weeds in a garden usually do not cause a problem. But different gardeners have different tolerance levels. Or take whiteflies. Many plants can withstand a relatively large number of adults, but some people are disturbed when they see a cloud of whiteflies (which looks like flying dandruff) take off from a plant. In general, try to live with some pests until they begin to cause significant damage. On the other hand, it makes sense to remove certain pests, such as the tomato hornworm, whenever you see one, since even one large caterpillar can eventually do a lot of damage.

Long-term Controls

You need some knowledge of plant and pest biology to develop the primary or long-term controls to keep pest density below the damage threshold. There are three basic classes of long-term controls: *genetic*, *biological*, and *cultural*.

Genetic Controls. Plant breeders try to identify genetic traits of garden plants that confer resistance to or tolerance of pest attack and incorporate them into plant varieties. Thus you can buy tomatoes that are resistant to root-knot nematodes or several fungi. Bean varieties with hairy stems and leaves resist aphids, because the hooked hairs rip the aphids climbing on them. Shrub roses and garden roses exhibit greater resistance to pests than do the hybrid teas, floribunda, and grandiflora roses. Such information sometimes appears on the label; other times you have to ask a knowledgeable salesperson.

If you have problems with a particular pest (especially diseases) attacking a certain plant, call your local garden store, Cooperative Extension agent, or the horticultural department of your state's land-grant university and ask about varieties of that plant that are resistant to or tolerant of the particular pest.

Biological Controls. Biological controls use natural enemies (predators, parasitoids, and diseases) to control pests. Indeed, natural enemies are often the major factor for controlling insect pests. There are two main strategies: You can create conditions conducive to natural enemies already present in your garden, or you can buy natural enemies and release them when needed.

The range of natural enemies you can purchase is extensive; there are insects, mites, nematodes, fungi, bacteria, and viruses (see References and Resources). Whatever you choose, you need to become familiar with the local natural enemies of the pests in your garden. Vertebrate natural enemies, such as birds, toads, frogs, and snakes, are usually recognizable; the invertebrate natural enemies of garden pests are less so. To identify beneficial insects, spiders, and their relatives, use a field guide or other book (see References and Resources). Increase the numbers of beneficial insects by planting a variety of plants so

that some are flowering at any point in the season and/or by spraying commercially prepared food for natural enemies. Beneficial fungi, bacteria, and viruses are often very difficult to identify. However, any disease organism that attacks a pest usually can be considered beneficial. While monitoring for pests, also keep an eye out for natural enemies. The more abundant the natural enemies, the higher your action threshold can be.

Cultural Controls. Cultural controls involve modifying the environment to reduce a pest's numbers, make its natural enemies more effective, or decrease its breeding, feeding, or shelter habitats. These controls include habitat diversification, trap cropping, crop rotation, mulching, waste removal, and using solar energy to kill harmful organisms and stimulate beneficial ones (a process called *soil solarization*).

• *Diversify the habitat* inside your garden by planting a number of different crops or flowers, intermingling rows of different crops or flowers, or allowing some weeds to grow. Outside the garden proper, diversify the habitat by growing some wildflowers, allowing some weeds to grow, building a hedgerow, or planting trees and shrubs. Habitat diversification makes it harder for pests to find their host plants and provides living space and alternate food sources for various natural enemies, from birds to insects.

Diversify the habitat inside your garden by planting different crops or flowers, using mulch to prevent weeds from germinating, and cleaning up plant trash at the end of the growing season.

• *Use trap cropping* to lure pests away from your crop by planting something that will attract them elsewhere. For instance, eggplant is more attractive to Colorado potato beetles than are potatoes. If you plant a few eggplants in your potato patch, the beetles will attack the eggplants and leave your potatoes alone. Simply pull the beetle adults, larvae, and eggs off the eggplants and throw them in soapy water.

• *Employ crop rotation* by rotating host and nonhost plants so as to reduce buildup of pests, particularly those that overwinter or live in the soil. Growing tomatoes on the same plot of land year after year often leads to problems with nematodes, especially root-knot nematodes. Corn grown on the same plot year after year often develops problems with corn rootworm. Move plants around in your garden from year to year to prevent the buildup of soil-inhabiting pests.

• *Use mulch*, either organic or inorganic, to prevent weeds from germinating, prevent transmission of various fungal diseases, and interrupt the life cycle of certain insect pests, such as thrips. Organic mulch, coupled with organic fertilizer, helps create a healthy soil that increases populations of beneficial organisms and leads to vigorous plants. In some circumstances organic mulches can encourage pests, especially slugs and mice.

• *Clean up weedy debris* and plant trash at the end of growing season and dispose of it to reduce populations of the numerous diseases and insect pests that normally overwinter in such debris. However, don't leave the soil unprotected since this may cause soil erosion. Cover the soil with an organic or inorganic mulch or covering.

• *Solarize your soil* to control serious pests. Soil solarization produces high levels of heat and humidity that effectively pasteurize the soil, destroying harmful fungi, bacteria, some nematodes, insect larvae, and weed seeds. Paradoxically, solarization enhances the soil's ability to grow healthy plants, in part by enhancing certain beneficial fungi.

Solarization is easy. Loosen or turn over the top 4 inches of soil in the plot, level it, soak it heavily with water, and let it sit overnight. The next day, cover the whole plot with a thin clear plastic tarp and seal the edges of the tarp, preferably by burying them in the soil. The plastic acts like a greenhouse; heat builds up beneath it and "cooks" the soil. Do not pierce the plastic, and remove any water puddles after rains, to enhance the greenhouse effect. Depending on the weather, the bed may need to be covered for four to six weeks. Solarize your soil during the sunniest, hottest time of year—July and August. You can cut the solarization time down to a couple of weeks if you use two layers of plastic—one pulled tautly over the soil, the other looser. Double layers increase the soil temperature by 5 to 7 degrees. After solarization, you can plant anything. When you do plant, be careful not to turn over too much soil; only the weed seeds in the top 2 to 4 inches of soil are killed by solarization.

Short-term Controls

Although the long-term controls will generally keep pest numbers below your action threshold, occasional outbreaks may require immediate action. This can include physical, cultural, biological, and chemical tactics.

Physical Controls. Some methods are as simple as hosing down plants with a strong jet of water to knock off aphids. Handpicking caterpillars off plants, hand-pulling weeds, or using a small vacuum on a cool morning to remove adult whiteflies or beetles can often alleviate an outbreak.

Biological Controls. To check many pests you can purchase natural enemies and release them in your garden. Predatory mites control pest mites and fire ants; green lacewings, ladybugs, spined soldier bugs, and preda-

tory midges fight aphids, mealybugs, whiteflies, and caterpillars; parasitic wasps help control whiteflies and caterpillars; nematodes attack flea beetles, moths, fire ants, white grubs, and root maggot flies; predatory snails control garden snails. If you have a pond, you can buy fish to eat the mosquito larvae.

You can also spray microbial pesticides on your garden to control an infestation. The most widely used is the bacteria *Bacillus thuringiensis* (or *Bt*), which attacks a wide range of caterpillars. There are now strains of *Bt* that are effective against mosquitoes (sold commercially as *Bti*), and against Mexican bean beetles and Colorado potato beetles (M-One). Other bacteria work against crown gall disease, fungus gnats, various beetle grubs, or grasshoppers. Certain fungi control aphids (Vertalec), whiteflies (Mycotal), milkweeds (Devine), and northern joint vetch (Collego); there is a virus for bollworms or budworms (Elcar). *Caution:* Wear gloves and a dust mask when applying bacteria and fungi.

Last Resorts

Finally, there are the chemical pesticides, the most widely used short-term control measures. Most of these treatments are unnecessary, and they can exacerbate problems. However, if you must use a pesticide, try to choose a chemical that is specific to the target pest. If you choose a broader-spectrum pesticide, choose one that has the least negative impact on natural enemies, nontarget organisms, and the surrounding environment, and is the least persistent in the environment. (Products listed in later sections of this chapter meet any, if not all, of these criteria. Minimize your use of pesticides to decrease the likelihood that the pests will become tolerant of them.

Among the safer groups of chemicals are the fatty acid soaps, some botanical insecticides, insect growth regulators, pheromones, and sulfur. Sulfur compounds are successful against many fungal diseases, such as the various rusts, powdery mildew, and Fusarium. Fatty acid soaps are effective against many weeds and insects, particularly aphids, whiteflies, mealybugs, mites, and many caterpillars. Wear a dust mask when applying these substances, and do not spray on windy days.

Botanical pesticides, which come from plants, tend to break down more rapidly and are often (but not always)

less toxic than the synthetic organic pesticides. Particularly useful botanicals include pyrethrum, pyrethrins, and their synthetic analogs, and extracts from the neem tree (Bio-Neem). Do not use pyrethroids near water; they are highly toxic to fish and some aquatic invertebrates. *Caution:* Some people are allergic to the pulverized flowers that are the source of pyrethrum. Be sure to wear a dust mask or respirator if you are allergic to pollen and are spraying pyrethrum, or choose a different chemical.

The insect growth regulators, especially methoprene, hydroprene, and fenoxycarb, prevent reproduction in a range of insects, including cockroaches, fleas, mosquitoes, some ants, and leaf miners. Pheromones and other attractant chemicals lure gypsy moths, yellowjackets, cucumber beetles, pink bollworms, and apple, blueberry, and cherry fruit maggots to traps that contain chemicals to kill them.

WEEMS

Natural History

A garden is an ideal habitat for weeds. Annuals are particularly adept at invading patches of bare soil, then growing and reproducing quickly before other plants have a chance to take root. Common weeds produce huge numbers of seeds that can lie in the soil for long periods, waiting for the appropriate conditions to germi-

nate. An average lawn has about 5,000 weed seeds per square foot.

The general strategy for controlling garden weeds relies on physical controls such as pulling, digging, cutting, ground covers, or using a flamer and cultural controls such as mulching. IPM minimizes the need for chemical herbicides, or uses them only as a last resort. (For basic biology and ecology of weeds and weed types, see chapter 3.)

Make environmental conditions inhospitable for the weeds, then concentrate on physical and cultural controls. In general, suppress germination of weed seeds, remove or kill weeds in the early stages of growth, and prevent their reproduction. For those species that thrive on particular conditions such as compacted or waterlogged soils, correct these conditions.

When to Take Action

The necessary first step in developing an IPM system for weeds in your garden is deciding what plants you can tolerate, at what levels, and what plants you cannot tolerate. Remember that "weed" is not a biological category but a human one; one person's weed is someone else's wildflower. Many weeds also serve as alternative food for some insect pests or as habitats for insect pests' natural enemies. A more diverse habitat is more attractive to pests' natural enemies.

Once you've decided on your threshold, make the environment inhospitable to your selected weeds. You need to know a weed's identity, since the conditions that work

---

"Weed" is not a biological category but a human one; one person's weed is another person's wildflower.

---

against one weed can actually benefit another. Guides to weeds and wildflowers are found in the References and Resources section.

Long-term Controls

Two preventive measures, soil solarization and mulch, minimize the need to weed. If you can allow your garden bed to remain fallow for most of a season, solarize the soil to destroy weed seeds in the top 2 to 4 inches. If the garden is already planted, use a mulch.

Mulch. Organic mulch—particularly compost, rice or bran hulls, sawdust, tan bark, wood chips, and straw—helps build up soil health. In addition, mulch covers the soil and blocks out the light that weeds need to germinate. Apply mulch in the spring before the weeds sprout. Use 3 or 4 inches of mulch so that weed seeds that do germinate will use up their food reserves and die before reaching the top of the mulch.

Other mulch, mostly inorganic—newspapers, plastic sheets or pellets, gravel, boards, and new geotextile weed mats—covers the soil but obviously does not add nutrients. Black plastic sheets increase soil temperature and can help extend the growing season in northern climates. Mulch traps excessive moisture near the soil surface, which can exacerbate some fungal or bacterial diseases, certain insects, and slugs. Many mulches, particularly organic ones, can provide enough ground cover to attract rodents.

Geotextile weed mats, which consist of a synthetic woven fabric, cover (and shade) the soil but allow it to breathe, preventing moisture build-up. These are ideal for ornamental flower beds; try one if you have slugs or diseases.

Some perennial weeds, particularly those with long taproots or large storage organs, may have enough food reserves underground to push up through most mulch. Remove these weeds by hand.

Cover Crops. Instead of using mulch, you can eliminate bare soil in your garden with a cover crop. Clover is a good cover crop because it's a "living mulch" that also adds nutrients, especially nitrogen, to the soil. For landscaping areas, there are many low-maintenance plants such as spurges, bugleweed, English ivy, stonecrop, lily of the valley, and periwinkle. Your local soil conservation service office or county Cooperative Extension office can suggest plants suitable for your area. You can also call Native Plants, Inc. (see References and Resources), for suggestions of appropriate low-maintenance ornamentals.

Short-term Controls

Besides preventing weed emergence, you can also kill weeds in the early growth stages just after germination. Annual weeds are particularly susceptible at this time. A couple of weeks before planting your garden, turn over the soil with a rototiller or hoe and water it. Weed seeds brought to the surface will germinate. Wait 7 to 10 days, then cultivate the soil as shallowly as possible to kill these weed seedlings.

Instead of cultivation, you can use a flamer, which is basically a propane torch modified to kill plants. It's most effective if you lightly pass the flame over the weeds, rather than incinerating the plants. (Burning the leaves can actually stimulate the plant to grow back faster.) Flamers are particularly useful against young broad-leaf weeds. They are less effective against grasses. *Caution:* Keep flamers away from clothing and flammable materials. Keep the fuel out of children's reach. (See References and Resources for buying information.)

Once grasses are past the seedling stage, cultural and physical controls are the best bet. If your garden is small, hand weeding, cultivation, or mowing should control the weeds. In larger gardens or landscaped yards, where those measures may be too laborious, you may want to rethink your attitude toward these "weeds." Can you possibly live with them? Consider the benefits they provide— food and shelter for natural enemies, esthetics, and so on. If the weeds must go, and if cultural or physical controls such as mulching don't work, identify the weed and learn as much as you can about its biology.

Last Resorts

Use a chemical herbicide. The least toxic to other organisms and the environment is herbicidal soap. Wear a mask to prevent inhalation. If something stronger is needed, try glyphosate, but apply only to the surface of the weed using a small paintbrush or squirt bottle. Wear gloves.

Perennial Weeds

Of the various weed types, perennials are the hardest to control. These plants live for several years and usually spread or reproduce by vegetative means. Perennial grasses, such as crabgrass, can cover large areas. Perennial broad-leaf weeds cover less area, but can have large root systems that store food. In both cases, eradication can be difficult. Removing the aboveground portions of older plants, by mowing or cultivating, or using most herbicides, causes the plant to resprout. Indeed, cultivation often results in more plants sprouting once the roots are broken up into pieces. If the plants are widespread or the individual patches are large, look for an indication of an underlying condition that favors the weed. If you find one, change it. For example, large patches of yellow nutsedge usually indicate waterlogged and/or compacted soil.

Cultivation

With smaller weed patches or more isolated plants, the best strategy is usually to make the plant use up its stored food. You can do this by repeatedly killing or removing the aboveground portion. One way is to cultivate (i.e., turn over the soil) in the area, leave it fallow, then cultivate again when the weeds return. Re-

---

OUTDOOR CAUTIONS: Pesticides

1. Wear protective clothing if necessary, and do not spray pesticides during windy weather.
2. Refrain from applying pesticides to a lawn or garden if there is a creek, river, lake, or pond nearby.
3. Do not smoke while applying or spraying pesticides.
4. Refrain from using granular pesticides on the lawn where birds can consume them.
5. Never use poison baits unless they are in a covered bait station.

---

peat until the weeds stop returning. For broad-leaf perennials, use a flamer and repeat as often as needed. With either technique, do not allow the plant to grow back or it will make more food and store it; this usually means repeating the treatment every couple of weeks, three or four times in all.

Herbicides

In an IPM program, herbicides—even the least toxic ones—are a last resort. There is an herbicidal soap, composed of fatty acids that are also found in the soil, that works against a range of plants and has few known adverse environmental side effects. The soap kills only the aboveground vegetation, so perennials may continue to resprout. If this happens, repeat applications until the plant dies. For aggressive perennial weeds that herbicidal soap does not control, use glyphosate (Roundup), a systemic herbicide. *Caution:* Apply glyphosate only with a paintbrush or small squirt bottle because it significantly damages virtually any plant it touches. Wear protective gloves and a dust mask when applying.

To minimize the environmental impact (as well as the cost) of chemical herbicides, apply them only at the most appropriate time during the season, and only to the weed you want to kill. For most weeds, but particularly annuals, apply the herbicide before the plant flowers. Use a wick applicator, which looks like a small mop with a hollow handle, to put the herbicide exactly where it is needed and nowhere else.

Although we generally advise avoiding chemical herbicides, there are some exceptions. Poison ivy and poison oak are prime candidates for chemical control around the garden, particularly if you are highly susceptible to skin irritations from them. For nonchemical control, you can usually pull up small plants by hand. Be sure to wear gloves. With larger plants, use an herbicide containing glyphosate. Since it is a potent nonselective herbicide, use it sparingly and apply it only to the leaves of the target plant. Use a spray bottle, brush, or wick applicator to apply. Since it is a systemic plant poison, you do not need to apply it to all leaves in order to kill the entire vine. Apply it first to the youngest leaves, which are still growing, and then to slightly older leaves, to maximize the amount of poison that is transferred to the roots. Glyphosate is transported from leaves to roots, where it kills the plant. Younger leaves do a better job of absorbing the glyphosate than do older leaves.

INSECTS

Judging from the amount of pesticides applied against them, insects are the most troubling garden pests. Insects cause three general types of damage: they eat foliage; they suck plant juices; and they transmit plant diseases. Biting or chewing insects, such as caterpillars, beetles, and carpenter bees, create holes in plants.

Some insects that suck plant juices have toxic saliva that causes plant damage beyond the feeding. Insects transmit diseases either indirectly, when plant damage permits the disease organism to enter, or directly, when sucking insects actively transmit the bacteria or virus. Attack by tomato fruitworms, which bore holes into tomatoes, or fruit flies, which tunnel into fruit, is often followed by a fungal or bacterial disease, and many aphids transmit a number of viral diseases.

Aphids

Natural History

Aphids are about the size of a pinhead, soft-bodied and pear-shaped, with long antennae and two tubelike appendages protruding from the rear of the abdomen. Most you will encounter are green, but some are pink, brown, or black. They have piercing mouthparts that they use to suck plant juices from leaves, stems, flowers, or fruits, although a few species, such as the corn root aphid, live in the soil and feed on underground plant parts. They tend to congregate on the growing tips of plants in the early spring and fall.

Their short life cycle allows aphids to increase quickly. Aphids often overwinter as eggs. In early spring the egg hatches and a plump aphid, called the *stem mother*, emerges. The stem mother gives birth to live daughters, which, in turn, grow up and give birth to more daughters. This asexual process requires no mating, and no eggs are laid. The daughters, granddaughters, great-granddaughters, and successive generations all cluster around the stem mother, forming one big colony that keeps growing until something stops it. At the end of the growing season (usually during fall),

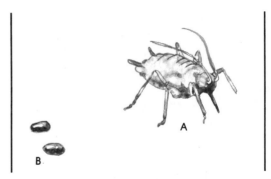

APHID LIFE CYCLE

In spring, a stem mother (A) hatches from an overwintering egg (B). She reproduces asexually, giving birth to female aphids, which in turn give birth to more female aphids within a week or two. In the fall, special aphids—males, which are frequently winged, and egg-laying females—are produced; they mate and produce the overwintering eggs.

the aphids produce a sexual generation—composed of winged males and females—which mate, disperse, lay eggs (which overwinter), and die.

Aphid damage can cause leaves to turn yellow, curl, or pucker, while the plant becomes weaker and weaker. Aphids can also transmit viral diseases such as cucumber, tomato, or

---

Aphid damage can cause leaves to turn yellow, curl, or pucker, and aphids transmit viral diseases such as cucumber, tomato, or tobacco mosaic.

---

tobacco mosaic. Aphids secrete a sugary water solution called honeydew, which attracts ants and can lead to growth of a black sooty mold. Some ant species like honeydew so much that they protect aphids from natural enemies. Protected colonies often build up to very high levels and cause significant plant damage.

An aphid colony's growth rate depends on the quality of its food source, temperature (aphids grow more slowly at low temperatures), and the rate at which aphids are being killed either by wind and rain or natural enemies.

Warning Signs

Start monitoring in the spring. Inspect the growing tip and the tops and bottoms of young leaves. Monitor the plants every week to ten days. Monitor more frequently when you discover aphids. If you find just a few on a plant, do not take immediate action. Delay action and monitor the plant more frequently, to give natural enemies some time to do their work. Look for ladybugs and other natural enemies of the aphid. Any aphids that appear shiny, hard, and a darker color contain parasitoids.

The dark color comes from a parasitic wasp developing inside the aphid, which has simply become a shell. Wait a few days to see if the parasitoids successfully control the aphids.

When to Take Action

Take action if the number of non-parasitized aphids starts rising rap-

To control aphids, reduce use of nitrogen fertilizer and use an organic nitrogen fertilizer, such as composted manure or regular compost.

idly over the course of a couple of sampling periods.

Long-term Controls

Cultural Controls. Modify your fertilizer use and schedule. High nitrogen levels stimulate aphid reproduction. Reduce nitrogen fertilizer and spread it out over a number of feedings. Avoid soluble nitrogen fertilizer. If you want to use a synthetic fertilizer, use ammonium- or urea-based formulas. There are also synthetic fertilizers, such as urea-based Ozmacote, that slowly release lower levels of nitrogen over a longer period. Even better, use an organic nitrogen fertilizer, such as composted manure or regular compost, which releases nitrogen slowly over the whole growing season.

Spread out fertilizer application over the whole growing season, and use a fertilizer that has a relatively low ratio of nitrogen to phosphorous. Fertilizers with a low N/P ratio tend to emphasize flower and fruit production over foliage growth, as well as limit soluble nitrogen that aphids need.

Biological Controls. Aphids are vulnerable to natural enemies, in part because they are soft-bodied, live in fairly exposed habitats, and have few defenses. A wide range of both parasitic and predaceous insects attack aphids, including, for example, spiders, lacewings, ladybug beetles, sryphid flies, and big-eyed bugs. In addition, one whole family of parasitic wasps specializes in parasitizing aphids.

Insect natural enemies are more susceptible to pesticides than the aphids are; therefore, use of pesticides often causes aphid outbreaks. It is particularly important not to use pesticides early in the season. The earlier natural enemies start attacking aphids, the less likely it is that aphid numbers will reach outbreak levels.

Encourage aphids' natural enemies by planting wildflowers and allowing a number of weedy species to grow at the border of the garden; encourage diversity of flowering plants in the area. Natural enemies are attracted both by the flowers themselves and by the aphids on those weeds, both of which provide food for predators. Plants with small, more easily accessible flowers but significant nectar production, such as wild carrot, alfalfa, oleander, white clover, and yarrow, are particular favorites of parasitic wasps.

If you want to buy natural enemies to release, the best bets are green lacewings or aphid midges. Adult lacewings feed on nectar, pollen, and honeydew; the larvae (commonly called aphid lions) are voracious predators of small insects and eggs. A green lacewing larva can eat up to 60 aphids per hour. You can get lacewings either as eggs or larvae, although larvae are easier to handle. For smaller plants, put on one larva, or a few eggs, per plant. The larvae of aphid midges prey heavily on aphids and are sold by a number of mail-order companies.

We do not recommend ladybug beetles or praying mantises. Ladybug beetles usually leave the garden quickly. Driven by instinct, they act as though they have just come out of hibernation and, as in the nursery rhyme, "fly away home." With a little luck, ladybug beetles returning naturally to your garden from their overwintering sites will arrive in time to keep aphids in check. Praying mantises are general predators that eat just about any insect they can catch—including other praying mantises. They kill natural enemies just as readily as pests, and do not necessarily concentrate on eating aphids, even if the aphid population grows.

Short-term Controls

Physical Controls. Physical controls are particularly appropriate early in the season, before natural enemies are abundant. Hose down plants with water to kill aphids and knock them off the plant, where they can then be eaten by spiders, beetles, and other ground-dwelling predators. Depending on the severity of the infestation and the type of plant, you may have to spray the plants twice a week for a couple of weeks.

You can also physically remove the

aphids, either by crushing them or brushing them off with your hands (wear gloves), or by pruning badly infested plant parts. If there are ants among the aphids, the ants may be defending them, and you need to get rid of the ants to allow natural enemies to attack. If you have woody plants such as roses, you may be able to prevent ants from climbing the stems by applying a band of sticky stuff (Tanglefoot, Stickem, or Tack Trap) around the base of the plant. Otherwise, you may have to take other means to control the ants (see chapter 6).

Last Resorts

Chemical Controls. If physical and cultural controls and natural enemies haven't done the job, try an insecticidal soap or light horticultural oil. Soft-bodied aphids are susceptible to them. Apply the soap only to infestations you can see, because some natural enemies are affected by the soap. Light horticultural oil can be used on a variety of plants to smother the aphids.

Whiteflies

Natural History

Whiteflies are not flies, but plant-sucking bugs related to scale, aphids, and mealybugs. A number of species attack garden plants and ornamentals. The adults, as their name implies, look like tiny white flies; the females are slightly bigger than the

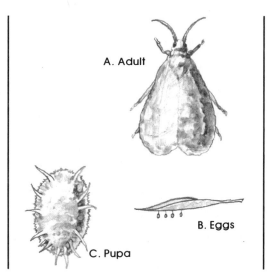

WHITEFLY LIFE CYCLE

The female (A) lays eggs (B) in a small circle on the underside of mature leaves. The eggs hatch into mobile larvae, which settle on the underside of the leaf and lose the ability to move. The insect is small, round, translucent (with a white, cream, or greenish tint); some species have a fringe of hairs. Larvae pupate (C) on the leaf and develop a cocoonlike protective cover; the adult emerges within a week.

males. They are attracted to the color yellow and cluster on the yellowish green parts of a plant—invariably the youngest leaves and the ends of stems. The adults don't fly very far; when disturbed, they usually fly off the plant in a small cloud and then land again on the young foliage. The female lays her eggs on the underside of older leaves. The eggs, laid in a small circle, look like greenish white pointed tubes hanging from the leaf

The larvae of whiteflies settle down on the underside of the leaf and suck plant juices.

by a short stalk. The eggs hatch into mobile larvae that look like translucent small moving spots. The larvae quickly settle down on the underside of the leaf, and start sucking plant juices. At this point the insect is small, round, and translucent (with a white, cream, or greenish tint); some species have a fringe of hairs. Larvae pupate on the leaf and develop a cocoonlike protective cover. Normal pupae are light-colored, while parasitized pupae are black.

As with scale, mealybugs, and aphids, populations in the wild are primarily controlled by keeping soluble nitrogen levels low in host plants and by the same natural enemies that attack aphids. Being immobile, the larval and pupal stages are particularly susceptible to natural enemy attack.

Warning Signs

Begin monitoring plants for whiteflies every week or two weeks, beginning in spring. Check the youngest foliage for the adults, which are easy to spot. If you notice any adults, check the oldest leaves (usually near the bottom of the plant) for eggs, larvae, and pupae. The larvae are pale and often hard to see, so look carefully—a hand lens is useful.

When to Take Action

Only the larvae damage the plant; adults do not pose a problem, as long as there are few larvae. If the number of leaves infected with larvae and the number of larvae and pupae per leaf start rising rapidly, take action. If a third or more larvae or pupae on the leaves is dark—an indication of parasitization that has already killed the organism—wait a while to see if the natural enemies that will emerge from the parasitized larvae will control the population.

Long-term Controls

Cultural Controls. As with aphid control, reduce use of nitrogen fertilizers; use organic fertilizers, ammonium- or urea-based synthetic ones, slow-release formulations (such as Ozmacote), or ones that emphasize phosphorous over nitrogen. Spread out fertilization over the whole season by using a larger number of smaller doses.

Biological Controls. Many of the natural enemies (such as lacewings and parasitic wasps) that control aphids also work against whiteflies. Follow the strategies described in detail in the aphid section; avoid insecticides, especially early in the season. Plant a diversity of wildflowers to

Parasitic wasps can be used to control whiteflies.

provide both food and sources of alternate prey.

Short-term Controls

Physical Controls. Wipe larvae and pupae off the plant with a gloved hand. Early in the season, when the populations are low, vacuum the adults off the leaves with a hand-held vacuum. This is best done in the morning when the bugs are sluggish.

Sticky traps are also effective. You can make the traps easily yourself. Paint a board with yellow paint (whiteflies particularly like Rustoleum yellow No. 659) and cover it with a thin layer of Stickem, Tack Trap, or Tanglefoot. Or you can make your own sticky solution by mixing one part petroleum jelly or mineral oil and one part dishwashing liquid. This homemade material is easy to wash off the boards, so that they can be reused; the same is not true of the commercial sticky material.

The size, shape, and location of the boards depends on the plant you are protecting. Since adult whiteflies are found on the youngest leaves and do not fly far, squash on a trellis would require a wide, tall board, while viny squash might require a long, narrow board laid close to the ground. Try to orient the board so that the sticky side faces away from the sun. You will need a trap for every few plants. Shake the plants to get the adults to fly; some will be attracted to the yellow and try to land on the board. Periodically check the trap and replace it when it fills up with insects and dirt. Since the traps can also catch natural enemies, take them down when the larval populations start declining, an indication that the natural enemies are working. You can store the traps and reuse them later in the season.

Last Resorts

Chemical Controls. Spray an insecticidal soap or light horticultural oil, but only on the undersides of older leaves, which contain larvae and pupae, or directly on the adults. Wear a dust mask when applying these substances.

Mites

Natural History

Mites are tiny arachnids (like ticks and spiders), about the size of a grain of pepper. There are a large number of species, many of which are beneficial and serve as predators or decomposers. The pest species come in yellow, red, green, or brown, and attack virtually any garden or ornamental plant. Spider mites, broad mites, and cyclamen mites cause the bulk of the

---

Mites are tiny arachnids (like ticks and spiders) the size of a grain of pepper. They can be yellow, red, green, or brown, and attack virtually any garden or ornamental plant.

SPIDER MITE

problems. Spider mites, the most common mite pest, live in colonies and most species spin small webs on the undersides of leaves and on new growth before beginning feeding. Broad and cyclamen mites are about one-quarter the size of spider mites, making them virtually impossible to see, even with a hand lens. They primarily attack flowers such as begonias, cyclamens, gerberas, and African violets. In general, mites love hot, dry conditions.

Mites have piercing mouthparts to suck sap from plants. They can feed on flowers, the blossom ends of fruits, and bulbs and corms, although they prefer the undersides of leaves, particularly along the main rib. A number of species also inject toxins into the plant, causing discoloration and distortion of new growth. The feeding damage first appears as tiny, needlelike punctures on the undersides of leaves. Silver or yellow stippling or red spots then appear on the upper sides of leaves. As the damage progresses, the punctures become brown and sunken, and leaves start turning yellow and may eventually curl and fall off. Both broad mites and their feeding damage are too small to see; often the first indication that you have these pests is the distortion of new growth caused by the mites' salivary toxins.

Warning Signs

Start monitoring for mites in the spring, just as buds are opening. Initially, you probably have to monitor the plants only once every 10 days to two weeks. When conditions become hotter and drier, monitor at least once a week. Look for the mites or signs of their damage—stippling on leaves, webs on the new growth and the undersides of leaves, or small moving dots (the mites themselves) on the undersides of leaves. A hand lens or small magnifying glass helps. You can also tap some leaves or branch tips against a piece of white paper and look for mites on the paper. To see the webs of spider mites more readily, turn the leaf over and expose it to sunlight at an angle, or mist the leaf underside so that the sun glistens in the water drops clinging to the web. Early in the season, concentrate your search along the midribs on the undersides of middle-aged leaves, where the initial infestation often occurs.

When to Take Action

If you find only a couple of mites on a plant, hold off treating them to see if natural enemies exert control.

Natural enemies usually control spider mites, with outbreaks occurring only if plants have been repeatedly sprayed with pesticides. If mite numbers steadily increase, or if environmental conditions are hot and dry, take action.

Long-term Controls

Cultural Controls. Keep your plants from getting hot and dry. If possible, plant them where they can get some shade during the heat of the day. Use organic mulch on your garden soil to increase the humidity near the plants. Mist your plants on particularly hot days. Mites often travel from plant to plant via overlapping leaves, so space plants so that they do not touch. Since mites can also be spread via tools or hands, wash your hands frequently while working with the plants, and dip your tools in alcohol between uses, especially if your plants are infested with mites. For shrubs and small trees, spray a horticultural oil over the whole plant as a preventive measure in late fall or early spring while there are no leaves on the plant. The spray smothers overwintering adults.

Biological Controls. Mites have numerous natural enemies, including predatory mites, green lacewings, predatory thrips, predatory midges, big-eyed bugs, and ladybug beetles. First, preserve and augment natural enemies already in your garden. Eliminate, or at least minimize, use of pesticides to avoid killing off natural

To avoid killing off natural enemies of mites, eliminate or minimize the use of pesticides. Plant a variety of wildflowers, particularly ones from the carrot family, to attract and support the enemies.

enemies. Plant a variety of wildflowers to attract and support the enemies.

Second, you can buy and release predatory mite species against spider mites and broad mites. To determine which predatory mites, and how many, to buy, you need the following information: the identity of the pest mite (collect some and have them identified by a scientist at your county Cooperative Extension agent's office); the number of plants, or the area, to be protected; the number of infested plants and/or the average number of mites per leaf. Follow the supplier's instructions for how to apply the mites.

To maximize the impact of the predatory mites, don't spray any insecticide for about a month before releasing them, because some pesticides leave a residue that could kill them. For greatest effect, release the predatory mites while pest mite numbers are low. If pests are abundant, spray the plants with an insecticidal soap about three days prior to release of the predatory mites. Release the mites on the middle or upper foliage of all plants. (The predatory mites are often shipped in sawdust, so just sprinkle a little sawdust on each plant.) Keep some predatory mites and look at them with a hand lens to learn how to recognize them.

For best results, release three batches of predatory mites: the first two batches two weeks apart, the third batch one month later. Monitor your plants every week and keep track of the number of pest and predatory mites. Finally, while using the predatory mites, don't use pesticides at all.

Short-term Controls

Physical Controls. As soon as you notice mites, hose down the plants with water, which knocks off mites and sometimes kills them. Make sure you spray the undersides of the leaves. Do this once a day for three or four days.

Last Resorts

Chemical Controls. Use a chemical spray, but only to reduce severe mite infestations. Spray one of the light superior horticultural oils on woody plants and on some ornamentals. Varieties of light horticultural oil have recently become available for use on many kinds of garden plants. Insecticidal soaps also work against mites. For both the oil and the soap, make sure you first test them on a few leaves to see if they damage the foliage; any damage will be evident within a day.

Dust or spray the undersides of infested leaves with sulfur. With all sprays, make sure to coat the undersides of only the infested leaves. Spray the infested plants every four to five days for a couple of weeks, and continue to monitor the plants to see if mites persist. Wear a dust mask if you use a powder; wear rubber gloves if you use a liquid.

You can also use a pyrethrum- or pyrethroid-based product, although it kills natural enemies as well as pest mites. Do not spray pyrethroids around water; they are highly toxic to fish and aquatic invertebrates. *Caution:* Some people are allergic to the pulverized flowers that are the source of pyrethrum. Wear a dust mask or respirator if you are allergic to pollen and are spraying pyrethrum, or choose a different chemical.

Thrips

Natural History

Thrips are tiny insects less than ¼ inch long, and you may have a hard time seeing them without a hand lens. They can be pale or dark, and are very thin. To the naked eye, they look like tiny squiggly lines. Some thrips eat other thrips, mites, and small aphids; others feed on plants and can cause significant damage. For example, large infestations of pear thrips on maple trees in the Northeast, particularly in Vermont, have severely hurt maple syrup production. Pest thrips feed on leaves, flowers, and fruits, al-

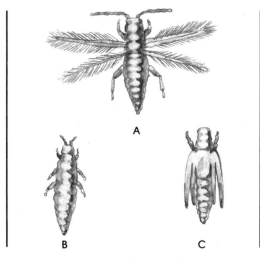

THRIP LIFE CYCLE

Depending on the species, the adult female thrip (A) lays her eggs on leaves, fruits, and stems. The eggs hatch into wingless larvae (B), which feed on sap. In most thrip species the resting, or pupal, stage (C) is spent in the soil, while some species pupate on the plant itself. The adults emerge from the pupae.

Large infestations of thrips lead to brownish or silver flecks on the leaf and scarred fruit.

stages, prepupa, pupa, and adult. The larval and adult stages do the damage; the prepupa and pupa are resting stages. In some species the resting stages remain on the plant, but usually the resting stages are spent in the soil. Most species overwinter as adults in the ground or in cracks and crevices in trees or the woody parts of bushes. Adults can fly, but the larvae cannot. Females lay eggs on leaves, fruits, and stems. Thrips seem to do best under dry conditions.

Warning Signs

Monitor every two weeks beginning in late spring or early summer, depending on where you live. The farther north you live, the later the thrips arrive. Look for signs of thrips or their damage on leaves and flowers. If you think flowers are infested, shake them over paper and look for the larvae or their fecal pellets (small, dark-colored specks, which are often more noticeable). Look on the undersides of leaves, particularly along the veins.

When to Take Action

If you find only a few infested flowers or leaves, begin monitoring more frequently and wait to see if natural enemies control the problem. Other-

though light-colored flowers seem to be particularly attractive to them.

Thrips scrape the surface of the plant tissue and then suck up the resulting juice. They feed on the underside of the leaf, near the veins. Badly damaged flowers can become discolored and disfigured. Damaged growing tips can become distorted. Large infestations lead to brownish or silver flecks, with the leaf eventually drying up. Sufficient damage can also scar fruits. In addition, some thrips transmit viral diseases.

The thrip's life cycle includes six developmental stages: egg, two larval wise, take action if thrip numbers steadily increase over three or four sampling periods.

Long-term Controls

Genetic Controls. Certain plant varieties are more resistant to thrip damage than others. For example, gardeners in the Northeast often have thrips on cabbage. Red Danish, Danish Ballhead, and Early Jersey Wakefield cabbage resist them well. Ask your garden store, or county Cooperative Extension Service for resistant or tolerant varieties.

Dry conditions favor thrips. Mist your plants during dry weather, use organic mulch, and water your plants to increase moisture.

Cultural Controls. For thrips that pupate in the soil, aluminum or aluminum polyethylene mulches appear to disorient the thrips and may prevent them from finding their way underground. Similarly, paper, plastic, or other impermeable mulches help prevent pupation. A 4-inch-thick organic mulch also works, possibly by increasing fungi or nematodes.

Biological Controls. Preserve thrips' natural enemies by avoiding the use of pesticides. Keep the vegetation around your garden diverse, to provide a habitat and alternate food sources for the natural enemies. Plant some wildflowers or a hedge in the area, or even leave a few weeds around. Plant trees, bushes, and hedges that are attractive to birds that prey on adult thrips.

Thrips have three kinds of natural enemies: predatory mites, lacewings, and nematodes. The mites prey mainly on the eggs and larvae. Since the adults are long-lived, you may continue to see them for a while, even though the egg and larval stages are under full control. However, you should notice a reduction in larvae and damage. Sometimes, though, environmental conditions are not amenable to these mites, or they are not effective against the species of thrip that is the source of your problems.

Green lacewing larvae feed voraciously on thrips. One particular benefit of lacewings is that they prey on thrip species that pupate on the plant and that can't be controlled with mulch. When buying lacewings, try to purchase larvae rather than eggs (it can be difficult to find someone who will ship larvae), which don't work as well. Ask that they be packed individually; otherwise they may eat one another. Apply one larva per plant; on bigger plants you can put two larvae.

Nematodes, particularly *Heterorhabditis heliothidis*, can help control those thrip species that pupate in the soil or in which the second larval stage and/or the resting stages occur in the soil, but not those species that

Paper, plastic, or other impermeable mulches help prevent pupation of thrips.

pupate on the plant. Apply the nematodes at the recommended rates, and irrigate the soil after applying them.

Short-term Controls

Physical Controls. For low to moderate thrip infestations on flowers, crush the thrips between the petals with your fingers. Pick and burn badly infested buds and flowers.

Last Resorts

Chemical Controls. Use chemicals only on infested plants, which usually occur in a "hot spot," and only against thrip species that attack leaves. Thrips that attack flowers are often down between the petals, where chemicals have a hard time reaching. For woody plants, spray a superior (less viscous) light horticultural oil over the entire plant; this will smother any thrip adults or larvae. Light horticultural oils are now available for most garden plants. Or spray an insecticidal soap on the underside of infested leaves every four or five days for two weeks. For both oils and soap, make sure you first test them on a few leaves to see if they damage the foliage; any damage will be evident in a day. Also, wear a mask when spraying oils or soap.

Butterflies and Moths

Natural History

We all enjoy butterflies flying gracefully through the yard, stopping to visit flowers. The same can't be said for many butterfly and moth caterpillars (frequently called "worms"). Caterpillars are voracious, consuming more than their own weight in food every day. They attack all parts of garden plants, as well as the green parts of trees.

Thousands of butterfly and moth species exist, but only a relatively small number of them become significant pests. Though specific aspects of their biology differ, all have the same basic life cycle. Eggs give rise to larvae (or caterpillars), which shed their skin a number of times and then go into a resting pupal stage, from which an adult butterfly or moth emerges. The adults usually visit flowers and feed on nectar and/or pollen, while larvae feed on plant material, doing the damage.

Larvae can be grouped into two broad categories: exposed and hidden. Exposed larvae, such as the tomato hornworm, tomato fruitworm, and inchworm, constitute the bulk of the species. They feed in the open and hide when they are not feeding. Most feed during the day, although some species, including the cutworms and armyworms, feed at night. The hidden larvae are not exposed while feeding, either because they have burrowed into a plant part such as the stem (stem borers), leaf (leaf miners), or fruit (corn earworm), or have fashioned their own shelters by rolling up leaves (leaf rollers) or fastening them together, often around growing tips, or tying flower heads closed with the use of silk. Exposed larvae are usually easier to locate than hidden larvae, although evidence of some hidden larvae—such as rolled-up leaves—is easy to spot. Control measures such as sprays are far more effective against exposed larvae.

Warning Signs

Begin monitoring in the spring or as soon as you notice any damage. Look for eggs, larvae (caterpillars), or feeding damage. Eggs can be almost anywhere on a plant, although they are frequently laid near the growing tips. Exposed larvae often feed on the plant, or hide, usually near the base. If you suspect nocturnal caterpillars, check your plants with a flashlight after dark. For hidden larvae, look for signs of their presence such as rolled-up or folded leaves, tunneling in the leaf, or wilted branches or runners (for stem borers). The feeding damage of most caterpillars is obvious, consisting of chewed sections and/or holes in leaves and other plant parts.

When to Take Action

Action thresholds depend on the type of damage (direct or indirect) and when it occurs. Direct damage occurs to the "valuable" part of the plant, usually the fruit or vegetable, or flowers and leaves in the case of let-

The feeding damage of most caterpillars is obvious—chewed sections and/or holes in leaves and other plant parts.

tuce or cabbage. Indirect damage occurs to the "nonvaluable" plant part.

The action threshold is much lower for direct damage. Thus a corn plant can withstand a high level of leaf damage without any adverse effects on yields, whereas virtually any damage by the corn earworm, which bores into a developing ear of corn, causes problems. Similarly, a tomato plant can withstand a good deal of attack from tomato pinworm, but the tomato fruitworm attacks fruit as well as leaves, so its action threshold is lower.

Timing of the damage is also important. Plants are most vulnerable to caterpillars during the seedling stage and while they are flowering, so thresholds are lower then. Contact your county Cooperative Extension agent or entomologist at your local land-grant university or state agricultural college to see if action thresholds exist for the caterpillar pests plaguing you.

Long-term Controls

Genetic Controls. Some varieties of plants are less susceptible genetically to caterpillar attack than others. Ask your local Cooperative Extension office, land-grant university, or botanical garden for suggestions.

Cultural Controls. Diversify the habitat to attract natural enemies. A ground cover such as clover or mulch can increase populations of predatory ground beetles as well as toads; wildflowers and many weeds, particularly those with tiny flowers, attract and nourish numerous predatory and parasitic insects (you can order by mail special collections of seeds of flowering herbs and/or cover crops); bushes, trees, and hedgerows attract insect-eating birds.

Plant *trap crops*, plants that are more attractive to the pest than the plants being protected. Some people plant eggplants near potatoes because the Colorado potato beetle prefers eggplant. To use trap crops effectively, you must know a good deal about the biology of the pest you want to control.

Lightweight polypropylene row covers that let sunlight, air, and water through, but not insects, can be helpful in preventing butterflies and moths from laying eggs on your plants.

Physical Controls. Try lightweight polypropylene row covers. These let sunlight, air, and water—but not insects—pass through. Put these covers over your plants early in the season to prevent butterflies and moths from laying eggs on your plants. The new super-lightweight covers transmit 95 percent of sunlight and can be used for the entire growing season. Remove the covers during blossoming to permit pollination of crops that need it.

Short-term Controls

Physical Controls. In most cases, simply handpicking caterpillars off plants suffices. This works best with the exposed larvae. For hidden larvae, you may have to remove the infested plant part. With the squash stem borer, for example, find the caterpillar's entrance hole in the vine, slit open the vine at that point, and then remove the caterpillar and drop it in a bucket of sudsy water.

Biological Controls. A couple of natural enemies can be purchased and released into your garden. A tiny parasitic wasp, *Trichogramma*, lays its eggs inside other insect eggs, killing the pest before it has a chance to hatch. The species *T. pretiosum* attacks a variety of lepidopteran pests and prefers to remain within 5 feet of the ground, so it is best for vegetable gardens. The adults feed on nectar from small flowers, particularly those in the daisy and carrot families. Release these wasps in two or three small batches at two-week intervals—rather than in one big batch—to catch newly laid lepidopteran eggs. For best results, use *T. pretiosum* in conjunction with *Bt* in order to attack both the eggs and larvae at the same time.

You can also use spined soldier bugs against cabbage looper and imported cabbageworm (as well as Mexican bean beetles). In the case of broccoli worm, use these bugs, sold as larvae, only if you have a significant infestation; they will die if not enough food is available.

Chemical Controls. The most appropriate pesticide for use against caterpillars is the bacterial insecticide *Bt.* One variety of *Bt—Bacillus thuringiensis kurstaki, Btk* for short—is specific to caterpillars and is sold under brand names including Caterpillar Killer, Dipel, Javelin, Thuricide, Topside, and Worm Attack. Apply *Btk* as you would any pesticidal spray. *Btk* is particularly appropriate for exposed caterpillars, but very ineffective for hidden caterpillars. *Caution:* Wear a dust mask when spraying *Btk.*

Last Resorts

Chemical Controls. Use fatty acid soaps, an extract of the neem tree (called BioNeem), pyrethrum/pyrethrins, and synthetic pyrethroids. Apply them only to infested plants, and only to infested areas of the plant. Do not use pyrethroids near water; they are highly toxic to fish and some aquatic invertebrates. *Caution:* Some people are allergic to the pulverized flowers that are the source of pyrethrum. Wear a dust mask or respirator if you are allergic to pollen and are spraying pyrethrum, or choose a different chemical.

Cutworms and Armyworms

Natural History

This group comprises a number of moth species that all share some larval traits. The larvae usually eat plants near or below ground level, although some species do climb plants.

Most feed at night and hide in the soil, or under plant trash, during the day. Cutworms cut off young seedlings around ground level. Some species of cutworm do not even eat the seedlings; they just cut them off and move on, acting like lawnmowers. Armyworms get their name from the fact that the larvae often move army-like across a field, eating all the tender plants in their path.

Warning Signs

Monitor your garden most closely in early spring, when cutworm and armyworm damage is at its worst. Look for lawnmower type damage (horizontal cuts, not diagonal ones, which are characteristic of rabbits), often with the seedling lying next to the cut stem. Go out after dark and look for the caterpillars themselves. During the day, dig in the soil or plant mulch around seedlings and look for caterpillars that are usually curled up in the shape of a C.

When to Take Action

Take action at the first evidence of cutworm or armyworm damage or, if you have problems every year, take preventive action to protect seedlings.

Long-term Controls

Use a ground cover or mulch to attract natural enemies such as predatory ground beetles, snakes, or toads. A diversity of vegetation also enhances natural enemy protection.

In the fall clean up plant trash and weeds, where the adult cutworm moth lays its eggs. The larvae in the North or pupae in the South overwinter in the soil.

Short-term Controls

Physical Controls. Focus on protecting seedlings and transplants from attack. You can erect barriers around individual plants or entire plots. If there are not too many plants, you can put barriers such as milk cartons, paper cups, or paper collars around each plant. Or put cornmeal and bran in a spiral leading away from the plant. When the cutworm or armyworm eats the mixture, it expands in its gut and the worm dies. You can also mix the bran with sawdust and a bit of molasses. Cutworms feed on this mixture, get caught in it, and then die when the molasses hardens.

A layer of diatomaceous earth (composed of the skeletons of microscopic aquatic creatures called *diatoms*) around the plant cuts the caterpillars' skin, causing them eventually to dehydrate and die. Diatomaceous earth doesn't work when it's wet, so you need to reapply it after a rainstorm.

You can also go out in your garden at night and remove the cutworms and armyworms by hand.

Biological Controls. Apply predatory nematodes to the soil. Be sure to water profusely when applying these, because nematodes need a film of water in order to attack their hosts.

Last Resorts

Chemical Controls. Drench the soil with an insecticidal soap, or with an insecticidal soap/*Bt* mixture.

Yellowjackets and Hornets

Natural History

Yellowjacket is the common name for a number of species of wasps, including hornets, found throughout the United States that have similar biologies. Yellowjackets are social wasps; they live in colonies that build large, multiple-layered "carton" nests. Most species build their nests in the ground (often in an old rodent burrow), but some build aerial nests that hang from trees or inside or on houses or tool sheds. Virtually all species are predators that feed on both live and dead prey. It is during the periods of extensive scavenging that yellowjackets become pests. Otherwise they are beneficial insects that prey heavily on other insects, particularly caterpillars.

Yellowjacket colonies are started each spring by a single fertile female (the queen). She feeds on floral nectar and insects and then searches for a suitable nest site. The queen builds a small nest and begins producing sterile offspring (workers). The colony grows throughout the summer, and can contain thousands of individuals. By late summer the colony has reached peak size and begins reproducing. After reproduction, the colony begins to decline. The workers become more aggressive, begin desperately searching for food, and often switch their diet from protein to more sugary food. It is at this point that yellowjackets usually become a problem. Yellowjackets also vigorously defend their nest and attack any intruder.

Warning Signs

Look for the yellowjackets themselves or for their nests. You will begin to notice them during the summer, particularly in August and September. Yellowjackets, particularly those from declining (or dying) colonies, often forage aggressively at picnics or barbecues, searching for protein or sugar.

Yellowjackets often forage aggressively at picnics, searching for protein and sugar.

You can use traps to monitor yellowjackets. Simple cardboard or plastic traps are available at garden supply stores, to which you add your own bait, such as cat food, early in the season, and jelly or syrup later on. A trap sold by Sterling International comes with a chemical attractant, and can be used for many, but not all, yellowjacket species.

When to Take Action

Take action when the yellowjackets exceed your nuisance threshold.

Long-term Controls

Physical Controls. To reduce the number of yellowjackets around your house, make sure that there is no steady source of protein available, such as pet food left in a bowl outside or an open garbage can. If the yellowjackets discover food early in the season, they may return again and again. Keep garbage in a can with a tight-fitting lid. Use a plastic liner to keep the can clean and free of odors. Feed your pet indoors, or, if you must feed it outdoors, promptly remove uneaten food. To minimize problems at picnics and other outdoor gatherings, make sure all soft drinks have lids on them, do not leave food exposed for any period of time, and keep all refuse covered.

Although yellowjackets are beneficial insects, there are some situations where they should be eradicated—for instance, when a huge nest is in a wall of your house, in your lawn, or hanging from a tree near your front door. Nests in the lawn are particularly dangerous; if you walk over one, the workers will attack you.

Physical removal is appropriate for underground nests. It is done with a

---

To reduce the number of yellowjackets around your house, eliminate any steady source of protein, such as pet food left in a bowl outside, or an open garbage can.

---

vacuum and requires a professional who is used to handling bees and has the appropriate protective clothing. Don't try to do it yourself.

Chemical Controls. For nests in the walls or voids of houses, nest removal is usually impractical; a chemical approach is the better choice. Observe the yellowjackets and find out where they are entering the house or wall. Next, blow an insecticidal dust containing silica gel or diatomaceous earth, pyrethrins, or a synthetic pyrethroid spray through the entrance hole and onto the nest if possible, and then plug the hole with steel wool that has been dusted or sprayed. Yellowjackets not immediately killed by the spray, trying to leave or enter the hole will chew on the steel wool and die. Use of a repellent insecticide, such as resmethrin, will keep the yellowjackets away from the hole. *Caution:* Wear a dust mask and protective clothing. If you are allergic to bee or wasp stings, hire a professional to do the job.

Short-term Controls

Environmental Controls. Minimize your chances of getting stung. Scented materials, such as perfume, hairspray, suntan oil, after-shave, and cologne, as well as brightly colored or patterned clothing, attract yellowjackets. Don't carry drinks or meat in open containers. If confronted with a yellowjacket, stay calm. The yellowjacket is looking for food and has no reason to sting you unless you give it one. Slowly brush it off you or calmly wait until it leaves. Do not swat at it. If you're using a wet towel, examine it carefully before sitting down on it or picking it up, to ensure that a yellowjacket is not trying to drink water from it.

Physical Controls. At picnics and other outdoor functions, minimize incursions of yellowjackets by using traps baited with pet food or something sweet. Buy a yellowjacket trap, or buy a fly trap and put bait in it. Use the traps only while you're having the picnic.

Because of the generally beneficial nature of yellowjackets, do not use bait traps or commercial poison baits on a long-term basis. Aim for reducing yellowjackets' nuisance value, rather than exterminating them. Some traps have the drawback that, if left out for days or weeks, the wasps learn how to escape from them.

Spiders

Since spiders only feed on other organisms (primarily insects), they should be considered allies in your struggle with garden pests, and as many of them as possible should be tolerated. In spite of this spiders may become a nuisance at a given time of year. For example, along the East Coast, some homeowners may notice an explosion of tiny spiders indoors and outside in early to late spring. This occurs when spider egg sacs hatch. If there are any insects for the baby spiders to eat they will live; otherwise, they will die rather quickly.

In either case, you should benefit. If possible, leave them alone. If not, try sweeping or vacuuming them up.

 # SLUGS AND SNAILS

Natural History

Slugs are basically snails without shells. Both are mollusks, related to clams, octopuses, and oysters. These animals have soft bodies covered with slime; Snails have protective, coil-shaped shells. Both leave slime trails wherever they go.

Although many snails and slugs live on land, they require a good deal of moisture and like cool conditions. Some live underground, others dwell at the surface; some prefer alkaline soils, others prefer acidic soils. Because they can easily dry out, slugs feed at night and hide during the day. Slugs and snails also come out on cloudy and rainy days. All snails and slugs have the same basic life cycle: eggs are laid underground or under rocks and need a high level of moisture to hatch. The young look like miniature adults.

The most pestiferous slugs are *Deroceras*, *Arion*, and the garden slug. Slugs from the genus *Deroceras* are usually under 1 inch long, cream to flesh to gray in color, and attack a wide range of vegetables, especially seedlings. Slugs from the genus *Arion* can reach 3 to 4 inches in length, although most average 1 to 2 inches;

they are orange or yellow and are very destructive. The garden slug, as the name implies, wreaks havoc in gardens.

Snails are not as important a pest as slugs, although they cause severe problems in parts of the West Coast. Two pestiferous species stand out: the brown garden snail and the cold-season snail. The brown garden snail is a problem in the West and throughout the South, where it preferentially attacks ornamentals, although it will eat fruits and vegetables. The brown garden snail was initially introduced from Europe as escargot, but later escaped and became a pest. The cold-season snail is found primarily in California, Colorado, New England, Tennessee, Virginia, and Wisconsin.

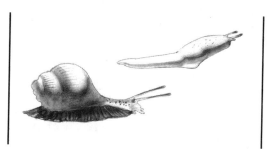

SLUG AND SNAIL

Warning Signs

Start looking for slugs and snails in early spring. It is best to look at night or in the rain. Look for the slugs and snails themselves, or signs of their presence such as slime trails on leaves or soil or large, ragged holes in leaves, fruits, and stems. Slug damage starts at the bottom of the plant and moves up. Slugs prefer damaged plants or rotting leaves, but will eat almost anything. Be sure to check the undersides of leaves.

A variety of traps works for slug and snail populations. For slugs, anything that produces a cool, shady environment works best. Flat boards,

---

Slime trails on leaves or soil, or large, ragged holes in leaves, fruits, and stems indicate the presence of slugs and snails.

---

sacking, old shingles (smooth side down), carpeting, garbage bags, grapefruit rinds, and cabbage leaves have all been used. Put the traps out on bare soil (if possible) in the garden, check them early in the morning while it is still cool, count the number of slugs, and dispose of them in sudsy water or vinegar. Check the traps every day. If they are left too long without checking, they may increase the local slug population by offering a snug hiding place. If put out early in the season, cover traps serve both as a useful monitor to let you know when slugs first invade your garden and as a control.

An overturned clay flowerpot placed on the shadiest side of a plant, or flat boards (especially painted green or red) placed on 1-inch risers attract snails. For both slugs and snails, a commercial trap/bait station, which creates a cool, dark, moist microenvironment, also works.

When to Take Action

Action thresholds have not been developed for garden slugs or snails. Small numbers do not cause appreciable damage. If you put traps out early in the season, snails and slugs will show up in the traps before you see any signs in your garden. Take action when you begin to notice significant damage.

Long-term Controls

Biological Controls. Mulch, particularly the organic kind, often exacerbates slug and snail problems, since they like moist conditions. Remove mulch if you have a serious slug problem. Modify your garden to make it more attractive to the wide range of slugs' natural enemies such as ground beetles, rove beetles, firefly larvae, flies, toads, salamanders, snakes, and some birds. Plant a ground cover such as white clover to attract the predatory beetles. Diverse vegetation and ground cover are appealing to snakes and toads. Put cracked corn or sunflower seeds around susceptible plants to attract birds.

In addition, you can purchase a predatory snail, the decollate snail, for use against the brown garden snail. (Check with your county Co-operative Extension agent to see if the decollate snail can be released in your area. It is sometimes against the law.) Also learn to recognize the native predatory snails living in your area so that you won't accidentally kill them along with the pest snails. Whenever you see a new kind of snail, get it identified (again, your county Cooperative Extension Service or local land-grant university can help). Cover traps can help provide extra living habitat for these good snails.

Short-term Controls

Genetic Controls. Remove or decrease the numbers of plants to which snails and slugs are attracted. You will notice which plants are attacked most frequently.

Physical Controls. Handpick the snails or slugs off plants at night or during moist, overcast days (you might want to use oven mittens if you are squeamish) and drop them in a pail of sudsy water or vinegar. If you have a large population of brown garden snails, consider eating them rather than throwing them away. After all, this is the same snail eaten in France as escargot. You can also use a hand-held vacuum to remove slugs. If you do, you should put the vacuum bag in the freezer to kill them before you empty it. Physical removal is most effective when combined with

Barriers about 1 inch wide and 2 inches high, surrounding individual plants or trees or the entire garden, can physically prevent slugs or snails from reaching plants.

other techniques, particularly traps and barriers. Rototilling the garden usually kills most slugs in the ground.

Barriers about 1 inch wide and 2 inches high, surrounding individual plants or trees or the entire garden, physically prevent the slug or snail from reaching plants. One class of barriers has an irritant or drying action; slugs and snails do not like to crawl over them. Irritants include wood ashes, diatomaceous earth, sawdust, and ammonium sulfate. These substances work only when dry. Reapply them after rain, or cover them with plastic strips or a board during a shower. Ammonium sulfate has a dual use; it's also a commercial plant fertilizer.

A second class of barriers is made of copper or zinc, in the form of bands or screening. These tend to be most effective. The slime reacts with the metals to cause a weak electrical charge. The slug or snail usually crawls about an inch before turning around, so make the barriers at least 2 inches tall.

A third method is to bury screening at least 4 inches into the soil, leaving at least 2 inches above ground. Bend the upper edge of the screening away from the plant or garden to form a flange. Unless you surround the entire garden with a barrier in early spring, you will need to destroy the mollusks that are already in the garden. Handpick them or use a trap.

Traps are useful for controlling as well as monitoring snails. They can be used with or without baits. Unbaited traps include boards, inverted flowerpots, garbage bags, or old carpet. Beer is a popular bait. Put a dish of beer at ground level. Slugs will enter it, become anesthetized, and drown. You may need to cover the bait to prevent rainwater from diluting it. Beer gets unpredictable results, perhaps because different species are attracted to different ingredients used in different brands. Other baits, which you can add to the cover trap, are potato, lettuce, rotting leaves, and crushed snails.

Last Resorts

Chemical Controls. Use a molluscicide, such as methiocarb or metaldehyde, only in conjunction with a covered bait station also called a snail trap. Exposed bait can be eaten by dogs, birds, or children. Some manufacturers sell secure bait stations (see References and Resources).

 # DISEASES

All gardeners have lost plants—sometimes whole crops or gardens—to diseases. Once a disease infects a plant, control can be difficult. Consequently, you need to focus on prevention, particularly if you want to avoid using synthetic fungicides. Prevention requires the ability to diagnose disease damage and to identify what's causing the disease.

Diagnosis

Diseases can be divided into two categories, noninfective and infective. Noninfective diseases result from some environmental problem, most often temperature, light, water, soil chemistry, or location. Infective diseases result from a causal organism such as fungus, bacteria, virus, or nematode. Fungal diseases are often stimulated by environmental conditions, especially weather; their symptoms usually develop over the course of a few weeks. Bacterial and viral diseases usually develop in a couple of days.

The first step in developing an IPM program for diseases is to monitor your garden weekly. To do this effectively, you must recognize general as well as specific symptoms of diseases. General symptoms include yellowing of leaves, dead or brownish areas on leaves, rotten spots on leaves and/or fruits, water-soaked or greasy appear-

🐛 *Some insects can cause damage similar to that caused by diseases. Search your plants to try to determine which is causing the problem.*

ance, wilting, abnormal plant growth, or a mysterious death.

Some insects can cause damage similar to that caused by diseases; for example, wilted squash vines are often caused by the squash stem borer rather than by a disease. Search your plants for insects. They may be the culprits. In addition, although some overlapping occurs, each disease can usually be recognized by a general set of symptoms.

Color photographs of the prominent diseases in your area may be available from your county Cooperative Extension office or gardening store. There are also a number of good books that contain pictures of typical disease damage (see References and Resources).

After diagnosing the symptom(s) of disease, you need to determine the specific disease and its causal agent. Most problems are with noninfective diseases; among the infective diseases, fungi are the most common agents. These two factors are often hard to separate. The fungi and bacteria that cause plant disease are usually present in the environment, but a healthy soil can keep them at levels

low enough that they are not a problem.

Susceptibility to disease is also important, and resistance to given diseases has been bred into many varieties of garden plants. The key to the presence of a disease is usually the existence of conditions conducive to the disease. Plant diseases can be fostered by bad weather or microclimate, over- or underwatering, over- or underfertilizing, changes in soil chemistry, or pesticides that decrease antagonists or enemies of the disease-causing organism. To treat a disease, correct the source of the problem, not just the symptoms.

Prevention

Correct any problems with soil fertility, chemistry, compaction, or drainage. Minimize or discontinue use of fertilizers and pesticides, which kill off beneficial soil organisms. If possible, solarize the soil. Use organic fertilizers such as compost, green manures, organic soil condi-

🐛 *Plant diseases can be fostered by bad weather, over- or underwatering, over- or underfertilizing, changes in soil chemistry, or pesticides that decrease antagonists or enemies of the disease-causing organism. To treat a disease, correct the source of the problem.*

tioners, or Clandosan to build soil health. Mulching also stops the spread of many fungal and bacterial diseases. Seaweed extract sprays applied to the younger leaves help fertilize plants as well as provide potential protection against fungal diseases.

Rotate your crops to minimize soil-borne diseases. In addition to these cultural controls, use varieties resistant to the most prevalent diseases in your area.

Noninfective Diseases

The environmental disorders that cause noninfective diseases are the easiest to correct, once you identify the cause. The most common symptoms are weakening of the whole plant; color changes in the leaves, particularly yellowing; and dead areas on leaves. Yellowing on different parts of the plant often indicates different environmental disorders. Yellowing on all leaves can indicate a general lack of nutrients, high temperatures, or too much sun; on older leaves it can indicate soil compaction or lack of nitrogen or potassium; on younger leaves, it can mean the plant is not getting enough light, iron, or manganese; on leaf edges, not enough magnesium and potassium. Cold water can cause yellow or dead spots on the leaves. Dead spots on the leaf edges can indicate not enough potassium or water, or too much fluoride, boron, heat, or cold. A water-soaked appear-ance of the leaves can be caused by too-high or too-low temperatures or cold water on the leaves.

A few specific examples of noninfective diseases include tomato leaf roll, sunscald, and blossom-end rot. In tomato leaf roll, the edges of the older leaves roll up, because of some kind of stress, particularly high temperatures, drought, waterlogged soil, or a heavy fruit. Big Boy, Beefsteak, and Floramerica are especially susceptible to leaf roll.

Sunscald, which attacks peppers and tomatoes, is characterized by a light, blistered area that becomes sunken, covered with a papery surface, and can eventually become infected with rot. Direct sunlight on the developing fruits can lead to sunscald. (If necessary, cover the fruits with a light netting to shade them.) Plants that have lost a lot of leaves to diseases or are heavily pruned while fruits are still green are most susceptible to sunscald.

Blossom-end rot attacks tomatoes, peppers, cucumbers, squashes, and melons, and is characterized by a water-soaked spot on the blossom end of the fruit, which turns dark brown and leathery and may become infected with rot. It frequently occurs when plants grow rapidly at the start of the season and then set fruit during dry weather. A deficiency of calcium causes blossom-end rot, particularly when environmental conditions prevent the root from absorbing enough calcium. Large fluctuations in soil moisture or too much salt (often the result of synthetic fertilizer use) inhibit calcium uptake by the roots. To control blossom-end rot, use a mulch to conserve water during hot weather, and water deeply and uniformly throughout the season.

Infective Diseases

Fungal Diseases

Fungi cause the bulk of infective disease problems. Fungi often cause distinctive symptoms, as indicated by their names: rust (rust-colored spots), leaf spot (yellowish spots that slowly darken), downy mildew (pale patches on the upper leaf surface and hairy white or purple mold on the lower leaf surface), powdery mildew (powdery white growth on the upper leaf surface). Other symptoms include sudden death of small seedlings and water-soaked patches on leaves or stems. Fungal diseases are usually spread by wind, rain, and human activity, and can often be eliminated by proper yard care.

Environmental Controls. Use mulch or drip irrigation to prevent splashing and spread of spores. Water before noon, so that the plant dries out before evening. Increase spacing between plants to improve airflow and reduce moisture levels. Remove infected tissue and dispose of it. Do not work in the garden when plants are wet, as you can spread fungal spores. Disinfect tools with alcohol to prevent the accidental spread of fungal spores.

Genetic Controls. Plant a variety with genetic resistance to the particular fungal disease. If not listed on the seed package, ask your local garden or nursery store, Cooperative Extension agent, or local botanical garden which varieties show resistance to the disease that plagues you.

Chemical Controls. A number of compounds, both organic and synthetic, combat fungal diseases. The organic ones tend to have lower toxicity and a lower environmental risk. Most of these compounds, particularly the organic ones, prevent infection of the plant, rather than cure an infection, so they must be applied before infection occurs. Find out the environmental conditions that favor a given fungal disease, and then apply the compounds prophylactically.

Antitranspirants—sprays composed of biodegradable materials that form a thin film over the leaves—protect ornamental flowers from mildews by both repelling the fungi and preventing the film of water needed for germination of the fungal spores. To protect against powdery mildew, spray plants weekly with 1 teaspoon

Use mulch or drip irrigation to prevent splashing and spread of fungal spores.

baking soda dissolved in 2 quarts water.

Bordeaux mixture (a mixture of copper and lime) burns fungal spores and is effective against anthracnose, black rot, fire blight, and leaf spot. Copper sulfate, copper oxychloride, and cuprous oxide work against anthracnose, downy or powdery mildew, early and late blight, and leaf spot. Don't use Bordeaux mixture or the fixed-copper compounds during cool, wet weather, as it can damage the plants.

Lime sulfur burns germinating fungal spores and is useful for controlling leaf spot, powdery mildew, scab, and cedar-apple rust. Apply it after a rain, and do not use it on flowering shrubs and trees or when temperatures exceed 80 degrees, since leaves and flowers may be damaged. Sulfur compounds are available in powder and liquid form. *Caution:* Wear a mask, goggles, and protective clothing if you are spraying large areas.

Sulfur prevents spore germination and works against anthracnose, rusts, leaf spot, and powdery mildew.

Bacterial Diseases

Bacterial diseases are characterized by rotting plant parts, irregularly shaped swelling on stems or leaves near the soil level, or wilted leaves. These diseases are spread via soil, water, insects, and tools, and are virtually impossible to cure. Immediately remove infected plants and destroy them or throw them out. Do

> *Bacterial diseases are characterized by rotting plant parts, irregularly shaped swellings near the soil level, or wilted leaves.*

not keep seeds from infected plants, and do not put them in compost piles. As a preventive action, do not work in a wet garden, and disinfect tools, particularly those used for cutting and grafting. Also, rotate crops. Finally, to control insect-vectored bacterial diseases, you need to control the insects that spread them. For example, bacterial wilt, spread by the cucumber beetle, attacks cucumbers, muskmelons, pumpkins, and squash. To control this beetle, cover young plants with a floating row cover, dust plants weekly with an insecticidal dust, or use traps baited with the chemical eugenol. *Caution:* Wear a mask when applying powder. Be sure bait traps are covered.

Viral Diseases

Viral diseases cause sudden death of plants; stunted foliage or blossoms; yellow leaves, often combined with

> *To prevent the spread of bacterial diseases, do not work in a wet garden, and disinfect tools.*

leaf curling or excessive branching; yellow and green mottling on leaves, stems, or blossoms; dead patches on leaves; or rolled, puckered, or extremely narrow leaves. Viruses are spread by insects—particularly sucking insects such as aphids and leafhoppers—garden tools, hands, and tobacco smoke.

Wash your hands and don't smoke in the garden to prevent the spread of tobacco mosaic virus, which can attack tomatoes.

Call your county Cooperative Extension Service or land-grant university to help with disease identification. Destroy infected plants. Buy resistant varieties or disease-free seed; otherwise, try to prevent the spread of the disease. For diseases spread by aphids, control the aphids (see pages 77–80). For diseases spread by tools, disinfect the tools with alcohol or bleach when working around infected plants. Wash your hands and don't smoke in the garden to prevent the spread of tobacco mosaic virus, which can attack tomatoes.

Nematode Diseases

Nematodes are microscopic worms that often live in the soil; some species are crop pests and others are beneficial. Nematode disease symptoms continue for the life of the plant, as is the case with infective diseases.

Parasitic nematodes often invade a plant's roots, although stems and buds can be attacked. Nematode damage is hard to diagnose. In general, infested plants look sickly, wilted, or stunted, with the leaves turning yellow or bronze. Plants with damaged root systems often recover poorly from drought.

Although there are many pest species, a common one is the root-knot nematode. This species attacks a number of plants, including tomatoes; it lives in the soil and invades a plant's roots. The young worms feed on the root tissue and inject toxic saliva and bacteria, which cause rot and swelling on the plant in the area around the nematodes. You can monitor these nematodes by pulling up a suspect plant and looking for swelling on the roots.

General monitoring for nematodes is difficult, since they are always present in the soil. One method consists of taking two samples of soil from near the suspected infested plants. Freeze one sample to kill any nematodes and then plant radishes or cucumbers, both sensitive to nematode damage, in the two samples. A week to ten days later, look at the plants. If the plants grown in the soil that was frozen look healthier than those in regular soil, you probably have nematode problems.

If you have severe nematode problems, consider solarizing the soil and buying resistant plant varieties. Mold

compost, such as from rye and timothy grasses, releases fatty acids that
suppress nematode populations. A
healthy soil with plenty of organic
matter usually harbors many nematode predators such as beneficial
fungi and springtails.

Clandosan, a product composed of
chitin, a waste product of the shellfish
industry, suppresses pest nematode
populations, in part by stimulating
predatory nematodes.

For root-knot nematodes, plant
French marigolds, which release a ne-
matocide from their roots. If your tomato garden is badly infested with
root-knot nematodes, leave it fallow
for a year and grow marigolds instead. Fallow periods in general decrease nematodes. Finally, clean tools
and garden shoes that have come into
contact with infested soil with bleach
or alcohol.

CHART 4-1
Products to Control Garden Pests

The products listed below are a fraction of those pesticides marketed to control garden pests. These products were selected for listing based on CU's judgment that they can be effective when used in the context of an Integrated Pest Management strategy, and that they pose the least risk to humans, pets, or the environment, based on the active ingredients they contain.

Some products not listed here contain one or more of the same active ingredients as these products and may be substituted for them. But many widely available products are not listed because they contain active ingredients that, in CU's judgment, pose greater potential risks to health or the environment than the ingredients of products listed. In our view, effective pest control does not require use of more toxic pesticides, and we have chosen not to list products that contain them. Products are listed below in alphabetical order.

Recommended products listed below contain the least hazardous active ingredients including one or more of the following: *Bacillus thuringiensis kurstaki* (BTK) or *sandiego* (BTS), fatty acids (FA), copper (CU), neem (or azadirachtin [AZ]), nematodes (*Steinernema feltiae*, or SF), petroleum (or paraffinic) oil (PO), sulfur (SU), pyrethrins (PY), or pyrethroids, such as allethrin (AL), phenothrin (PH), resmethrin (RS), or tetramethrin (TM).

Other products listed below also contain synergists, such as MGK 264 (M2) or piperonyl butoxide (PB), rotenone (RO), and/or petroleum distillates (PD), whose chronic health effects remain unknown.

Code for pests controlled: aphids (A), whiteflies (W), mites (M), caterpillars (C), cutworms (CW), thrips (T).

BRAND NAME	ACTIVE INGREDIENT(S)	PESTS
Recommended		
Ace Hardware House & Garden Bug Killer II	PH, TM	A, M, W, T
BioLogic SCANMASK	SF	CW
Bonide Dipel .86% W.P.	BTK	C, CW
Dexol Aphid, Mite, and Whitefly Killer	PH, TM	A, M, W, T
Dexol Worms Away	BTK	C, CW
Dipel Worm Killer	BTK	C, CW
Gro-Well Horticultural Oil Spray	PO	A, W, M, T
Gro-Well Thuricide-HPC Insect Control	BTK	C, CW
Gro-Well Wettable Sulfur Dust or Spray	SU	M
Hot Shot House & Garden Bug Killer	PH, TM	A, W, M, C, T

Products to Control Garden Pests (*continued*)

BRAND NAME	ACTIVE INGREDIENT(S)	PESTS
HWI Hardware Rose & Garden Insect Killer	PH, TM	A, W, M, C, T
Hydro-Gardens Guardian	SF	CW
Ortho Dipel Caterpillar Spray	BTK	C, CW
Ortho FLOTOX Garden Sulfur	SU	M
Ortho Garden Sulfur Dust or Spray	SU	M
Ortho Horticultural Spray Oil	PO	W, M, T
Ortho Insecticidal Soap	FA	A, W, M, C, T
Ortho Volck Oil Spray	PO	W, M, T
Perma-Guard Plant & Garden Insecticide	DE	C, A, CW
Raid Multi-Bug Killer D39	AL, RS	A, W, M, C, T
Real-Kill House & Garden Bug Killer	PH, TM	A, W, M, C, T
Safer BioNeem	AZ	A, W, M, C, T
Safer Bt Caterpillar Attack	BTK	C, CW
Safer Fruit & Vegetable Insect Attack	FA	A, W, M, C, T
Safer Insecticidal Soap for Fruits and Vegetables	FA	A, W, M, C, T
Safer Leaf Beetle Attack M-One	BTS	
Safer Rose and Flower Insect Attack	FA	A, W, M, C, T
Safer Tree & Shrub Insect Attack	FA	A, W, M, C, T
Safer Vegetable Insect Attack	BTK	C, CW
Safer Yard and Garden Insect Attack	FA, PY	A, W, M, C, T, CW
SA 50 Citrus & Ornamental Spray	PO	W, M, T
SA 50 Thuricide-HPC	BT	C, CW
Spectracide Lawn & Garden Insect Control 1	PM	A, M, W, T, C

Brand Name	Active Ingredient(s)	Pests
Other		
Ace Hardware Flower & Vegetable	PB, PD, PY	A, W, M, C, T
Acme Garden Guard	RO	A, W, M, C, T
Black Flag House & Garden Insect Killer	AL, PB, RS	A, W, M, C, T
Dexol Rose & Flower Insect Killer	PB, PY	A, W, M, C, T
Gro-Well Flower & Rose Spray	PD, PH, TM	A, W, M, C, T
Gro-Well 1% Rotenone Dust	RO	A, W, M, C, T
Gro-Well Organic Insecticide	PB, PD, PY, RO	A, W, M, C, T
Gro-Well Rose & Flower Insect Killer	PB, PD, PY	A, W, M, C, T
Gro-Well Vegetable & Tomato Insect Killer	PB, PD, PY	A, W, M, C, T
Gro-Well Vegetable Tomato Dust	RO, CU	C
Meijer's Rose & Flower Insect Killer	PB, PY	A, W, M, C, T
Ortho Home & Garden Insect Killer Formula II	PB, PY, TM	A, W, M, C, T
Ortho Rose & Flower Insect Killer	PB, PY	A, W, M, C, T
Ortho Rose & Flower Insect Spray	PB, PY, RO	A, W, M, C, T
Ortho Rotenone Dust or Spray	PB, PY, RO	A, W, M, C, T
Ortho Tomato & Vegetable Insect Killer	PB, PD, PY	A, W, M, C, T
Raid House & Garden Formula 8	AL, RS	A, W, M, C, T
Raid House & Garden Formula 11	PB, PY, TM	A, W, M, C, T
Spectracide Rose & Garden Insect Killer	PB, PY	A, W, M, C, T

C H A R T   4 - 2
Products to Control Snails and Slugs

The products listed below are a fraction of those pesticides marketed to control snails and slugs. These products were selected for listing based on CU's judgment that they can be effective when used in the context of an Integrated Pest Management strategy, and that they pose the least risk to humans, pets, or the environment, based on the active ingredients they contain.

Some products not listed here contain one or more of the same active ingredients as these products and may be substituted for them. But many widely available products are not listed because they contain active ingredients that, in CU's judgment, pose greater potential risks to health or the environment than the ingredients of products listed. In our view, effective pest control does not require use of more toxic pesticides, and we have chosen not to list products that contain them. Products are listed in alphabetical order.

Code for active ingredients: carbaryl (CB), methiocarb (MB), metaldehyde (ME).

BRAND NAME	ACTIVE INGREDIENT(S)
Cooke Slug-n-Snail Granules	CB, ME
Corry's Liquid Slug & Snail Control	ME
Corry's Liquid Slug, Snail, & Insect Killer	CB, ME
Corry's Pellets	ME
Corry's Slug & Snail Death	ME
Corry's Slug, Snail & Insect Killer	CB, ME
Deadline Slug & Snail Bait	ME
Dexol Snail & Slug Pellets	ME
Ortho Bug-Geta Liquid Snail & Slug Killer	ME
Ortho Slug-Geta Snail & Slug Bait	MB
Snarol Snail & Slug Killer Pellets	ME

CHART 4-3
Products to Control Garden Diseases

The products listed below are a fraction of those pesticides marketed to control garden diseases. These products were selected for listing based on CU's judgment that they can be effective when used in the context of an Integrated Pest Management strategy, and that they pose the least risk to humans, pets, or the environment, based on the active ingredients they contain.

Some products not listed here contain one or more of the same active ingredients as these products and may be substituted for them. But many widely available products are not listed because they contain active ingredients that, in CU's judgment, pose greater potential risks to health or the environment than the ingredients of products listed. In our view, effective pest control does not require use of more toxic pesticides, and we have chosen not to list products that contain them. Products are listed in alphabetical order.

Recommended products listed below contain the least hazardous ingredients, including one or more of the following: Bordeaux mixture (BM), chitin (CH), copper (CU), lime sulfur (LS), or sulfur (SU).

BRAND NAME	ACTIVE INGREDIENT(S)
Acme Bordeaux Mixture	BM
Acme Lime Sulfur Spray	LS
Acme Wettable Dusting Sulfur	SU
Clandosan	CH
Earl May Lime Sulfur Spray	LS
Gro-Well Bordeaux Mixture	BM
Gro-Well Vegetable Tomato Dust	CU, RO
Gro-Well Wettable Sulfur	SU
Ortho Dormant Disease Control Lime-Sulfur Spray	LS
Ortho FLOTOX Garden Sulfur	SU
Ortho Garden Sulfur Dust or Spray	SU
SA 50 Liquid Copper Fungicide	CU
That Liquid Sulfur	SU
Top Cop Liquid Copper	CU
Top Cop with Sulfur	CU, SU

5

ROSE
PEST
CONTROL

Roses are notoriously difficult to grow. Books on growing roses invariably say they require a large amount of pesticides, particularly fungicides, because they are susceptible to disease. But a good IPM program focusing on physical, cultural, and biological controls can dramatically reduce or even eliminate the need for pesticides.

The many different types of roses are divided into classes and subclasses based on species, bloom type, and growth habit. Each group serves a specific landscape use. There are two overall classes: older and modern.

Among the older types are *Rosa majalis* (cinnamon rose), *R. rugosa* (bush rose), *R. multiflora* (multiflora rose), and *R. wichuraiana* (memorial rose), which include old garden varieties, shrub, climbing, and miniature roses. Among the modern types are the hybrid tea, grandiflora, and floribunda roses, specially bred for their brilliant colors and spectacular flowers. The modern varieties are susceptible to diseases and adverse environmental conditions. The older varieties tend to be hardier and more resistant to pests and, consequently, are gaining popularity again.

🌹 *The older varieties of roses tend to be hardier and more resistant to pests than modern varieties.*

Roses grow better in slightly acid (pH 6.0–6.5), rich, well-drained, but moist soils that receive a full sun and at least occasional breezes, which dry out dew and rain, thereby reducing disease problems. In general, the old garden and shrub roses can tolerate cold winters, while most of the modern varieties cannot. Within groups, however, there is great variability in cold-hardiness.

Roses also need five to six hours of strong sunlight each day. The plants can survive with less sun or filtered shade, but they tend to have more disease problems. Ideally, the plants should receive sun in the morning and shade during the hottest part of the day. They need substantial amounts of water, particularly during flowering, to attain peak performance so don't let the soil dry out. Most roses thrive on one major feeding a year, just after early-spring pruning. Additional "snacks" of di-lute liquid fertilizer once a month during the flowering season sustain health. Spray a mixture of one or two tablespoons of brewer's yeast per gallon of water on the plants during mid-season, for greener foliage, sturdier growth, and improved flowers. Peat moss mixed with bone meal applied as a mulch also improves the general health of the plant.

Identifying the Problem

To decide what steps to take to protect and treat your roses, you need first to determine whether the symptoms you observe result from local environmental conditions (such as sun, water, and soil), insect infestations, or disease.

🌹 *To protect and treat your roses, you need to determine whether the symptoms you observe result from local environmental conditions, insect infestations, or disease.*

 ENVIRONMENTAL CONDITIONS

Temperature Problems

Do your roses open halfway and then stop, a phenomenon known as *balling*? Cool nights and overcast damp days cause balling. In the short term, cut off the balled flowers and wait for the weather to improve. When it does, the problem will correct itself. If you have this problem year after year, or if you have cool, foggy summers, try a rose variety that has fewer petals. The fewer the petals, the less likelihood of balling.

Early-fall freezes can do more damage than severe winters, since they occur before the plant has hardened itself for the cold weather. Fall frosts can damage rose canes, with new growth suffering the most. To prevent these problems, reduce or discontinue fertilization and watering during late summer or early fall. Also cover your plants on nights when the temperature is predicted to fall below freezing.

Canes need chilling for normal growth. If the winter is not cold enough, your roses may grow irregularly in the spring, with certain canes showing no evidence of growth either of leaves or of side buds. The only way to deal with this problem is to prune the affected canes.

Light Problems

If your plants appear spindly and produce few or no flowers, there is probably insufficient sunlight. The only solution is to move your plants to a sunnier location.

On the other hand, if your plants get too much sunlight, particularly during the summer when the temperature exceeds 90 degrees, they will wilt. On such hot days, provide shade for your plants. If the plants are small enough, you can use a beach umbrella.

Water and Soil Chemistry Problems

If your roses are growing very slowly, they may not be getting enough water. Keep the soil around the base of the plant moist.

If the leaves turn yellow and then die, check the soil pH. Very alkaline soils (pH greater than 7.5) cause such damage. If the pH exceeds 7.5, add flowers of sulfur to help acidify the soil. If new leaves are stunted, off-color, and then die, your soil may have too many salts, a possible result of excessive fertilizer use. Drench the area with water (to dissolve the salts), and then water heavily for the next few days in an attempt to wash away the excess salts.

DISEASES

Both *infective* diseases, caused by fungi, bacteria, or viruses, and *non-infective* diseases, caused by environmental problems, attack roses.

Infective Diseases

The top three infective diseases, in terms of prevalence and destructiveness, are black spot, powdery mildew, and rust. Fungi cause all three diseases, and all three have similar life cycles. Each fungus overwinters as a spore on infected canes, buds, and fallen leaves. In the spring, the spore germinates and infects new buds; to do so, it must generally move from its overwintering site to a new bud. During the summer, leaves and flowers become infected by airborne spores produced from the earlier infection.

Most strategies for disease control focus on one or more interventions: preventing the spores from overwintering; preventing overwintered spores from reaching susceptible tissue; preventing spore germination; and killing germinated spores. An IPM approach focuses on the first three strategies; a chemical approach is used for the last.

Black Spot

Natural History

As its name suggests, black spot is characterized by small circular black spots, from ⅟₁₆ to ½ inch in diameter, with ragged edges and surrounded by yellow. The fungus can appear on both surfaces of the leaves. As the disease progresses, the spots can grow and connect to form irregular patches. In severe cases, the leaves turn yellow and fall off, with complete defoliation occurring by midsummer. The disease can also infect canes.

Black spot spores overwinter on fallen leaves and infected canes, as well as inside leaf buds. In the spring, the spores are spread by rain, wind, and insects (the spores are sticky and adhere to any insect that comes in contact with them). If spores reach immature buds and the environmental conditions are appropriate, they germinate and infect the immature leaf tissue. Appropriate environmental conditions include temperatures in the eighties, high humidity, and enough moisture to keep the spores wet for at least seven hours.

Black spot is most severe in areas with warm, wet summers, especially the East and some midwestern states. Light-colored roses seem to be more susceptible to black spot than darker colored varieties.

Warning Signs

Start monitoring for black spot as soon as temperatures reach the mid-sixties in the spring and there is extensive humidity and rain. Roses are most susceptible to disease during their most rapid growth phases, which occur in spring and early summer, and sometimes fall. At these times monitor your plant at least twice weekly, especially during rainy,

Black spot is characterized by small circular black spots, from ⅟₁₆ to ½ inch in diameter, with ragged edges surrounded by yellow.

hot weather. Look for the small black specks, initially on leaves close to the ground and then on buds, leaves, and flowers at the top of the plant.

When to Take Action

Take action at the first appearance of the spot. Immediately remove the infected tissue and dispose of it, preferably by burning; do not put it in a compost pile. Carefully monitor and act quickly whenever signs of infection appear, to prevent a full-blown attack. If the disease worsens in spite of your efforts, take some of the steps described below.

Long-term Controls

Genetic Controls. If black spot infects your plants severely every year, consider switching to a variety that is resistant to, or tolerant of, the disease. In general, the older garden and shrub roses are more resistant, but a few tea, floribunda, and grandiflora varieties also show some tolerance to black spot. Ask your local nursery, rose society, municipal rosarian, or botanical garden for help in locating resistant varieties. Since black spot prefers roses with light-colored flowers, plant varieties with darker flowers.

Cultural Controls. To decrease the humidity that encourages the disease, plant your roses in full sun and increase the space between them to three to four feet to promote air circulation. Pruning away the center of the plant also improves air circulation and reduces humidity there.

> *To decrease the humidity that encourages black spot, plant your roses in full sun and increase space between them.*

To minimize overwintering of spores, rake up fallen leaves and debris in the fall. At this time, or during the early winter, prune any canes with dark blotches on the wood (we assume you've already removed all infected leaves and flowers). To minimize transport of spores to susceptible tissue, spread mulch (compost, wood chips, paper, weed mat) around the bases of the plants after the fall cleanup and again in early spring to cover any overwintering spores. Keep the plants mulched during the spring and summer also. If the weather is very humid, try a landscape mat of the type used to combat weeds.

Chemical Controls. An antitranspirant (also called an antidesiccant) spray can prevent spores from reaching plant tissue. Antitranspirant sprays form a clear, flexible, air-permeable film over the surface of a plant. They keep the plant from losing water, and also repel water, which spores need to germinate. Spray the antitranspirant over your plants in late fall while the temperature is above 40 degrees. The spray should protect your plants from exposure to spores during any warm weather during the winter. It has the added bene-

CAUTION: All Pesticides Have Some Degree of Toxicity

1. Store all pesticides in their original containers in areas where children and pets cannot get at them.
2. Read the label thoroughly before using the product.
3. Minimize your exposure to the pesticide.
4. Wear protective clothing, including goggles, hat, long pants, long-sleeved shirt, rubber boots, and unlined rubber gloves when necessary.
5. Always wear either a dust mask or a respirator mask when spraying pesticides.
6. Thoroughly wash application equipment, hands, and clothing after using pesticides.

fit of preventing your roses from drying out during severe or windy winters.

Some experts also suggest using antitranspirant sprays during the spring and early summer. Although these sprays protect a number of ornamental plants from fungal diseases, there is no data on roses. Since the newer rose varieties grow and bloom continuously from spring to fall, it is not clear how often they would have to be sprayed. (The film will thin or crack, allowing entry points for spores.) Try spraying new growth with the antitranspirant once a week.

Short-term Controls

Cultural Controls. Since the spores require continuous contact with water for seven hours, keep the leaves dry. Use drip irrigation or gently water the ground beneath the plant. Water the plants in the morning on sunny days so that the plants have time to dry out before nightfall. To reduce spread of the spores, do not work in your garden when the plants are wet; you may accidentally transfer spores. Disinfect your tools between use and immediately after working near an infected plant by dipping them in alcohol or bleach and then rinsing in clean water.

Last Resorts

Chemical Controls. Surfactants, fatty acid salts similar to those used for insect control, inhibit fungus infection and kill germinated spores. You can make your own garden surfactant solution by adding 1 to 2 tablespoons of detergent to a gallon of water. You can also use inorganic fungicides, including sulfur-based fungicides or Bordeaux mixture.

Do not use the sulfur-based fungicides when the temperature exceeds 85 degrees, because they may damage

Disinfect your gardening tools between uses and immediately after working near an infected plant by dipping them in alcohol or bleach and then rinsing in clean water.

OUTDOOR CAUTIONS: Pesticides

1. Do not spray pesticides during windy weather.
2. Refrain from applying pesticides to a lawn or garden if there is a creek, river, lake, or pond nearby.
3. Do not smoke while applying or spraying pesticides.
4. Refrain from using granular pesticides on the lawn where birds can consume them.
5. Never use poison baits unless they are in a covered bait station.

---

plants. Apply both surfactants and sulfur-based fungicides during hot, rainy weather—but before you have seen the first evidence of black spot. Continue spraying weekly when your roses are growing rapidly—in spring, early summer, and possibly fall.

Rust

Natural History

Rust is characterized by small red-orange or yellow pustules on green plant parts. The pustules usually first appear on the undersides of leaves and are easy to overlook. Eventually the pustules appear on the upper surfaces of leaves and then on the stems. At the end of the growing season, black overwintering spores appear on these pustules. The spores are spread by rain or wind. Infected canes exhibit dark, corky blotches or scars at the sites of infection.

Although the disease occurs throughout the West, it causes severe problems only on the Pacific Coast, where the conditions are ideal for its development—moderate temperatures and ample water in the form of heavy dew, rain, fog, or cloud cover. Low temperatures in the winter and high temperatures in the summer prevent this disease from moving much farther east.

Warning Signs

Start monitoring for the pustules in the spring as soon as new leaves begin to appear. Initially, focus your search on the undersides of leaves.

When to Take Action

Take action at the first sign of pustules.

Long-term Controls

Genetic Controls. Controls are similar to those for black spot. For recurrent rust problems, switch to a resistant or tolerant rose variety. Ask your local nursery or rosarians which varieties are most suitable.

Cultural Controls. Since rust, like black spot, thrives in high humidity, improve air circulation and decrease humidity: Plant in the sun, increase spacing between plants, and prune the interior of the plant. To minimize overwintering and dispersal of spores, clean up debris in the fall; prune infested canes; add a layer of mulch in late fall and early spring; use antitranspirants during the winter; water soil directly and avoid splashing water on the leaves.

Short-term Controls

Physical Controls. When you find pustules, immediately remove the infected tissue and burn it or throw it out in the trash. Do not compost infected plant parts. Removing infected tissue as soon as it appears may keep the disease under control. If the disease starts to spread, try some of the long-term controls.

Last Resorts

Chemical Controls. Use a flowable sulfur-based fungicide and/or a surfactant (see page 115 for a recipe). When the temperature is mild and there is heavy moisture in the form of dew, rain, or fog, apply these low-toxicity fungicides every week for preventive purposes until the environmental conditions are no longer optimal. You may also want to use an antitranspirant at this time. *Caution:* Wear gloves and a dust mask when spraying sulfur-based fungicides.

Powdery Mildew

Natural History

Powdery mildew attacks succulent young growth, but rarely infects mature foliage. It usually first strikes

Powdery mildew attacks succulent young growth and appears as small raised blisters that cause the leaves to curl up.

young leaves, appearing as small raised blisters that cause the leaves to curl up, exposing the undersides. As the disease progresses, infected leaves become covered with what looks like a grayish white powder. Infected flower buds that become covered with the powder may never open.

Spores overwinter in leaf buds, particularly those just below infected flowers, as well as in infected canes and fallen leaves. In the spring, with days in the mid-sixties, cool nights, high relative humidity, no rain, and low light levels, the spores germinate, are dispersed by the wind, and infect young leaves. Interestingly, although the spores do best in conditions of high humidity and low light (owing to lots of fog or heavy cloud cover), a film of water over the leaves actually inhibits spore germination and growth. The disease creates the greatest problems when the plant is growing most vigorously—in spring, early summer, and possibly fall.

Powdery mildew occurs in many parts of the United States, but it is most serious on the West Coast.

Warning Signs

Begin monitoring in spring as soon as the temperature reaches the sixties.

Search the newly opened buds and young leaves for raised blisters and curling leaf edges. As the season progresses, start looking at all new growth, including stems.

When to Take Action

Take action at the first sign of blisters. Immediately remove the infected leaves and dispose of them in the trash or by burning, but do not throw them in a compost pile. If this preventive pruning does not stop the disease, try one of the measures discussed below.

Long-term Controls

Genetic Controls. If you consistently have problems with powdery mildew, switch to a resistant or tolerant variety of roses.

Cultural Controls. To minimize the numbers of overwintering spores, rake up the fallen leaves and trash from around your plants and discard them. Spores can overwinter in leaf buds, and buds below infected flowers are likely to become infected. Since infected leaf buds show no outward signs of infection, prune off all the buds just below flowers that were infected at the end of the season. Also prune away any infected stems. Severely prune plants that were badly infested during the previous season, because they may contain overwintering spores. Prune during the late fall when you are hardening your plants for the winter.

Chemical Controls. Spray your plants with an antitranspirant in late fall to protect them from desiccation and to prevent powdery mildew spores from infecting the plant during uncharacteristic warm weather during the winter.

Changing fertilizing patterns can also influence the disease. Powdery mildew primarily attacks succulent young foliage. Heavy fertilizer use produces lush but weak growth that is susceptible to attack. Reduce use of nitrogen fertilizers. Use a slow-release ammonium fertilizer, mulch, or compost in late fall and early spring.

Short-term Controls

Physical Controls. Spray roses with water. The water physically removes both germinated and ungerminated spores from the plant and prevents the spores that remain from germinating. When you have cool nights, warm days, and high humidity but no rain, begin vigorously hosing down your plants, making sure to get both sides of the leaf wet. Do this once or twice a week, preferably in the early

To reduce powdery mildew, spray roses with water. The water physically removes both germinated and ungerminated spores from the plant and prevents the remaining spores from germinating.

afternoon so that your plants have time to dry before nightfall. Allowing them to remain wet encourages other fungal diseases such as rust, black spot, blight, or leaf spot.

Last Resorts

Chemical Controls. When the environmental conditions are right, start weekly prophylactic sprays of surfactant, a flowable sulfur-based fungicide, or Safer's Fungicidal Soap (see page 115). *Caution:* Wear gloves and a dust mask when spraying these substances.

You can also try using a solution of 2 to 6 teaspoons of baking soda per gallon of water; spray at the first sign of pustules and whenever new symptoms appear, or spray weekly.

 # WEEDS

The weed problems encountered in a rose garden are no different from those encountered in a regular garden (see chapter 4).

If you have enough time before planting, consider solarizing the soil in your rose bed or covering it with newspapers for a full year. In either case, you must start such treatment the summer before planting. Newspaper forms an impermeable layer, effectively stopping weed germination. It is also easy to lay down and remove a year later. Solarizing the soil takes more work, but has the added benefits of killing many fungal diseases and stimulating beneficial fungi and bacteria. If you don't have the lead time of a year, use mulch for a shorter time. Finally, there is the old-fashioned alternative—elbow grease and weeding by hand or hoe.

To minimize weeds, solarize the bed or cover with newspapers or mulch.

 # INSECTS

The first step in controlling insects is to monitor your roses regularly and look for leaves with chewed sections, holes, discolored spots, wilting and discoloration, or curling. If you find any of these symptoms, look for and identify the insect that is causing the damage.

Spray horticultural or dormant oil to help control certain insect pests that attack roses.

Since roses are woody plants, you can use horticultural or dormant oil to help control certain insect pests. Spray it on your rose bushes in the late fall, after all the leaves have fallen off, and in the early spring, before buds open. Do not spray the oil when leaves are present, as it can damage them. The oil covers the stems and suffocates aphids, scale, thrips, spider mites, and mealybugs. The fall spray also smothers most overwintering eggs or larvae. If you mix some flowable sulfur fungicide with the horticultural oil, you help control many diseases by killing spores on the plants and providing a barrier to new spores landing.

Normal horticultural oils are restricted to use on deciduous woody plants when they have no foliage. However, a new class of horticultural oil has been developed that has a much lower viscosity and does not damage foliage. This oil, usually called superior horticultural oil, can be used on woody ornamentals during the growing season as a weapon against numerous insect pests. Use it in a 2-to-3-percent solution. Since the foliage of some plants may be sensitive to even the superior horticultural oil, spray a small portion of the plant and wait for a couple of days to see if the leaves turn yellow or die. If not, it can be used against a wide range of sucking and chewing pests. *Caution:* Wear a mask when spraying horticultural oils.

Aphids

Natural History

For general information on aphids, see page 77. Aphids suck juices from rose plants and tend to occur on the undersides of leaves, although flowers and stems can be attacked. If the infestation is heavy enough, the leaves turn yellow or brown and start to curl or pucker, and the flowers become malformed.

There is also an aphid species that attacks rose roots. Root aphids cause similar symptoms, and in addition the leaves may wilt in bright sunlight and the plants grow poorly because the root damage interferes with the flow of water and nutrients. The effects are most severe in young, recently transplanted bushes.

Warning Signs

Aphids' natural enemies should be present unless you have been regu-

Aphid infestation can turn leaves yellow or brown, cause curling and puckering, and result in malformation of flowers.

larly spraying insecticides. Ordinarily, aphids are a problem in the early spring and late fall, before their enemies arrive and after they leave.

Begin monitoring your plants in the spring, as soon as the first buds open. At first, focus on the buds and growing tips, where aphids tend to congregate; then look on the undersides of older leaves.

Monitoring for root aphids is more difficult. If you see signs of aphid damage, but there are no aphids on the leaves, check the roots. Damaged roots are scarred or knotted, and you can see aphids there. If you notice ants in the soil at the base of the plant, there is a strong possibility that they are tending the root aphids.

When to Take Action

Carefully observe when the aphids first arrive and when their natural enemies arrive, and then decide whether you can tolerate the damage during the lag time. An experiment in northern California found that roses could tolerate up to 10 aphids per bud without suffering significant damage. Determine your tolerance level.

Long-term Controls

Cultural Controls. Follow the fertilization guidelines described on page 78. Avoid highly soluble nitrogen fertilizer; use less soluble synthetic or organic fertilizers, and spread the applications throughout the growing season. Add some rock phosphate or bone meal. Conserve aphids' natural enemies by minimiz-

 Plant wildflowers to help protect against aphids.

ing or eliminating pesticide sprays, especially early in the season when natural enemies become established. Plant wildflowers, particularly those with numerous small flowers, such as carrot, Queen Anne's lace, fennel, dill, oleander, alfalfa, and clover to attract parasitic wasps and flies. Don't bother to buy ladybug beetles or praying mantises to release in the garden.

Chemical Controls. If you have had problems in the past, spray your roses with a regular horticultural oil in the early spring as a general preventive measure.

Biological Controls. For root aphids, drench the soil around the bases of infected plants with parasitic nematodes, which can be applied mixed with water in a watering can.

Short-term Controls

Physical Controls. Wipe or brush the aphids off the leaves, and prune the growing tip if the infestation is bad enough. For light infestations, spray the plants with water in the afternoon twice a week for a couple of weeks. If ants are tending the aphids, place a 1-to-2-inch band of sticky material (Tanglefoot, Stickem, Tack Trap) around the base of each cane to prevent the ants from crawling up the plant. When using the sticky material, make sure that none of the branches or other parts of the plant touches the ground or any other plant; otherwise the ants will use these alternative pathways to get to the aphids.

Last Resorts

Chemical Controls. For light to moderate infestations, spray a light superior horticultural oil. Test the plant for susceptibility to the oil by spraying a few leaves and then waiting for a day or two to see if the leaves become discolored. An insecticidal soap, applied in the morning, can help stop medium to heavy infestations. For heavy infestations, and as a last resort, use a neem-, pyrethrum-, or pyrethrin-based product.

With all chemical sprays, spray only infested plants and only those parts of the plant that contain aphids. Do not use pyrethroids near water; they are highly toxic to fish and some aquatic invertebrates. *Caution:* Some people are allergic to the pulverized flowers that are the source of pyrethrum. Wear a dust mask or respirator if you are allergic to pollen and are spraying pyrethrum, or choose a different chemical.

Beetles

Natural History

Many beetles, especially Japanese beetles, rose chafers, rose curculios, rose leaf beetles, fuller rose beetles, and goldsmith beetles, can cause significant damage to roses. The adult beetles eat leaves or flowers and leave large holes. The larvae of some species feed on the roots. Large numbers of adults can destroy flowers and reduce the leaves to bare skeletons. To be sure of the identity of any beetles causing damage to your roses, collect some and take them to your county Cooperative Extension Service (or call and describe them over the phone), or look in one of the pest identification books listed in the References and Resources section.

The adult rose chafer is about ¼ inch long, has a tan body and long legs, and feeds on flowers. The larvae are 1 to 1½ inches long, C-shaped, plump, and white with brown heads. The adults breed in sandy soils. The larvae feed on turfgrass roots and are usually found in lawns near unused farmland or open fields. They are a major problem in the Northeast, but are found as far west as Colorado.

Fuller rose beetles are weevils about ⅓ inch long, gray-brown with a cream-colored stripe on either side of the body. The adults hide during the day and come out at night to feed on leaves. The larvae are yellowish with brown heads; they feed on rose roots and may girdle stems. They are found in the South and in California.

 Heavily chewed leaves or flowers may be signs of beetles.

Adult rose curculios are small weevils (a type of beetle), about ¼ inch long, with a red body and a long black snout. The adults chew holes in rosebuds, while the white larvae feed on seeds and flowers. The larvae drop to the ground to pupate and overwinter. The rose curculio is found throughout the United States, but is usually a problem only in the colder northern regions, especially in North Dakota.

The goldsmith beetle is about 1 inch long and has a hairy, lemon yellow body. The adults feed on the leaves, while the larvae, which are white grubs, feed on the roots of roses and other ornamentals. They cause problems throughout the East and Southwest.

The rose leaf beetle is ⅛ inch long, oval-shaped, with a metallic green or blue sheen. The adults bore into the flower buds, but also eat young shoots and foliage. The larvae feed on the roots of rose plants. They are found throughout the United States.

For a description of the Japanese beetle, see chapter 3.

Warning Signs

Look for the adult beetles themselves, or signs of their damage, such as leaves or flowers that are heavily chewed.

When to Take Action

Take action when the level of damage to the leaves or flowers exceeds your personal threshold, or when you see more than two or three adults per plant.

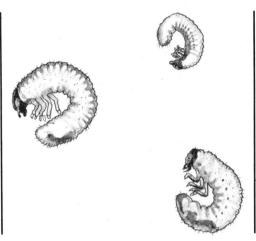

WHITE GRUBS

The larvae of many different beetles, called grubs, attack turfgrass. Though they vary from ¼ inch to 2 inches in length, all have a similar appearance.

Long-term Controls

Cultural Controls. To combat rose curculio, remove any dry, damaged buds, which can harbor larvae. Also clean up garden debris around the plants in the fall to remove any overwintering pupae.

Biological Controls. Predatory nematodes, sold through mail-order and gardening supply houses (see References and Resources), can control the larvae of the rose chafer and the fuller rose, goldsmith, Japanese, and rose leaf beetles. Mix the recommended number of nematodes with water and spray or sprinkle them where the beetle larvae are. For larvae of fuller rose, goldsmith, and rose leaf beetles, apply the nematodes around

Predatory nematodes, sold through mail-order and gardening supply houses, can control the larvae of the rose chafer and the fuller rose, goldsmith, Japanese, and rose leaf beetles.

the base of the rose plant. For goldsmith beetle, also apply the nematodes to the soil around any other ornamentals that are attacked. For both Japanese beetle and rose chafer, apply the predatory nematodes to the lawn. After application, water the area thoroughly because the nematodes need a film of water to attack the larval grubs.

Milky spore disease controls larvae of the Japanese beetle, rose chafer, goldsmith beetle, and rose leaf beetle. Follow label directions when applying it to the soil, and water the area after application to help move the milky spore disease into the soil. Milky spore disease takes one season or more to work fully. *Caution:* During application, wear a dust mask to avoid inhaling spores.

Short-term Controls

Physical Controls. Physical controls are most appropriate for the adults. You can use a hand-held vacuum to suck adults off plants. Otherwise, handpick or shake the adults off the plants and drop them into a bucket of soapy water. Do this early

in the morning, while it is still cool and the beetles are sluggish.

To combat the fuller rose beetle, apply a band of sticky material to the trunk of the plant to prevent adults from climbing up.

Last Resorts

Chemical Controls. Adults of all species can be controlled with chemicals. Use a pyrethrum or pyrethrin-based product. *Caution:* Some people are allergic to the pulverized flowers that are the source of pyrethrum. Wear a dust mask or respirator if you are allergic to pollen and are spraying pyrethrum, or choose a different chemical. If you absolutely must use something stronger, use a synthetic pyrethroid product. All of these pesticides kill natural enemies, so spray only the infested tissues and spray during the heat of the afternoon, when natural enemies are least active. (Since adults of fuller rose beetles are active at night, spray them about two hours after dark.) Neem tree extract (BioNeem) has antifeedant/repellent properties; try spraying roses with it to see how well it repels these beetles. *Caution:* Wear a dust mask when spraying neem extracts.

Caterpillars

Biology

Caterpillars, the larval stage of a number of moths and butterflies, attack rose plants, feeding on leaves, buds, and sometimes flowers. Although a large number of species feed on roses, the most common ones are the rose budworm, fall webworm, bristly rose slug, and various leaf rollers. Although all caterpillars have the same basic appearance, leaf rollers have a distinctive behavior: they roll the leaves into tubes, which they stick together with silk, and then feed on the rolled-up leaf. For information on caterpillar pests, contact your county Cooperative Extension Service or the entomology department of your state's agricultural college.

Heavily chewed leaves, silk nests, or rolled-up leaves may be signs of caterpillar damage.

Warning Signs

Look for the caterpillars themselves, signs of their damage—chewed leaves—or signs of their presence, such as silk nests (fall webworm) or rolled-up leaves (leaf rollers). Start monitoring in early spring and continue every week or two throughout the summer.

When to Take Action

Take action when you see caterpillars or when they begin to cause significant damage to leaves or flowers.

Long-term Controls

Cultural Controls. Encourage caterpillars' natural enemies by keeping a diverse array of plants growing in and around your garden. Try to have plants that flower throughout the season. Minimize use of synthetic pesticides, which kill natural enemies.

Short-term Controls

Physical Controls. First, handpick any webworm caterpillars and nests off the plants. Remove rolled-up leaves, which will contain leaf rollers, and dispose of them.

Biological Controls. Spray the plants with *Bt* once a week until all the caterpillars and symptoms disappear. *Caution:* Wear a dust mask when spraying *Bt.*

Last Resorts

Chemical Controls. Spray plants with an insecticidal soap or neem oil product at regular intervals. For leaf rollers, spray a horticultural oil in the early spring before the buds open; this will smother the eggs. *Caution:* Wear a dust mask when spraying these substances.

Leafhoppers

Natural History

Leafhoppers are small wedge-shaped insects that hold their wings in a rooflike position above their bodies. As their name implies, leafhoppers hop when disturbed. Both adults and young nymphs suck juices from leaves, buds, and stems and may leave behind small white spots. Some species also secrete honeydew, which can give rise to a sooty mold. If leafhopper numbers are high enough, the leaves shrivel up and die. A number of leafhopper species can be found on roses, but the rose leafhopper is a particular pest. The rose leafhopper's eggs hatch in May, and the nymphs feed on the undersides of leaves.

Warning Signs

Start monitoring your plants in May. Look for white spots on leaves and for the leafhoppers on the undersides of leaves. Monitor the plants every week or two. If you find leafhoppers, monitor twice a week.

White spots on leaves may be a sign of leafhoppers.

When to Take Action

If you find just a few leafhoppers, monitor them carefully. If the numbers increase and they spread, begin control measures.

Long-term Controls

Physical Controls. Cover your rose plants with floating row covers in early spring to keep leafhoppers from colonizing the plants.

Cover your rose plants with a floating row cover in early spring to keep leafhoppers from colonizing.

Short-term Controls

Chemical Controls. Spray an insecticidal soap or light superior horticultural oil, but only on infested areas of a plant. Test the oil on a few leaves to see if it damages them, which will usually be evident after one or two days. Use a dust mask when spraying insecticidal soap.

Last Resorts

Chemical Controls. Use a neem-pyrethrum or pyrethrin-based product. Spray the undersides of the leaves, where most of the leafhoppers are found. Do not use pyrethroids near water; they are highly toxic to fish and some aquatic invertebrates. *Caution:* Some people are allergic to the pulverized flowers that are the source of pyrethrum. Be sure to wear a dust mask or respirator if you are allergic to pollen and are spraying pyrethrum, or choose a different chemical.

Mites

Natural History

Mites are tiny, spiderlike arthropods (they are not insects) about the size of a grain of pepper. They can be found in a range of colors, including yellow, red, green, and brown. One group of mites, the spider mites, causes particular damage. They spin small webs on the undersides of leaf surfaces and on new growth, and then begin feeding. Feeding damage consists of yellow, red, or brown stippling. As the damage continues, the leaves may curl up and fall off the

Webs on new growth, or yellow, red, or brown stippling on leaves may indicate mites.

plant. Some varieties of floribunda roses are very sensitive to mite damage and prematurely drop their leaves. (See also the discussion in chapter 4.)

Warning Signs

Start monitoring for mites in the spring just as buds are opening. Look for webs on the new growth and for small moving dots on the undersides of leaves. If you find only a couple of mites on a plant, wait for natural enemies to exert control.

When to Take Action

Take action when mite numbers increase steadily over the course of a couple of weeks.

Long-term Controls

Genetic Controls. If mites are a continual problem and you have susceptible floribundas, consider switching to a more tolerant or resistant variety.

Chemical Controls. As a preventive, spray on a horticultural or dormant oil in early spring before any buds open.

Short-term Controls

Physical Controls. As soon as you notice mites, spray the plants, especially the undersides of the leaves, with water. This removes the mites from the plants and kills them. Do this once a day for three or four days. If mite damage continues, consider a chemical spray.

Last Resorts

Chemical Controls. Use a light superior horticultural oil, but first test the oil on a few leaves to see if it damages the foliage. (Wait a day or two for any damage to appear.) With either spray, coat the undersides of the leaves. Spray the plants every three to four days for a couple of weeks. Continue to monitor the plants to see if mites persist. You can also use a pyrethrum- or sulfur-based product. *Caution:* Wear a dust mask when spraying these substances. Some people are allergic to pyrethrum and must not inhale the chemical.

CHART 5-1
Products to Control Rose Pests

The products listed below are a fraction of those pesticides marketed to control rose pests. These products were selected for listing based on CU's judgment that they can be effective when used in the context of an Integrated Pest Management strategy, and that they pose the least risk to humans, pets, or the environment, based on the active ingredients they contain.

Some products not listed here contain one or more of the same active ingredients as these products and may be substituted for them. But many widely available products are not listed because they contain active ingredients that, in CU's judgment, pose greater potential risks to health or the environment than the products listed. In our view, effective pest control does not require use of more toxic pesticides, and we have chosen not to list products that contain them. Products are listed in alphabetical order.

Recommended products listed below contain the least hazardous active ingredients, including one or more of the following: *Bacillus thuringiensis*, either *kurstaki* (BTK) or *sandiego* (BTS), fatty acids (FA), neem (or azadirachtin [AZ]), nematodes (*Steinernema feltiae*, or SF), petroleum (or paraffinic) oil (PO), sulfur (SU), pyrethrins (PY), or pyrethroids, such as allethrin (AL), phenothrin (PH), resmethrin (RS), or tetramethrin (TM).

Other products listed below also contain synergists, such as MGK (M2) or piperonyl butoxide (PB), rotenone (RO), and/or petroleum distillates (PD), whose chronic health effects remain unknown.

Code for pests controlled: aphids (A), beetles (B), beetle larvae (BL), caterpillars (C), leafhoppers (L), mites (M).

BRAND NAME	ACTIVE INGREDIENT(S)	PESTS
Recommended		
Ace Hardware House & Garden Bug Killer II	PH, TM	A, B, C, L, M
BioLogic SCANMASK	SF	BL
Bonide Dipel .86% WP	BTK	C
Dexol Aphid, Mite, and Whitefly Killer	PH, TM	A, B, C, L, M
Dexol Worms Away	BTL	C
Dipel Worm Killer	BTK	C
Gro-Well Horticultural Oil Spray	PO	A, M, L
Gro-Well Thuricide-HPC Insect Control	BTK	C
Gro-Well Wettable Sulfur Dust or Spray	SU	M
Hot Shot House & Garden Bug Killer	PH, TM	A, B, C, L, M
HWI Rose & Garden Insect Killer	PH, TM	A, B, C, L, M

Brand Name	Active Ingredient(s)	Pests
Hydro-Gardens Guardian	SF	BL
K-mart House & Garden Bug Killer II	PH, TM	A, B, C, L, M
Ortho BioSafe Soil Insect Control	SF	BL, C
Ortho Bt Biospray	BTK	C
Ortho Dipel Caterpillar Spray	BTK	C
Ortho FLOTOX Garden Sulfur	SU	M
Ortho Garden Sulfur Dust or Spray	SU	M
Ortho Horticultural Spray Oil	PO	A, C, L, M
Ortho Insecticidal Soap	FA	A, C, L, M
Ortho Volck Oil Spray	PO	A, C, L, M
Perma-Guard Plant & Garden Insecticide	DE	C
Raid Multi-Bug Killer D39	AL, RS	A, B, C, L, M
Real-Kill House & Garden Bug Killer	PH, TM	A, B, C, L, M
Safer BioNeem	AZ	B, C
Safer Bt Caterpillar Attack	BTK	C
Safer Fruit & Vegetable Insect Attack	FA	A, C, L, M
Safer Leaf Beetle Attack M-One	BTS	B
Safer Rose & Flower Insect Attack	FA	A, C, L, M
Safer Tree & Shrub Insect Attack	FA	A, C, L, M
Safer Vegetable Insect Attack	BT	C
Safer Yard and Garden Insect Attack	FA, PY	A, B, C, L, M
SA 50 Citrus & Ornamental Spray	PO	A, C, M
SA 50 Thuricide-HPC	BTK	C
Spectracide Lawn & Garden Insect Control 1	PM	A, B, C, L, M
Other		
Ace Hardware Flower & Vegetable	PB, PY	A, B, C, L, M
Ace Hardware House and Garden Bug Killer II	PB, PD, PY	A, B, C, L, M

Products to Control Rose Pests (*continued*)

Brand Name	Active Ingredient(s)	Pests
Acme Garden Guard	RO	A, B, C, L, M
Black Flag House & Garden Insect Killer	AL, PB, PD, RS	A, B, C, L, M
Dexol Rose & Floral Insect Killer	PB, PY	A, B, C, L, M
Dexol Vegetable Insect Killer	PB, PY	A, B, C, L, M
Gro-Well 1% Rotenone Dust or Spray	RO	A, B, C, L, M
Gro-Well Organic Insecticide	PB, PD, PY, RO	A, B, C, L, M
Gro-Well Rose & Flower Insect Killer	PB, PY	A, B, C, L, M
Gro-Well Vegetable & Tomato Insect Killer	PB, PD, PY	A, B, C, L, M
Gro-Well Vegetable Tomato Dust	RO, CU	A, B, C, L, M
Meijer's Rose & Flower Insect Killer	PB, PY	A, B, C, L, M
Ortho Home & Garden Insect Killer Formula II	PB, PY, TM	A, B, C, L, M
Ortho Rose & Flower Insect Spray	PB, PY	A, B, C, L, M
Ortho Rose & Flower Insect Killer	PB, PY	A, B, C, L, M
Ortho Rotenone Dust or Spray	RO	A, B, C, L, M
Ortho Tomato & Vegetable Insect Spray	PB, PY, RO	A, B, C, L, M
Ortho Tomato & Vegetable Insect Killer	PB, PY	A, B, C, L, M
Raid House & Garden Formula 8	AL, PD, RS	A, B, C, L, M
Raid House & Garden Formula 11	PB, PY, TM	A, B, C, L, M
Spectracide Rose & Garden Insect Killer	PB, PY	A, B, C, L, M

CHART 5-2
Products to Control Rose Diseases

The products listed below are a fraction of those pesticides marketed to control rose diseases. These products were selected for listing based on CU's judgment that they can be effective when used in the context of an Integrated Pest Management strategy, and that they pose the least risk to humans, pets, or the environment, based on the active ingredients they contain.

Some products not listed here contain one or more of the same active ingredients as these products and may be substituted for them. But many widely available products are not listed because they contain active ingredients that, in CU's judgment, pose greater potential risks to health or the environment than the ingredients of products listed. In our view, effective pest control does not require use of more toxic pesticides, and we have chosen not to list products that contain them. Products are listed in alphabetical order.

Recommended products listed below contain the least hazardous active ingredients, including one or more of the following: antitranspirant (AN), baking soda (BS), Bordeaux mixture (BM), and sulfur (SU).

Code for diseases controlled: black spot (BK), powdery mildew (PM), and rust (R).

BRAND NAME	ACTIVE INGREDIENT(S)	DISEASE
Acme Bordeaux Mixture	BM	BK
Acme Wettable Dusting Sulfur	SU	BK, PM, R
Arm & Hammer Baking Soda	BS	PM
Gro-Well Bordeaux Mixture	BM	BK
Gro-Well Wettable Sulfur	SU	BK, PM, R
Nature's Touch Leaf Cote Clear	AN	BK, PM, R
Ortho FLOTOX Garden Sulfur	SU	BK, PM
PBI-Gordon Transfirm	AN	BK, PM, R
Precision Laboratories Preserve	AN	BK, PM, R
Safer Plant Protectant	AN	BK, PM, R
That Liquid Sulfur	SU	BK, PM, R
Wilt-Pruf	AN	BK, PM, R

PART THREE

INDOOR PESTS

6

HOUSEHOLD INSECT CONTROL

We expect to see bugs outdoors, but all it takes is one spider indoors to make us overreact and spray poisons indiscriminately all over the home. Every household has had problems with insect pests at one time or another. Termites, carpenter ants, clothes moths, and pantry beetles can cause significant damage to the house itself or its furnishings. Flies, ants, mosquitoes, cockroaches, and spiders cause more annoyance than real damage. In any case, there is no reason to panic. Instead, remember the principles of IPM and proceed with caution and moderation.

 ## STRUCTURAL PESTS

Virtually all wood damage is caused by either fungi or insects. Discolored wood is usually a sign of fungi; holes, tunnels, galleries, sawdust, powder, or piles of tiny black fecal pellets are signs of insects. Wood-damaging insects include termites, carpenter ants, carpenter bees, and various beetle species. Here we discuss only termites and carpenter ants, which cause the bulk of the serious damage.

The first step in an IPM program to control structural pests is an annual inspection of your home, both inside and out, for wood damage. To determine which areas in and around the home to inspect, get a checklist from your county Cooperative Extension Service, pest control company, or the federal government (see References and Resources). Search these areas systematically and mark the areas where you find insects or damage on a small map of your house.

In general, look first for the conditions that attract insects, then for damaged wood, and finally for signs of the insects themselves. Most structural pests like damp wood, although dry, cracked wood can attract drywood termites, which are common in the western United States. If the wood feels soft or crumbly, look for piles of sawdust, tunneling in the wood, piles of fecal pellets, or other indications of insect presence or activity.

If you don't want to do the inspection yourself, hire a professional. Some inspection services use beagles to sniff out termites and carpenter ants. Inspection services that use dogs cost more, but dogs are far more effective than humans in detecting an infestation.

Consider monitoring and treatment as separate activities. Don't suggest to the firm you hire for the termite inspection that it will be the one you intend to employ for treatment. This will minimize the possibility of unnecessary treatment.

Termites

Natural History

Termites live in colonies that consist of one or more pairs of reproducing kings and queens and large numbers of workers and soldiers. The colonies vary in size from hundreds to millions, depending on the species and the habitat. All termites have the same generalized life cycle. New colonies are formed when old colonies produce large numbers of winged members capable of reproduction, a process called *swarming.* After swarming, males and females pair off, find a suitable nesting site, and drop their wings. They establish a small colony that gradually grows.

Broadly speaking, termites can be classified into two major groups: subterranean and nonsubterranean. The former live in the ground or must have contact with it; the latter can live solely in the wood they infest.

The first step in an IPM program to control structural pests is an annual inspection of your home for wood damage, both inside and out.

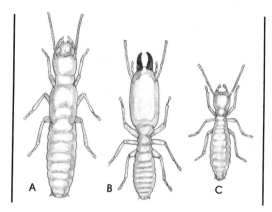

EASTERN SUBTERRANEAN TERMITES

Supplementary queen (A), soldier (B), and worker (C) castes.

Subterranean termites cause most of the damage to houses.

Subterranean termites are found throughout the United States, particularly in moist soils. Two types generally cause the problems: *Reticulitermes* and *Formosan*. *Reticulitermes* are small to medium-size softbodied insects that are very sensitive to drying out. They must therefore live in moist soil and can attack only wood that is within reach of the soil. They build mud tubes that protect them from predators (mainly ants) and desiccation, and serve as passageways between the soil and the wood.

Formosan termites are able to attack a wide range of materials and create serious damage in a short time. Present control methods are relatively ineffective against them. They are less susceptible to desiccation than other subterranean termites, and so can build aboveground nests as

well as subterranean ones. Colonies are large, frequently containing 2 to 3 million members. They are found in Alabama, Florida, Hawaii, Louisiana, Mississippi, South Carolina, and Texas.

Three types of nonsubterranean termites cause problems: *drywood*, *dampwood*, and *powderpost*. Drywood termites are found primarily in southern and coastal areas and can establish themselves in fairly dry sites, entering wood through cracks and crevices. Drywood termite damage may be found close to the ground or above it, such as in roofs, sills, and posts. Dampwood termites are found in the West, primarily in California, in damp, decaying wood. Powderpost termites are found in southern California, Florida, Louisiana, and Hawaii, and usually inhabit furniture or other wooden items imported from subtropical areas.

Warning Signs

Look for damaged or chewed wood, swarming insects near an indoor foundation, mud tubes, sawdust, powder, tiny black or brown fecal pellets, or a small pile of insect wings.

Pellets and wings can indicate the presence of either termites or ants. Collect specimens of the different types you see for identification. If the insects are ants, see the section beginning on page 142.

Damaged wood contains holes or tunnels and doesn't sound or feel solid when hit with a hammer or

TABLE 6-1

Character	Ants	Termites
■ DIFFERENCES BETWEEN ANTS AND TERMITES ■		
antennae	elbowed	not elbowed
waist	narrow	broad
wings (when present)	forewing larger than hindwing	forewing and hindwing same size
young vs. adults	young (larva, pupa) do not resemble adults; young not mobile	mobile young (nymphs) resemble adults, only smaller

poked with a screwdriver. But not all damaged wood is currently inhabited by an active colony. Mud tubes indicate subterranean termites, while fecal pellets indicate nonsubterranean termites. Piles of wings found in the spring usually indicate subterranean termites, while those found in the summer and fall usually indicate drywood termites.

❧ *Fecal pellets and discarded wings can indicate the presence of either termites or ants. Collect specimens of the different types of ants and termites you see for identification.*

When to Take Action

When you discover damaged wood, determine whether it contains an active termite infestation, and, if so, how large. Since termites gener-

ally take a long time to create damage, you don't need to take immediate action. Find out what your options are and decide what you want to do. Always take nonchemical and least toxic measures before moving on to strong pesticides. In general, take action whenever you find an active termite infestation. When using chemicals, remember to treat only the infested area.

Long-Term Controls for Subterranean Termites

Physical Controls. Modify the habitat to ensure that wood does not come into contact with the soil. If possible, remove wooden parts of the house that are in contact with the soil, or wood that is damaged, and replace them with chemically treated wood or a barrier such as a concrete or metal shield. Wood that is within 1½ feet of the soil should also be protected with a concrete or metal barrier. If repair or replacement is not feasible, treat all wood close to or touching the soil with a preservative. Buy chemically pretreated wood or treat the wood yourself. Preservatives containing copper naphthenate, zinc naphthenate, polyphase, or borate as the active ingredients are the least toxic. *Caution:* Wear rubber gloves, protective clothing, goggles, and a respirator mask when handling these substances.

When building a house, keep wood about 1½ feet away from the soil or use redwood, cedar, cypress, or oak, which have some resistance to termites. Coarse sand (ranging in size from 10 to 16 mesh) under or around a foundation, or in crawl spaces, makes a good barrier because most subterranean termites cannot penetrate it. Remove food sources such as tree stumps, construction debris, form boards, and buried wood from underneath the house or near the foundation, so that termites don't use it to build up their population before attacking the house itself.

Finally, minimize the moisture in the wood, particularly wood close to the ground. Repair leaks on the exterior or interior of the house; relocate or modify vents or drains; install

---

When building a house, keep wood about 1½ feet away from the soil or use redwood, cedar, cypress, or oak, which have some resistance to termites.

---

vapor barriers, such as a sheet of plastic, to prevent moisture in the soil from coming into contact with wood; install sump pumps. Use a moisture meter to tell you if you've actually reduced the moisture level in the wood. Have downspouts deliver water several feet from the house.

Biological Controls. Formosan subterranean termites require more drastic action than other termites because of their huge colony sizes, aggressive foraging, ability to live aboveground, and higher tolerance to chemical pesticides. Take action against both the underground and aboveground populations. If the underground colony is not large, try nematodes (particularly *Steinernema feltiae*) to control it. Mix the nematodes with water at the concentration recommended by the supply house, and spray them into exposed portions of the nest. Soak the nest with water to enable the nematodes to attack the termites.

Chemical Controls. It is very difficult to kill an entire underground colony; you may have to settle for treating the soil underneath the house (via a slab injection) with a pesticide, in the hope of creating an unbroken layer of pesticide between soil and house that prevents termites from crossing. Pyrethroids, either permethrin or cypermethrin, repel Formosan termites; the organophosphate and carbamate insecticides registered for termites do not. (Of the two pyrethroids, permethrin is more repellent and cypermethrin is more toxic to termites.) Slab injection must be done by professionals. Request that they use these compounds. See chapter 9.

If there is an aboveground nest, locate the source of moisture the termites are using to maintain the nest and eliminate it; it may be something as simple as a leak in the roof or a pipe. If you get rid of the moisture, the colony will dry up and die. If you cannot find the source of moisture, look for the nest itself (we recommend hiring an inspector with a specially trained beagle) and destroy it with heat, cold, electricity, or chemicals (see below on short-term control of drywood termites).

DRYWOOD TERMITES

Physical Controls. Seal cracks in wood or breaks in the outer skin of the house. Look carefully at corners, wall edges, siding-chimney or siding-roof contacts, exposed beam ends. Remove moisture sources that the termites need. Since drywood termites attack from aboveground, pay attention to wood in the upper areas of the house. Outdoors, for evidence of moisture damage or cracks leading to the interior, look on balconies and landings, windowsills, shingle roofs, eaves, roof overhangs, fascia boards, and gutters. Indoors, look for leaky faucets, pipes, or other areas of condensation, particularly around sinks, floor drains, vents, showers, tubs, and toilet bowls. Check ceilings and walls in rooms beneath bathrooms or showers, areas in contact with glass, interior walls beneath windows, and the attic.

Chemical Controls. In addition to sealing as many outside cracks as possible, dust the attic with a pesticidal or desiccating dust containing boric acid, diatomaceous earth, or silica gel just prior to the swarming season. Ask your county Cooperative Extension Service when that is for the type of termites you have. Attics are frequently the point of entry for drywood termites. Use a bulb duster or hand duster for smaller areas and rent a dust machine for large areas. When using a bulb or hand duster, shake it to agitate the dust inside to produce electrostatically charged particles, which cling to surfaces more readily. Cover exposed wood in the attic with a *thin* layer of dust. Be sure to wear a respiratory mask and goggles when applying these dusts.

DAMPWOOD TERMITES

Physical Controls. Locate the source of water infiltration and eliminate it. Replace rotted wood.

Short-term Controls for Subterranean Termites

Physical Controls. If possible, locate and destroy the nest. If the nest is small and accessible, dig it out.

Biological Controls. Destroy any mud tubes and, if possible, follow the mud tubes back to the nest and dig it out or expose it. This makes the termite colony vulnerable to their major natural enemy, ants, which may kill the colony, if it is not too big. Or use nematodes (particularly *Steinernema feltiae*). Mix them with water and sprinkle them in the soil near, or on, the termite nest.

Chemical Controls. Wood injection is particularly appropriate if there is a well or cistern beneath the house, if the water table is close to the surface (in such instances a soil termiticide could contaminate groundwater), or if there is a high population of termites, particularly Formosan termites, infesting the wood. Inject insecticide directly into damaged wood. Only one injection should be necessary, but check the wood in a few months to see if it has been reinfected. You may have to hire a professional pest control operator to do the injection, if you cannot find the equipment.

Try an insecticidal soap, boric acid, and pyrethroids, particularly permethrin, cypermethrin, and fenvalerate, or have a professional pest control operator treat unfinished wood with borate (BoraCare or Borid). It is absorbed throughout the wood to give total protection against both insects and fungi.

If you cannot destroy the nest, or if the infestation is too large, try a slab injection. In this technique, a pesticide is injected underneath or around the foundation to create a layer of pesticide between soil and house. (Slab injection must be performed by a licensed pest control operator.) Pyrethroids, particularly permethrin and cypermethrin, are as effective as, or more effective than, any of the registered organophosphate or carbamate termiticides. Find a professional who uses one of these ingredients. Foam works better than regular liquid for slab injection, giving a more uniform layer.

---

*C*AUTION: *All Pesticides Have Some Degree of Toxicity*

1. Store all pesticides in their original containers in areas where children and pets cannot get at them.
2. Read the label thoroughly before using the product.
3. Minimize your exposure to the pesticide.
4. Wear protective clothing, including goggles, hat, long pants, long-sleeved shirt, rubber boots, and unlined rubber gloves when necessary.
5. Always wear either a dust mask or a respirator mask when spraying pesticides.
6. Thoroughly wash application equipment, hands, and clothing after using pesticides.

DRYWOOD TERMITES

Physical Controls. Destroy indoor nests and infestations with cold, heat, or electricity. You must hire a professional for the cold, heat, and electricity options. The cold procedure involves injecting liquid nitrogen into the infested area, which freezes the termites. The heat procedure involves using a portable heating unit to heat the infested area to a temperature that kills the termites. Heat works well, but only a few pest control firms offer the service. For the electricity method, the professional passes an "electro gun" over the infected surfaces and basically electrocutes the termites in their galleries. The electro gun works best on smaller infestations that can be remedied by spot treatment.

Chemical Controls. If you have to treat the problem chemically, inject boric acid, borate, desiccating dusts, pyrethrins, or a synthetic pyrethroid directly into the wood. You will probably need to hire a professional to perform this procedure.

Carpenter Ants

Natural History

Carpenter ants are black, about ½ inch long. They nest in moist or rotting wood, and are found throughout the United States. In many areas, the ants live in single colonies with only one nest. In the Pacific Northwest, the Northeast, and other high rainfall areas, carpenter ants may live in bigger, more diffuse colonies, consisting of a main or parent colony (usually located in a log, stump, or tree) and from 1 to 10 satellite colonies, each with a small nest, located in a woodpile or a house. Trails usually connect the main colony and the satellites.

Warning Signs

Look for big black ants, often with hairs on the abdomen, that move quickly and appear able to notice movement a couple of feet away.

Look for dead ants, particularly winged ones, on windowsills or in light fixtures. Large numbers usually indicate that a nest is nearby indoors. Look for piles of finely shredded wood shavings, bits of soil, or dead ants, which are all expelled from the nest through slitlike openings. Finally, listen for rustling sounds inside wall voids, floors, hollow doors, or other suitable nesting sites, especially at night.

Ants are most active from 10:00 P.M. to 2:00 A.M. Look and listen for live ants during these hours and try to follow them back to their main nest,

> *Piles of finely shredded wood shavings, bits of soil, or dead ants expelled from the nest through slitlike openings may be signs of carpenter ants.*

which can be indoors or outdoors. Outdoors, nests occur wherever wood is moist enough—in firewood, shade trees, stumps, wooden decks, or buried wood. (On the East Coast, shade trees are the prime site for the main nest.) If there are satellite colonies indoors, there will be a trail of ants, which can be difficult to find, leading from the main nest to these colonies.

Indoors, carpenter ants nest in confined spaces with high moisture levels. Look for wood that is damaged, either by fungus or water. Pound lightly with a hammer on suspected damage areas to see if they sound hollow, or probe with an icepick to expose galleries. Favorite nesting sites are around or below dishwashers, washing machines, bathtubs, shower stalls, sinks, and toilets; in crawl spaces, wall voids, window trim, skylights, and hollow doors; behind attic insulation below chimneys and vents; and in outdoor steps, porches, and decks.

When to Take Action

The presence of carpenter ants indoors usually indicates that there is moist wood attracting them. Winged ants indoors usually indicate an indoor nest. Try to trace where the ants are coming from as soon as you spot one indoors.

Long-term Controls

Cultural Controls. Modify the habitat, making it less suitable for carpenter ants. Outdoors, remove stacks of firewood, stumps, or buried wood that contain the nest. Indoors, repair leaks, fix the roof, correct drainage, or do whatever is necessary to eliminate moisture near the infested area.

Physical Controls. Deny ants access to the house; this is particularly important if you cannot locate the outdoor nest or if the ants are coming from a woodlot close to your property. Prevent tree branches, which can be used as highways, from touching the house by cutting or tying them back. Plug entry points, often electrical lines or pipes, with caulk or other substances. If possible, replace damaged wood with pretreated wood.

Chemical Controls. Apply a dust (with boric acid, diatomaceous earth, or silica gel) or a liquid (a synthetic pyrethroid such as permethrin or cypermethrin) to the wires or pipes at the entry points. *Caution:* Wear a dust mask if you are applying these substances yourself. You could also hire a pest control operator to apply a borate solution (Timbor) to the affected wood.

Short-term Controls

Physical Controls. Destroy outdoor nests by removing the wood source, the nest itself, and any ants. Use a small vacuum to remove the ants. Carpenter ants will bite and some may sting; be sure to wear gloves, a long-sleeved heavy shirt, and long pants to ensure that you don't get bitten or stung.

Destroy indoor nests and infestations with heat or electricity (see page 142 on termites). The electro gun works best on smaller infestations where the ants are living in a gallery system within the wood. Ants living in a hollow space not of their making (wall void, hollow door) are not affected by the electro gun. The electro gun is also less effective in more humid conditions.

Chemical Controls. Inject boric acid, desiccating dusts, or pyrethrins directly into the galleries where the ants live. Blow diatomaceous earth, silica gel, or boric acid into all the hollow spaces.

For large outdoor colonies, spray a pesticide directly into the nest. Use a synthetic pyrethroid, unless there is a body of water nearby that may be contaminated. *Caution:* Wear a dust mask and goggles when using these sprays.

Last Resorts
Physical Controls. Use a sponge or mop and soapy water to remove individual ants.

Chemical Controls. If there are too many ants, or you feel you must use something stronger, spray them with pyrethrum/pyrethrins or a synthetic pyrethroid insecticide.

 # ANTS

Natural History
A number of ant species invade houses and become pests. Ants differ so greatly that no generalizations can be made. Nevertheless, most can be divided into two broad categories based on their food preference: sugar or protein.

Ants are similar to termites in that they are social insects that live in organized colonies. Adult workers do all the foraging for food and are usually the only stage seen. If you see the other stages, it means a nest is nearby. Most ants live in nests, located either underground—under boards, stones, or cement—or aboveground in trees,

twigs, or nests made of twigs, sand, and gravel. Indoors, certain species nest in cracks, crevices, cupboards, and wall voids.

Colonies reproduce in two ways. At certain times of the year, a swarm of winged queens and males leaves the main colony. The queens mate in the air, disperse, land, lose their wings, and then establish a new nest. In other species, a queen leaves a nest with a group of workers and establishes a new nest in a process called *budding.* Colonies may contain one or multiple queens. In general, it is easier to kill small colonies containing a single queen than colonies with

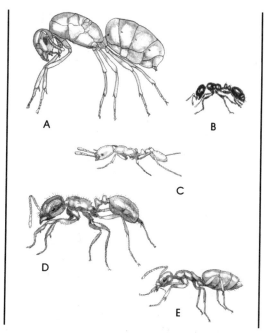

ANT PESTS IN THE HOME

(A) **Carpenter ant.** *A large (8 to 19 mm) ant that is black, brownish, or yellowish, depending on the species. (B)* **Little black ant.** *A tiny (.8 to 1.5 mm) jet black ant. (C)* **Pharoah ant.** *A small (1.5 to 2 mm) ant that varies in color from yellow to red. (D)* **Pavement ant.** *A small (2 to 3 mm), hairy, blackish brown ant with pale legs and antennae, and a black abdomen. (E)* **Odorous house ant.** *A small (2 to 3 mm) brownish to black ant that exudes an unpleasant odor when crushed.*

can become pests, the most common are the carpenter (see page 142), Argentine, pharaoh, thief, and odorous house ants.

Argentine ants are small ants, light to dark brown in color, that nest outdoors in multiple-queen colonies and invade houses in search of sweets. They are found on the West Coast, east of the Mississippi in warm climates, and in Arizona, New Mexico, and Texas.

Pharaoh ants are small yellowish-to-red ants that can nest almost anywhere indoors in a secluded spot. They are found throughout the United States, although they prefer warmth and moisture and so are often found near hot-water pipes in cooler parts of the country.

Thief ants are tiny yellowish ants that nest indoors in cracks, crevices, and cupboards. They prefer grease and high-protein foods and are found throughout the United States.

Odorous house ants are medium-size ants, dark reddish brown to black in color, that usually nest outdoors or in foundations. They give off a distinctive, strong odor when crushed. They prefer sweets and are found throughout the United States, particularly in the West.

Warning Signs

Look for the ants themselves. Although you can find them in any room, they are usually in the kitchen or bathroom, where they have access to food or water. You may notice a few ants wandering around by them-

multiple queens because you must kill all the queens to kill a colony effectively.

Ants cause problems throughout the United States, although they tend to be more of a bother in warmer, humid areas. Although many species selves, but most often they travel in larger numbers—either a trail or a swarm—around some food source, or come out of openings around pipes in the kitchen or bathroom. When you first see them, make sure they are not termites or carpenter ants.

When to Take Action

Ants can be a help in and around the house by feeding on the young of other household pests, such as flea and fly larvae, clothes moths, silverfish, and cockroaches. Outside, some ants attack subterranean termites. You may be able to tolerate a few ants (as long as they are not carpenter ants) indoors for their beneficial effects.

If the ants have wings, they could be coming from either an indoor or an outdoor nest. Identify the species by taking a specimen to your county Cooperative Extension Service or natural history museum. If they are nesting indoors, you need to take action. If they are nesting outdoors, then your problem could just be a hole in a screen.

Long-term Controls

Physical Controls. The primary control method is to modify the habitat to deny ants access to food, water, and shelter. Store all foods in the refrigerator or in plastic containers with tight-fitting snap-on lids and jars with rubber seals or rubber or plastic gaskets.

Wipe kitchen counters and stove tops and sweep or vacuum kitchen floors at least daily. Do not leave unwashed dishes in the sink or dishwasher; either wash them immediately, leave them in a tub filled with soapy water, or rinse them off before putting them into the dishwasher.

Carefully separate organic waste from other material and store it in a plastic container with a tight-fitting lid. Rinse out all items that come into contact with food before throwing them out. This includes plastic, glass, metal, and paper food containers, and food (especially meat) wrappings. Leave pet food out only while the animal is eating, or put the dish in a shallow pie pan or larger saucer of soapy water to create a moat that the ants cannot cross.

To deny ants access to the house or a room, find where they are entering and patch up the holes. This may entail using a silicone caulk on cracks and crevices, and caulk or plaster around pipes. If you are not sure where the ants are entering, place a desired food item (either a sweet or piece of tuna, depending on the ant species) near the ants so that they swarm over it. Then follow the trail that develops from the food to their entrance/exit.

Chemical Controls. Locate and destroy indoor nests. Spray silica gel, diatomaceous earth, or a boric acid dust into the cracks, crevices, and wall voids near the nest. Be sure to wear a mask when spraying. After spraying, caulk up the crevices near the nest.

For ant nests close to the house, either move them or destroy them. Because most ants are beneficial predators, try moving the nest first. To get ants to move their nest, flood it repeatedly with hot or cold water. Add soap or detergent to the water if you want to kill some of the ants. To destroy the nest, dig it up and drench it with an insecticidal soap. If this doesn't work, try a drench of pyrethrins or pyrethroids. Do not use pyrethrins or pyrethroids if a body of water is nearby.

Poison baits take advantage of the fact that ants share food. If the poison is slow-acting and is in a preferred food, the ants eventually feed it to the queen, who then dies. Boric acid and hydramethylnon are both ideal slow-acting poisons with low toxicity. Ingredients in many other ant baits are not slow-acting enough and are unnecessarily toxic to nontarget organisms.

Commercial baits may not prove sufficiently attractive to your ants, since they act on only those species listed on the label. You may have more success making your own bait according to the following recipe:

1. Mix boric acid with sugar water, cat food, or tuna fish in oil, depending on whether your ants prefer sweets or protein. (Watch the ants to see which food types they are eating.) Do not add too much boric acid; you want the ants to live long enough to spread the bait to other ants and the queen. You may have to do a little experimenting, but for starters try 1 teaspoon of boric acid or borax to 1 cup of sugar water or tuna packed in oil.
2. Put the bait into a few small screw-top jars and add cotton to soak up the bait.
3. Punch a few small holes in the lids and screw them on tightly.
4. Smear some bait on the outside of the jar, then place it in an area out of reach of pets and children, but where the ants will find it.

Check the bait every few days. If ants are not using it within a week or so, move it to another place or use another bait mixture.

Insect growth regulators, such as methoprene (Pharorid) and hydroprene (Combat Ant Control), are used in some baits. These compounds interfere with molting and prevent larvae and pupae from becoming adults. Both compounds are effective against ants but are slow-acting (since they prevent the birth of the next generation), and may take a few months to reduce colony size, particularly for ants with very large colonies or multiple queens. They are best used against the pharaoh ant, which is notoriously difficult to control by other means. Look at labels to see which species they can be used against. Most of these compounds come prepackaged in baits or traps. If you can find them in liquid form, mix them with your own homemade bait, which is usually more attractive to ants than prepared baits. To make them faster acting, add boric acid.

Short-term Controls

Physical Controls. Follow the trail or swarm of ants until you find both ends: their food source and their entry point. To help in the identification, note the kind of food the ants are swarming over. Pour soapy water over the ant swarm and wipe them up. Seal the entrance hole(s) with a caulk or petroleum jelly. If there are lots of ants or the entrance hole is big, spray boric acid or a desiccating dust in before sealing it up. Wear a respiratory mask when applying any dust.

Last Resorts

Chemical Controls. It is futile to kill individual ants or a small swarm; there are many more that can replace them. It is much more effective to block their entry into your house or destroy the nest. Use a drench or spray of insecticidal soap and/or pyrethrins or a pyrethroid to kill the nest. Use a spray combining silica gel and pyrethrum or pyrethrins (Drione, Revenge, Pursue) for crack and crevice treatment. Some sprays even come with a slender strawlike applicator to make crack and crevice treatment easier. Use a spray combining silica gel and pyrethrum or pyrethrins, or spray pyrethrins or synthetic pyrethroids for quick kills. *Caution:* Some people are allergic to the pulverized flowers that are the source of pyrethrum. Wear a dust mask or respirator if you are allergic to pollen and are spraying pyrethrum, or choose a different chemical.

 # PANTRY PESTS

Natural History

Although a range of insects attack stored grains and grain-based foods, beetles, moths, and mites are the primary culprits. The larval (or nymphal, in the case of mites) stages cause the bulk of the damage, although adult beetles and mites also feed on the same foods. All the beetle and moth pests have a similar life cycle: egg, larva, pupa, adult. The eggs are laid on the food; the larvae feed on the food itself. The larvae either pupate in the food or crawl away and pupate in nearby cracks and crevices. Adults emerge and search for food on which to lay eggs.

Common beetle pests include the granary weevil, the flour beetle, and the drugstore beetle. Common moth pests include the Mediterranean flour moth, the Indian meal moth, and the grain moth.

Warning Signs

Check stored food and grains for signs of infestation such as cast-off insect skins, chewed grain or dust,

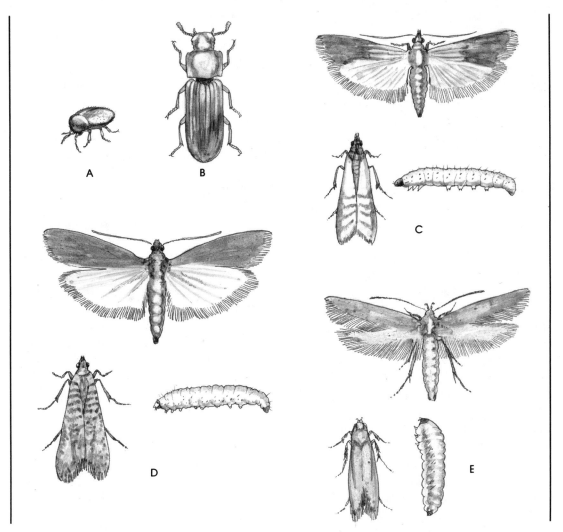

COMMON BEETLE AND MOTH PESTS OF STORED FOODS AND GRAINS

(A) **Cigarette beetle.** *(B)* **Confused flour beetle.** *(C) Adult and larvae of* **Mediterranean flour moth.** *(D) Adult and larvae of* **Indian meal moth.** *(E) Adult and larvae of* **Angoumois** *grain moth.*

silken tubes, and cocoons. Check the food packaging for small holes, which are usually made by adult beetles.

Look for the insects themselves. Moth and beetle larvae both look like small caterpillars, except that moth larvae have three pairs of legs while beetle larvae have more than three. Young mites look like miniature adult mites. While larvae will be confined to the food itself, adults may also be found flying or crawling in your kitchen cabinet or pantry. Adults are easy to tell apart: moths

❧ *Check stored food and grains for signs of infestation by grain moths or beetles, such as cast-off insect skins, chewed grain or dust, silken tubes, and cocoons.*

have wings with scales; beetles have a hard covering over membranous wings; mites are tiny eight-legged creatures that look like small dots. If the beetles have longish snouts, they are weevils.

If you find adults in some other area of your home, collect some for identification. They may be a different pest (clothes moth, powderpost beetle, or carpet beetle, for example), or they may indicate the presence of a food source you haven't been aware of: beanbag chairs or toys; jewelry, pictures, centerpieces, ornaments, or bottles made with beans or seeds; cork boards and backing; furniture stuffing; stuffed dolls or animals; hair or lint in cracks; dead animals, their nests, or food hidden by rodents in wall voids, attics, or crawl spaces.

Pheromone traps, used for monitoring purposes, are available for Mediterranean flour and Indian meal moths and some flour beetles. Place these little boxes or tubes in areas where you store food. They are highly attractive to the specific pest in question, and act as an early-warning system, alerting you to the presence of the pest.

When to Take Action

Control steps are warranted if you find infested food, adults crawling or flying in the food storage area or some other area of the house, or adults in pheromone traps.

Long-term Controls

Prevention is the best control. Since most problems are caused by bringing infested foodstuffs into the house, inspect all packaged food before buying. Look for holes or other damage to the wrapper that could permit pest entry. Check bulk food for pests or signs of their presence before you bag it up for purchase. At home, check all the food in the kitchen or storage area.

Physical Controls. Before storing cereals, crackers, beans, cookies, and other grain foods, put them in the freezer for four to seven days to kill any unseen eggs, larvae, or pupae. Heating grains or stored food to 150 to 175 degrees for at least 10 minutes also kills all pest stages. Be sure to dry any grain before storage.

After decontamination and drying, store foodstuffs in tightly sealed glass jars, or plastic tubs with tight-fitting lids, in a cool, dry place or in the freezer, to minimize chances of infestation. Periodically check food stored for any length of time for signs of infestation.

If you have a large infestation, thoroughly vacuum the kitchen and pantry. Be sure to vacuum the edges of shelves, behind counters, and other areas where food morsels might have fallen.

Short-term Controls

Physical Controls. For small infestations, throw out infested food. For large infestations, throw out infested food and clean the food storage area.

Last Resorts

Chemical Controls. If sanitary and physical controls don't work, spray the kitchen and pantry with an insecticidal soap or, as a last resort, a pyrethrin-based or synthetic pyrethroid-based product. *Caution:* Some people are allergic to the pulverized flowers that are the source of pyrethrum. Wear a dust mask or respirator if you are allergic to pollen and are spraying pyrethrum, or choose a different chemical.

For chronic problems, add diatomaceous earth, silica aerogel, or calcium carbonate to stored grains. These dusts kill moth and beetle larvae by causing them to desiccate. Read the labels for amounts to be added. For diatomaceous earth, be sure to use the amorphous type, *not* the type used in swimming pools, which is chemically treated.

Cockroaches

Natural History

Cockroaches are the most prevalent household pest in the United States. They are scavengers that feed on dead plant materials. They prefer starchy foods to protein, but will eat virtually anything if hungry. The bulk of species are nocturnal. Except when foraging, they stay in cracks and crevices. There are several pest species, which all have the same basic life cycle: egg, nymph, and adult. Of the 57 known species in the United States, the majority of problems are caused by the American, brown-banded, oriental, and German cockroaches.

The American cockroach is about 1¾ inches long, reddish brown, with a light-colored band behind the head on the back. It prefers warm, moist areas, and since it is a poor climber, tends to be found in the basement or first floor of the house. These roaches are common around warm pipes, such as those near furnaces or heating ducts. Upstairs, they are most likely to be found in the kitchen or bathroom. Egg cases are dropped in sheltered areas on the floor.

The brown-banded cockroach is about ½ inch long and tan to gold in color, with light V-shaped bands on the wings. The male is a good flier, but the female does not fly. They require less water than other cockroaches and prefer temperatures above 80 degrees. Although found throughout the United States, they are most abundant in the South. They like higher places, such as shelves, cupboards, and ceiling voids, but they also occur under furniture, in desk or dresser drawers, and in appliances that generate heat. Egg cases are most frequently glued to ceilings, in clos-

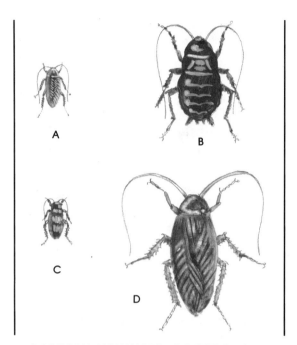

COMMON HOUSHOLD COCKROACHES

(A) **German cockroach.** *This medium-size roach, about ½ inch in length and brown with two dark streaks on the pronotum (the platelike structure on the back, behind the head) and thorax, is the most common roach in the United States. It prefers warm moist habitats and is most common in kitchens and bathrooms. (B)* **Oriental cockroach.** *This large roach, about 1¼ inches long, is dark red-brown to black and does not fly. It is most common in dark, damp basements, but is also found in crawl spaces or under washing machines, sinks, or refrigerators. (C)* **Brown-banded cockroach.** *This medium-size (about ½ inch long) cockroach, tan to gold, with light V-shaped bands on the wings, likes temperatures above 80 degrees. The male can fly. Most abundant in the southern United States, they favor high shelves, cupboards, and ceiling voids, but may be found under furniture, in desk or dresser drawers, and in appliances that generate heat. (D)* **American cockroach.** *This large roach (about 1¾ inches long) is reddish brown with a light-colored band surrounding the pronotum. It is a poor climber, and prefers warm, moist areas such as the basement or first floor around warm pipes.*

ets, under furniture, or in dark spaces.

The oriental cockroach, often called a waterbug, is about 1¼ inches long and dark red-brown to black. It does not fly. In part because they are poor climbers, tolerant of cold, and prefer damp porous surfaces, these roaches tend to be found at or below ground level in houses. They are most common in dark, damp basements, but are also found in crawl spaces or under washing machines, sinks, or refrigerators. However, they have been known to climb water pipes to the upper floors. They are slow-moving and do not disperse far from their hiding sites. Egg cases are put in food, debris, and other sheltered areas.

The German cockroach is about ½ inch in length, brown, with two dark streaks on the head and thorax. It is the most common roach in the United States. German roaches prefer warm, moist habitats and are most common in kitchens and bathrooms. Unlike other roaches, the female carries the egg case around until a day or two before it hatches, with the case becoming as large as the abdomen, and then drops the egg case anywhere.

Warning Signs

Look for adults or nymphs (smaller versions of the adults, without wings) or signs of their presence: small black fecal pellets (they look like large grains of pepper) or egg cases (small, leathery-looking striped pellets). Unless you have a large population, you are not likely to see a roach during the day (except for Oriental cockroaches, a problem in Florida). Often you see adults at night when you turn on a light and they run for cover. This is especially true of German roaches.

Use sticky traps to monitor roach populations. The many different brands all contain an adhesive, and some contain a chemical attractant. The traps differ in their effectiveness, depending on species. Identifying the roach species will help you determine which traps to use and where to place them. Put sticky traps in areas where you think roaches are a problem or where you've seen them or their fecal pellets. Most roaches stay in tight or closed spaces in warm, moist areas. They also prefer to move along the edges of areas. Put traps against walls on floors and counters; in drawers or cupboards; under refrigerators, stoves, or sinks; and in closets. Check the traps a couple of times a week. By keeping track of trap catches, you can determine where the highest infestations are. Focus your control efforts on these hot spots.

When to Take Action

Take action when numbers exceed your personal threshold.

Long-term Controls

Physical Controls. Permanent reduction in cockroach populations requires modifying their habitat to decrease or eliminate nesting sites, entry points, and sources of food and moisture. Remove clutter from the areas with the worst roach problems. With caulking material, plaster, or paint, seal cracks and crevices in walls, around pipes, behind and

Permanent reduction in cockroach populations requires modifying their habitat to decrease or eliminate nesting sites, entry points, and sources of food and moisture.

within cabinets, and underneath appliances. Sweep or vacuum the crevices before sealing to remove food, fecal pellets, and egg cases.

If significant numbers of roaches are coming from outside, caulk around pipes in the bathroom, put a fine-mesh screen over the drain in the basement, and trim shrubbery near air vents or other areas close to the house. Fix water leaks.

Improve garbage and food management to minimize or eliminate sources of food. Indoors, store food in tightly sealed containers. If you leave pet food standing in dishes, place the dish in a bowl or saucer of soapy water to create a moat. Keep kitchen counters clean and regularly sweep and vacuum floors, particularly in corners.

Chemical Controls. Treat hot spots (determined by trap catches) with a light dusting (roaches avoid thick coatings) of boric acid, diatomaceous earth, or silica gel. Places to dust include wall voids, cracks, crevices, and areas inaccessible to pets and children such as underneath appliances, behind cabinets, and under shelf paper in drawers and cupboards. Use a hand duster or bulb duster to help blow the dust into the appropriate areas. If possible, add a couple of coins or pebbles to the duster and shake it between applications. Shaking imparts an electrostatic charge to the dust, which helps it to cling to surfaces, including the roach's body. Wear a respiratory mask when applying a dust.

There are many brands of boric acid powder on the market. Roach Prufe and Roach Kill contain an anti-caking agent to protect against humidity, a dye to distinguish it from flour or sugar, and are electrostatically charged. You may find these brands most effective. *Caution:* Wear a dust mask when mixing or applying these products.

Short-term Controls

Physical Controls. For high populations, use a vacuum cleaner's corner or edge attachment to remove roaches and egg cases from cracks and crevices.

Chemical Controls. After vacuuming, use boric acid, diatomaceous earth, or silica gel, combined with a roach bait. Use a small squeeze bottle with a pointed tip to blow the dust into the cracks and crevices. The dusts take 10 to 14 days to kill roaches, and should be used only in areas not accessible to pets or children; baits usually kill roaches within a few days and can be used in areas where dusts cannot.

Baits minimize the amount of chemicals you use, and often come packaged in small plastic discs that make it much harder for pets and children to reach them. Use baits containing boric acid, hydramethylnon (Combat, MaxForce), sulfluramid, or abamectin, and place them where you put sticky traps. Boric acid baits are less likely to repel cockroaches than are most synthetic pesticides, and so are often more effective. Hydrame-thylnon baits work extremely well in controlling German and brown-banded cockroaches.

Some baits contain a roach phero-mone (periplanone-B), which attracts cockroaches. Abamectin, an extract from a soil bacteria, is effective against roaches, particularly those resistant to other insecticides, at very low doses. This product (with abamectin) has just appeared on the market and is now available from Black Flag.

Biological Controls. Insect growth regulators such as hydroprene (Gencor) and fenoxycarb (Torus) can also be part of a cockroach IPM program. These compounds do not kill roaches but rather act as a form of birth control, interfering with development and causing nymphs to mature into sterile adults. Adult roaches are not affected by these compounds. Since they do not kill roaches, these growth regulators are best used as part of an IPM program that includes habitat modification and use of boric acid or a desiccating dust. Apply the growth regulators to areas where you cannot use boric acid, such as on vertical surfaces and the undersides of shelves. Growth regulators take a number of weeks to work, so do not expect dramatic results overnight. *Caution:* Wear protective clothing and a respirator mask when applying these products.

Last Resorts

Chemical Controls. If you must use a general spray, use one with pyrethrins or silica gel (Drione, Revenge), diatomaceous earth (Diacide, Shellshock), boric acid, or a synthetic pyrethroid. *Caution:* Wear protective clothing and a respirator mask when applying these products.

 # HOUSEPLANT PESTS

Houseplants are plagued by a number of different problems that may be caused by bad horticultural practices, improper fertilization, disease, and insect and mite pests. Indeed, the last two problems are exacerbated by the first two, which can put great stress on the plant. In order to minimize problems with insects and mites, fertilize moderately and only during months when the plants are actively growing. Too much fertilizer can aggravate aphid, whitefly, mealybug, and scale problems. Finally, isolate new houseplants and search them carefully for signs of insect or mite pests before putting them with other plants.

Whiteflies

Natural History

Whiteflies are not flies but plant-sucking insects related to scale, mealybugs, and aphids. The most common species on houseplants is the greenhouse whitefly, which attacks begonias, coleus, fuchsias, poinsettias, salvia, verbenas, tomatoes, and cucumbers. All whiteflies have the same life cycle: egg, larva, pupa, and adult. Both adult and larval stages suck plant juices. The adults exude a sugary substance called honeydew, which can make leaves sticky and serves as the food source for a black sooty mold, which makes the leaves appear dirty.

The adults look like tiny white flies; females are slightly larger than the males. They are attracted to the color yellow, and cluster on the yellowest green parts of a plant—the youngest leaves and the ends of stems. The adults don't fly very far; when disturbed, they usually fly off the plant in a small cloud and then land again on the young foliage. The female lays her eggs on the undersides of older leaves.

Whiteflies are attracted to the color yellow and cluster on the yellowest green parts of a plant—the youngest leaves and the ends of stems.

The larvae go through a number of stages, called *instars*. The first larval instar is mobile, whereas later ones are not. The mobile larva finds a suitable place on the underside of a leaf, loses its ability to move the first time it sheds its skin, and starts sucking plant juices. Larvae pupate on the leaf and develop a cocoonlike protective cover. Normal larvae and pupae are light-colored. Adults emerge from the pupae and fly toward the yellow-green youngest foliage.

Warning Signs

Monitor plants weekly to once every two weeks, beginning in the spring. Check near the top of the plant or on the ends of branches for adults and the undersides of leaves near the bottom of the plant for eggs, larvae, and pupae. The eggs, usually laid in a small circle, look like greenish white pointed tubes hanging from the leaf by a short stalk. Larvae resemble small whitish or greenish translucent spots, with, in some species, a fringe of hairs. You may need a hand lens to spot the pale larvae. Pupae look like larvae with cocoonlike protective covers.

Signs of damage include a sticky residue on leaves, or sooty mold.

When to Take Action

Most plants are tolerant of whitefly damage. Adults do not pose a problem, so long as there are few larvae. Large numbers of larvae and pupae indicate a rapidly growing population. Take action if the numbers of infested leaves or the numbers of larvae and pupae per leaf start rising rapidly.

Long-term Controls

Control tactics differ for larvae and adults. Decrease use of nitrogen fertilizers; instead, use organic fertilizers, ammonium- or urea-based synthetic ones, slow-release formulations (Ozmacote), or ones that emphasize phosphorous over nitrogen. Fertilize only while the plant is actively growing, and spread out the fertilization by using more frequent smaller doses.

Biological Controls. If you have a large number of plants—such as in a small greenhouse, enclosed porch, or plant room—releasing tiny parasitic wasps (*Encarsia formosa*), available from supply houses (see References and Resources), may control your whitefly problem. These wasps are effective only against the greenhouse whitefly, so be sure to identify your pest species. The wasps work best against light to moderate whitefly infestations and take a couple of weeks to be effective. They attack third and fourth instar larvae. A parasitized larva is black, not its normal light color.

If whitefly populations are high, first knock them down with an insecticidal soap or superior light horticultural oil spray. Wasps need temperatures around 80 degrees and about 70 percent humidity. Release approximately 1 to 5 wasps per plant at 10-to-14-day intervals. Do this

three times so that successive generations will be susceptible to attack.

Make sure all windows and vents are covered with a fine-mesh screen to prevent the wasps from escaping. Don't use insecticides, since the wasp is more sensitive to them than the pests are. If environmental conditions remain warm and damp, the wasps may establish themselves and keep whitefly numbers down for an extended period.

Another biological control is the fungus *Verticillium lecani*, sold under the trade name Mycotal. Use it on medium to high levels of whitefly infestation. Mycotal works best when nighttime humidity exceeds 90 percent and the temperature is above 54 degrees. *Caution:* Wear a dust mask when applying this product.

Short-term Controls

Physical Controls. Isolate severely infested plants before treatment to reduce spread of whiteflies to uninfected plants.

Wipe adults, larvae, and pupae off the plant with a gloved hand or a moist towel. If mature leaves are severely infested with larvae, remove them.

If you have a small greenhouse or enclosed porch, vacuum the adults off the leaves. This is best done in the morning, when the adults are sluggish.

Sticky traps may also be useful. Since adult whiteflies are attracted by yellow, paint a small piece of cardboard, plastic, or wood with bright

ᛒ *Paint a small piece of cardboard,
plastic, or wood with bright yellow
paint and cover it with a thin layer of
a commercial sticky material to
trap whiteflies.*

yellow paint and cover it with a thin
layer of a commercial sticky material
(Stickem, Tack Trap, or Tanglefoot).
You can use a homemade variety (a
50-50 mixture of petroleum jelly or
mineral oil and dishwashing liquid)
instead. The sticky material washes
off easily, and the boards can be reused. Place the traps so that the tops
of the pieces extend a few inches
above the tops of your plants. You
will need a trap for every few plants.
Shake the plants; some adults will fly
off, be attracted to the yellow, and try
to land on the board. Replace the trap
when it fills up.

Last Resorts
Chemical Controls. Spray an insecticidal soap or a light superior horticultural oil on the infested areas of
the plant, but first test the soap or oil
on a few leaves to see if it damages the
foliage. Wait a day or two for any
damage to appear. Be sure to coat the
undersides of older leaves, which
contain larvae and pupae. If the insecticidal soap mixture doesn't work,
use a spray with pyrethrum/pyre-
thrins or synthetic pyrethroids. *Caution:* Some people are allergic to the
pulverized flowers that are the source

of pyrethrum. Wear a dust mask or
respirator if you are allergic to pollen
and are spraying pyrethrum, or
choose a different chemical.

Aphids
Natural History
Although there are more than
4,000 species of aphids, relatively
few are houseplant pests. For general
information on aphids, see page 77.

When to Take Action
If you find a few aphids on a plant,
monitor the plant more frequently. If
the number of aphids starts rising
rapidly, take action.

Long-term Controls
Chemical Controls. You may be ov-
erfertilizing. Reduce use of nitrogen
fertilizers; instead, switch to a slow-
release fertilizer with a moderate
level of nitrogen, such as liquid seaweed or fish emulsion. Dilute the fertilizer and apply it more frequently,
but in smaller doses—if the label says
to use 1 liter every 6 weeks, try ¼ liter
every 3 weeks. Fertilize only when
the plants are actively growing. Use a
fertilizer with relatively more phosphorous, which stimulates flower and
fruit production over foliage growth.

Biological Controls. Several predaceous insects eat aphids, including
lacewings and midges, but they are
appropriate for indoor use only if you
have a greenhouse or a large number
of plants.

ও *To help control aphids, reduce use of nitrogen fertilizers and switch to a slow-release fertilizer with a moderate level of nitrogen, such as liquid seaweed or fish emulsion.*

Adult lacewings feed on nectar, pollen, and honeydew, while the larvae are voracious predators of small insects and insect eggs. A green lacewing larva can eat up to 60 aphids per hour. You can buy lacewings as either eggs or larvae, although larvae are easier to handle. For smaller plants, one larva or a few eggs per plant should be sufficient.

Buy and release the larvae of midges a few times throughout the season to keep their numbers artificially high. You can also buy a food supplement (see References and Resources) to feed these predators when pest numbers are low.

Another natural enemy is a fungus, sold under the trade name Vertalec, which you can spray on aphid colonies. It works best under warm, humid conditions.

Short-term Controls

Physical Controls. Remove the aphids, either by crushing them or brushing them off with your hands (wear gloves) or by pruning infested plant parts. For smaller plants, cover the potting soil with a cloth to prevent it from falling out, hold the plant container upside down and shake the plant over a pail of soapy water, then rinse it with clear water. Take larger, sturdier plants outdoors and spray them with a hose. This knocks the aphids off, causing significant mortality. Using a soap spray rather than water helps kill the aphids not washed off the plant.

Last Resorts

See page 80.

Mites

Natural History

See page 82 for general information on mites. Spider mites, broad mites, and cyclamen mites cause the bulk of problems on houseplants.

Warning Signs

See page 83 for general information.

Both broad mites and their feeding damage are too small to be seen; the first indication that you have these pests often is the distortion of new growth caused by the mite's salivary toxins. Tap some leaves or branch tips against a piece of white paper and use a hand lens to look for mites on the paper.

When to Take Action

If mites increase, take action.

Long-term Controls

Cultural Controls. Try to keep your plants from getting hot and dry. If plants get full sun, shade them during the heat of the day. To increase the humidity, group plants together, put

To help control mites, mist your plants on hot days.

them on a layer of pebbles in a tray of water, and double-pot them, using moist sphagnum or peat moss between the pots. Mist your plants on particularly hot days.

Mites often travel from plant to plant via overlapping leaves; separate plants so that they do not touch. Since mites can also be spread from plant to plant via tools or hands, wash your hands frequently while working with the plants, and dip your tools in soapy water between use, especially if you know your plants are infested with mites.

Biological Controls. You can buy predatory mites that are effective against spider mites or broad mites. To determine which predatory mites you need, and how many, you must know the identity of the pest mite (collect some and have them identified by an entomologist at your county Cooperative Extension Service); the number of plants, or size of the plants' environment, to be protected; the number of infested plants and/or average number of mites per leaf. Unless you have an extensive plant collection or a greenhouse, it is not practical to buy and release predatory mites.

To apply the mites, follow the instructions of the supplier. To maximize the impact of the predatory mites, don't spray any insecticide for about a month before releasing them. Some pesticides leave a residue that can kill mites. For greatest impact, release predatory mites while pest mite numbers are low.

If pest mite numbers are high, spray the plants with an insecticidal soap about three days prior to releasing the predatory mites. Release the mites on the middle or upper foliage of all plants. (The predatory mites are often shipped in sawdust, so just sprinkle a little sawdust on each plant.) Keep some of the predatory mites so that you can look at them with a hand lens and learn how to recognize them.

For best results, release three batches of predatory mites: the first two batches two weeks apart, the third batch one month later. Monitor your plants every week and keep track of the number of pest and predatory mites. The number of pest mites should decline steadily. While using predatory mites, don't use pesticides at all.

Short-term Controls

Physical Controls. As soon as you notice mites, wash the plants with soapy water. You can shake plants over a bucket of soapy water, or hose down larger plants outdoors (the same procedure used against aphids—see page 159). Make sure you spray the undersides of the leaves. Do this once a day for three or four days. Prune infested areas.

Last Resorts

See page 85.

FLIES

Natural History

Flies make up one of the largest groups of insects, ranging from gnats and mosquitoes to horseflies, houseflies, and fruit flies. Although many are considered pestiferous (mosquitoes, for example), we focus here on *filth flies*—those that feed on decaying plant and animal matter.

Unlike other winged insects, flies have only two wings and four basic stages: egg, larva, pupa, and adult. Filth-fly larvae, called *maggots*, are grublike and play an important role in recycling dead plants and animals. They are sensitive to drying out, and so must live in a moist environment. Consequently, adults deposit eggs in moist decaying organic matter. Many of the differences among species rest on the larval food source. The most common pestiferous species are *houseflies, fruit flies,* and *drain flies.*

Houseflies are ¼-inch-long gray to black flies with four dark stripes on their thoraxes. The larvae feed on garbage and pet feces, either indoors or outdoors, while the adults frequently fly indoors.

Fruit flies are about ⅛ inch long, and are brownish with reddish eyes. The larvae feed on decaying and fermenting fruit and vegetables or any sweet substance. The adults hover over fruits and vegetables and are usually seen in kitchens.

Drain (or moth) flies, which resemble whitish moths, are tiny flies with hairy wings. Larvae are aquatic, and are frequently found in sewage sedimentation tanks and kitchen drains, where they feed on decayed organic matter. Indoors, adults are most frequently seen on windows above sinks.

Warning Signs

Adults are far more visible than the larvae, which often remain hidden. Capture some adults in a jar or net, put them in the freezer to kill them, then identify them yourself with the help of an identification guide, or take them to a specialist. Identification of the fly species will help you pinpoint the larval food source. If you have problems with flies outdoors, begin looking for exposed garbage or animal feces in the spring, or when flies are first seen. Check your compost pile to be sure flies are not breeding in it.

When to Take Action

Take action when flies become too much of a nuisance, or when you notice their numbers rising over time.

Long-term Controls

Physical Controls. Find the larval food source and get rid of it, or prevent egg laying by blocking adults' access to the food or making it unattractive to them. For fruit flies, look for rotting fruits or vegetables; for houseflies, rotting animal or plant material; for drain flies, sewage or drainage garbage.

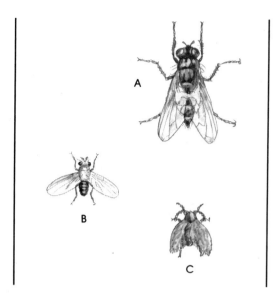

COMMON INDOOR FLY PESTS

*(A) **Housefly**. This medium-size (4 to 7.5 mm) fly is gray with four narrow black stripes on its thorax. It is the fly most commonly found in homes throughout the United States. (B) **Fruit fly**. This small (about 3 mm) fly has a tan-colored head and thorax, with a blackish abdomen and bright red eyes. It feeds on yeast in fermenting fluids and so is usually found in the kitchen, near rotting fruit and vegetables. (C) **Drain fly**. This small (less than 3 mm) fly has hairy wings and looks like a tiny whitish moth. Larvae can be found in sewage sedimentation tanks and kitchen drains. Indoors, adults are most frequently seen in kitchens, especially on windows above sinks.*

Determine whether the food source is indoors or outdoors. In general, garbage and waste management are the key to prevention and control. Keep organic waste dry to deny the

To control houseflies, try to find the larval food source and eliminate it, or prevent egg laying by blocking adults' access to food.

maggots the moist food source they need. Keep it in tightly sealed containers. Indoors, separate organic waste and, if possible, compost it. Otherwise, drain it and thoroughly wrap it in newspaper to dry it out. Dispose of the bundle in the trash.

Outdoors, use a trash or garbage can that has no holes in the sides or bottom, as well as a lid that will not come off if the can is knocked over by animals. You can buy special spring-attached lids or construct your own. Never leave the can uncovered; flies will enter and lay eggs. Apartment dwellers should make sure that the garbage pail remains closed. Scrub it with soap or detergent and let it dry thoroughly, to kill any developing maggots. Collect moist pet feces and flush them down the toilet, or wrap them in newspaper and throw them in the trash can.

Outdoors, if the source of your fly problem occurs on neighboring land, try to persuade the neighbor to change garbage and waste management. If you aren't successful, use a fly trap. Place an attractive bait in the trap, and keep the bait moist. The proper bait depends on the fly species, but try table scraps, sweet fermenting foods, beer, or syrup for starters.

Commercial baits contain chemicals attractive to flies. Experiment with different foods and locations to determine the most attractive ones for your purposes. Place the trap in the sun and upwind from the area you want to protect, but keep it away from entrances to the house, or locations used for eating or recreation. For large fly problems, ring an area with traps.

Each spring, repair holes or tears in screens; caulk cracks or gaps around windows, drains, or vents; and install weather stripping around doors. If drain flies are a problem, pour a gallon of hot soapy water down the drain or, if that doesn't work, install a fine screen over your drain to prevent the adults from emerging from the drain and entering the room.

Short-term Controls

Physical Controls. Kill adult flies with a flyswatter. Flypaper reduces fly numbers. Thoroughly scrub and dry infested garbage cans.

Last Resorts

Chemical Controls. If you can't wait for the long-term controls to work, or if there are so many adult flies that you must spray something immediately, use a product containing pyrethrum, pyrethrins, or a synthetic pyrethroid. *Caution:* Some people are allergic to the pulverized flowers that are the source of pyrethrum. Wear a dust mask or respirator if you are allergic to pollen and are spraying pyrethrum, or choose a different chemical. *Always* combine a spray with the long-term controls mentioned; otherwise the problem will recur.

 # MOSQUITOES

Natural History

Mosquitoes are among the most dangerous insect pests in the world because the adult females feed on human blood. Human-biting species can transmit serious diseases such as malaria, yellow fever, dengue fever, and encephalitis. (None of these diseases is a problem in the United States except for encephalitis.) In addition, proteins in mosquito saliva can cause severe allergic reactions in some individuals.

There are four stages to the life cycle of the mosquito: egg, larva (or wriggler), pupa (or tumbler), and adult. The young stages develop primarily in still or slowly moving water, while the adults feed on plant juices (usually nectar) and, in some species, blood.

Unlike most insects, mosquito pupae are active and move with jerking, somersaulting motions. Adults are small, have slender, elongated bodies, long legs, and two wings that

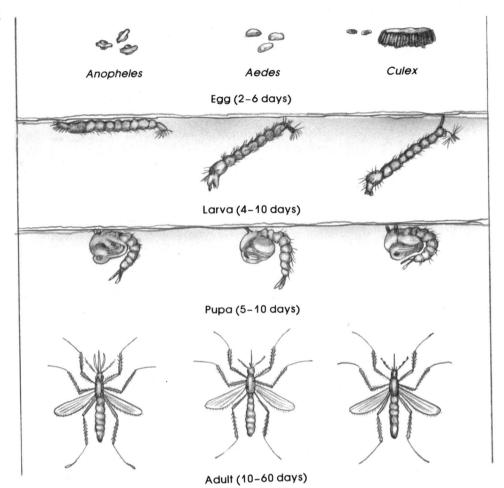

| Anopheles | Aedes | Culex |

Egg (2–6 days)

Larva (4–10 days)

Pupa (5–10 days)

Adult (10–60 days)

LIFE CYCLES OF THE THREE MEDICALLY IMPORTANT MOSQUITOES

The four stages of the life cycle of the mosquito are the egg, larva (wriggler), pupa (tumbler), and adult. A key to telling these three most common mosquito types apart is their egg-laying behavior. Anopheles eggs have finlike structures that help them float by themselves on the water surface. Aedes eggs are laid individually in holes drilled into the mud, hatching later when the mud is flooded. Culex eggs float in a mass on the water surface.

have scales on the veins and margins; the females have long, piercing mouthparts (proboscises).

Three kinds of mosquitoes cause the bulk of problems: *Anopheles, Culex,* and *Aedes.* The major differences are in egg laying and larval hab-

itat. *Anopheles* lays eggs singly on the water surface, preferably in shaded locations with dense vegetation. *Culex* is the most abundant household mosquito. It lays floating clusters of eggs, called *rafts,* in any standing body of stagnant water,

Mosquitoes are among the most dangerous insect pests in the world because the adult females feed on human blood and transmit many serious diseases.

from ponds to puddles to rain barrels, including polluted water. *Aedes* lays eggs singly out of the water, usually in or on soil next to a temporary water source. The eggs resist drying out, and so the larvae can be found in a wide range of habitats, including salt marshes, pastures, temporary pools, tree holes, containers, and discarded tires.

Adult females are fairly easy to tell apart. The *Anopheles* holds its body, head, and proboscis in a straight line and at an angle with the head pointing toward the resting surface, and has spotted wings. Both *Culex* and *Aedes* adults hold their bodies parallel to the resting surface, with their proboscises pointing downward, and have wings of uniform color. However, *Aedes* has a pointed abdomen, a silver thorax with white markings, and bristles around its air holes; *Culex* has a blunt abdomen and is often uniform in color.

Warning Signs

Almost everyone is familiar with the characteristic high-pitched buzz and itchy bite of the mosquito. Capture adults for identification. Do not confuse mosquitoes with fungus gnats, midges, and crane flies. These other flies superficially resemble mosquitoes but lack the proboscis and/or scales on the wing veins and margins.

After identifying the species, look for the larvae in the appropriate habitat. Collect larvae using a dip net and take them to a local museum, a Cooperative Extension Service, or the entomology department of the local university for identification to ensure that it is the same species of mosquito as the adult that is doing the biting.

Do not confuse mosquitoes with fungus gnats, midges, and crane flies. These other flies resemble mosquitoes superficially but lack the characteristic proboscis and/or scales on the wing veins and margins.

Many mosquito species do not feed on humans, so there's no need to get rid of them. Also, if the species is one that lives in salt marshes and can travel for distances up to 5 miles, getting rid of water sources on your property may not solve the problem; you'll need to get the local government or community involved to solve the problem.

When to Take Action

When the frequency of bites exceeds your tolerance threshold.

Long-term Controls

The only way to reduce adult populations is to find the larval population and destroy it. Prevent or eliminate standing water sources.

Physical Controls. Aedes is the most amenable to small-scale control measures because its larvae live in small pools of standing water. Search your yard and vicinity for any small containers of water: cans, jars, and other vessels, including dishes underneath plants; clogged drains and gutters; overwatered lawns; discarded tires; standing water in tire ruts; small ornamental ponds; wading pools; birdbaths; tree holes; puddles from evaporative cooler or air-conditioning drains; cesspools or septic tanks. Drain the water source or fix the problem. To drain cavities in trees, use a wicking material, such as cotton. The material should be long enough to reach from the bottom of the hole over the closest edge and down the tree trunk to a level below the cavity.

If water sources not on your property—flooded fields, marshes, drainage ditches, clogged storm sewers, puddles associated with large-scale construction, tree holes in wooded areas—are the source of your problem, you may need to enlist community assistance. Get in touch with your local mosquito abatement district office. Encourage them to map potential breeding sites and use nonchemical controls aimed at larval sources, rather than aerial fogging, to

Encourage your local mosquito abatement district office to use nonchemical controls aimed at larval sources, rather than aerial fogging, to kill adults.

kill adults. Help educate your neighbors on the need for controlling standing water on their property.

You can deal with adults in a number of ways. First, keep them from entering the house. Each spring, check windows and doors for tears in the screening or spaces around windows. Pay special attention to areas such as enclosed porches. Repair screens, caulk gaps around windows, and add weather stripping around doors. Check the house, especially the basement or laundry room, for puddles, to ensure that the mosquitoes aren't reproducing indoors.

During peak mosquito season, stay indoors in the early evening hours, when many mosquitoes are most active.

Biological Controls. If you can't drain the water, turn first to biological agents. For ornamental ponds and other small artificial bodies of water without drainage, add mosquito fish or goldfish; they devour the larvae and pupae. Mosquito fish do well in warm water with low oxygen content. Goldfish are hardier. Do not release these fish into any natural bodies of water containing native fish.

Strains of the bacteria *Bt*, called *Bacillus thuringiensis israelensis*, or *Bti* (Bactimos, Tecknar, Vectobac, Mosquito Dunks), attack mosquitoes. *Bti* kills mosquito larvae but not pupae. Mosquitoes differ in their sensitivity to *Bti*, although in general, *Culex* is most sensitive, *Aedes* requires higher doses, and *Anopheles* is the least susceptible. *Bti* is not persistent and breaks down rapidly. Most of these products consist of granules or solid rings; follow the label instructions and add them directly to the water.

Chemical Controls. Surface films clog the breathing apparatus of the mosquito and can be used only in relatively calm standing water. A chemical surface film called Arosurf can kill all stages it touches: larvae, pupae, emerging adults, some egg rafts, and adults landing on water to lay eggs or rest. *Caution:* Wear gloves when handling this product.

Short-term Controls

Physical Controls. Indoors, use a flyswatter to kill adults or flypaper to catch them. Outdoors, smoke coils or citronella candles repel them, and may work with low to moderate infestations. Wear densely knit long-sleeved shirts and long pants and use repellents. If mosquitoes are a severe problem and you must be outside, wear a head net similar to those worn by beekeepers, as well as suggested clothing.

Last Resorts

Chemical Controls. When going outdoors, apply insect repellents to your clothes. If absolutely necessary, put them on your skin. Stay away from repellents that contain concentrations of DEET greater than 50 percent. (Repellents can stain clothes and damage materials such as plastic wristwatch bands and glasses frames, so read the label carefully.) Some people have found that Avon Skin So Soft bath oil works quite well as a repellent.

 # FABRIC PESTS

Two types of pests attack fabric: clothes moths and dermestid (carpet and hide) beetles. Unlike many pests, clothes moths and dermestid beetles can digest keratin, the main component of fur. In nature, these moths and beetles also feed on a range of animal products including feathers, dead insects, and dried animal remains. For both types of pests, the larval stages cause the vast bulk of damage. Adult clothes moths cause

> *Clothes moths and dermestid beetles can digest keratin, the main component of fur and wool.*

no damage to fabrics, while adults of some dermestid species cause minor damage to fabrics or stored food.

Moths

Natural History

All the clothes moths have the same basic life cycle: The female lays eggs on a larval food source (such as woolen clothes, carpet, feathers, fur, hair, or dried animal remains); the larvae feed and then pupate on or near the food source.

Both larvae and adults shun light. If disturbed, they immediately try to hide, often in the folds of fabric. The larvae, which have white, naked bodies and dark heads, are fragile and prefer to feed in dark, undisturbed locations. Larvae cannot survive solely on clean, processed wool, but require stains of food, drink, urine, or sweat to supply the needed extra nutrients. Adults can be found flying in dark areas or walking on the food source.

The two most common moth pests are the webbing clothes moth and the casemaking clothes moth. The webbing clothes moth is most common, and is found throughout the United States. The adult is about ¼ inch long, golden buff to yellow-gray in color, and has upright reddish hairs on its face. Larvae sometimes construct a silken mat or tube over part of the fabric and feed underneath it, although you may find them wandering on the fabric itself. The casemaking clothes moth is primarily a problem in the South. The adult is brownish, about ¼ inch long, with three indistinct spots on the wings, and has light-colored upright hairs on its face. The larvae construct a silken tube, which they hide in and carry around as they feed. In addition to wool, they are particularly fond of feathers and down products.

Warning Signs

Woolen clothes and fabrics, or fabrics mixed with wool, that have been left undisturbed for a period of time, or carpeting where there is no foot traffic are most in danger of attack. Chewed fabric or carpet can contain all stages of the pests. Damage consists of a number of small holes. Look closely for larvae and adult moths. If you find only larvae, determine whether they are clothes moths or dermestid beetles. Moth larvae are white and hairless, with dark heads; beetle larvae have long hair or bristles, and are brownish. Look for silken tubes, mats, or a cell-like case.

> *Moth larvae are hairless, white, with dark heads; beetle larvae have long hair or bristles, and are brownish.*

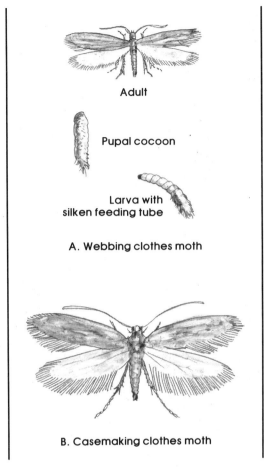

Adult

Pupal cocoon

Larva with
silken feeding tube

A. Webbing clothes moth

B. Casemaking clothes moth

LARVAE AND ADULTS OF THE COMMON CLOTHES MOTHS

*(A) The **webbing clothes moth**, about ¼ inch long, is golden buff to yellow-gray and has upright reddish hairs on its face. Larvae sometimes construct a silken mat or tube over part of the fabric and feed underneath it, but they may also be found on the fabric itself. (B) The **casemaking clothes moth**, a problem in the southern states, is about ¼ inch long, brownish, with three indistinct spots on the wings, and light-colored, upright hairs on the face. The larvae construct a silken tube, called a* cell *or* case, *which they carry around as they feed.*

You may or may not find adults near the larval food. Adult clothes moths are often mistaken for grain moths, but a number of differences distinguish them. Clothes moths fly in dark areas, usually closets or storage areas, while grain moths fly in lighted areas, such as the kitchen or pantry. Clothes moths intersperse short, fluttery flights with periods of running or walking; grain moths have a steady flight. Clothes moths have long, upward-pointing hairs on their heads, long legs, and move rapidly; grain moths lack these traits. If you find adult clothes moths in rooms that lack obvious food sources such as woolen clothes, carpets, or fabrics, suspect an animal nest in the walls of the house.

If you find adult moths in rooms that lack obvious food sources such as woolen clothes, carpets, or fabrics, suspect an animal nest in the walls of the house.

When to Take Action

Take action as soon as you see the pests or signs of their damage. Be certain that the moths are clothes moths and not grain moths.

Long-term Controls

Physical Controls. Practice preventive controls. Vacuum thoroughly to remove pet hair, lint, and organic debris from cracks and crevices in infested areas. Focus on closets where fur, feather, or woolen items are stored, and beneath furniture that's rarely moved. Remove old bird nests on or near the walls of your house.

Store susceptible *uninfected* clothing in *tightly sealed* containers (chests, boxes with sealable lids, or sealed plastic bags). Heat-sealed polyethylene bags are particularly good. Use cleaning or freezing to get rid of an infestation before storage. Dry cleaning kills the pests as well as removing the nutrients necessary for the larvae to mature. Freezing also kills these pests. Put susceptible clothes in sealed plastic bags, leave them in the freezer for a few days, then put them in storage.

Put susceptible clothes in sealed plastic bags, leave them in the freezer for a few days, then put them in storage.

Clothes moth larvae are fragile; every couple of months, expose susceptible unstored clothing to sunlight. Shake the garments out and vigorously brush them, particularly hidden areas such as cuffs, collars, or folds. This can kill the larvae and cause the eggs to fall off.

Temperatures in excess of 104 degrees kill fabric pests in a few hours. If your attic gets very hot during the summer months, store susceptible clothing or fabric there.

In rooms with woolen or wool-based carpets, periodically move heavy pieces of furniture to see if any damage has occurred.

Short-term Controls

Physical Controls. Throw out heavily infested items. Clean lightly infested items. For carpets, use steam cleaning; for clothing, dry clean or freeze.

Last Resorts

Chemical Controls. If you have an exceptionally severe infestation, and feel that you can't wait for the adults to die after you have removed all food sources, use a spray containing pyrethrum/pyrethrins or one of the synthetic pyrethroids. *Caution:* Some people are allergic to the pulverized flowers that are the source of pyrethrum. Wear a dust mask or respirator if you are allergic to pollen and are spraying pyrethrum, or choose a different chemical.

If you must use moth balls, crystals, or flakes to protect stored woolens from larval damage, use camphor rather than naphthalene or PDB (par-

----

If you must use moth balls, crystals, or flakes to protect stored woolens from larval damage, use camphor rather than naphthalene or PDB (paradichlorobenzene), both of which are more toxic than necessary.

----

adichlorobenzene), both of which are much more toxic than necessary, given the risk involved in using them. Scatter the camphor (ball, crystal, flake, or cake) both in the storage container and among the clothes themselves. Make sure the container can be tightly closed, or use duct tape to seal it.

Instead of camphor, you can spray the clothes with a pyrethroid. Permethrin-containing products also protect clothes from clothes moths. Be sure to wash or dry clean pesticide-treated clothes before wearing them. Finally, spray heavily infested carpets with *Bt*, permethrin, or another pyrethroid.

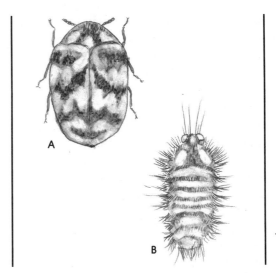

VARIED CARPET BEETLE

(A) Adult is quite small (about 4 mm). (B) Larva. Both adult and larva feed on fabrics.

Dermestid Beetles

Natural History
Dermestid beetles have the potential to cause far more damage than clothes moths because they have a broader appetite. In nature, dermestids are important decomposers, feeding on virtually any animal products and many plant products. A large number of species can become pests. They can be divided into two major groups on the basis of diet: those that feed solely on animal matter, and those that can feed on either animal or vegetable matter. Dermestids are also called carpet beetles or hide beetles; the carpet beetles feed mainly on carpets and fabrics, while the hide beetles feed on dried meat, dead animals in wall voids, etc.

All have the same basic life cycle. The adults are from 1/10 to 3/8 inch long, depending on species, and may be mistaken for small ladybug beetles. Adults lay eggs on the larval food source, and the eggs usually hatch within two weeks. Larvae are smaller than adults, slow-moving, hairy or bristly, and brownish in color. They roughly resemble maggots with long hairs or bristles. Larvae feed for various periods, depending on species and microenvironmental conditions. They may pupate within the food, or wander away and pupate elsewhere.

Warning Signs
Wool, leather, or fur clothing that has been left undisturbed for a period of time, or carpeting where there is

little foot traffic, is most in danger of attack. Chewed fabric or carpet can contain all stages of the pests. Damage over a large area of fabric or carpet is characteristic of dermestids. Look closely for larvae or adults. Collect any and identify them yourself or take them to your county Cooperative Extension Service, natural history museum, or the entomology department of your land-grant university for identification. If you find only larvae, determine whether they are clothes moths or dermestid beetles. Moth larvae are hairless and white with dark heads; beetle larvae have long hair or bristles and are brownish. Look for the characteristic cast skins of the dermestid larvae.

Adult dermestids are often first seen on windowsills, though you can find them in many places, depending on their food source. The adults of some species feed on pollen and may be found on cut flowers brought into the house.

When to Take Action

Take action as soon as you see the pests or signs of their damage. Be certain that the beetles are dermestids and not ladybug beetles.

Long-term Controls

Physical Controls. Prevention is the best form of treatment. Vacuum thoroughly to remove pet hair, lint, and organic debris from cracks and crevices in floors, baseboards, shelves, drawers, window molding, and trim. Pay particular attention to closets where fur, leather, or woolen items are stored; places beneath furniture that's rarely moved; around and in heaters, vents, and ducts. Trap rodents instead of using poison baits; poisoned animals often crawl away and die in a wall cavity, where their carcasses can serve as a food source for dermestids. Remove old bird, wasp, bee, or squirrel nests on or near the walls of your house.

Locate the source of infestation and correct it. Check stored clothing as well as carpets, furniture, boxes or bags of pet food, dried milk, dried meat, spices, and cereal or grains. Store all edible foods, including pet food, in mason jars, sealable plastic tubs, or glass or ceramic jars or tins. Search for dead animals or nests inside wall voids or close to an exterior wall. Check stuffed animals, cut flowers, or flowers close to doors or windows. Use good screening on doors, windows, and vents, particularly those closest to flower beds.

Short-term Controls

Physical Controls. Throw out heavily infested clothing or upholstered furniture. Clean lightly infested items. For carpet and furniture, try steam cleaning; for clothing, dry clean or freeze.

Last Resorts

Chemical Controls. If you have an exceptionally severe infestation, or you feel that you must use a pesticide, use pyrethrum/pyrethrins or a pyrethroid such as permethrin. Spray infested carpet or upholstery with a pyrethroid. *Caution:* Some people are allergic to the pulverized flowers that are the source of pyrethrum. Wear a dust mask or respirator if you are allergic to pollen and are spraying pyrethrum, or choose a different chemical.

 # WASPS

Natural History

Because of the threat of stings, wasps inside the home can be a problem. Virtually all wasps likely to get into your home belong to one family, Vespidae, which includes paper wasps, hornets, and yellowjackets. These species are all social, living in colonies that build paper or mud nests. Adults are predators and/or scavengers on other insects and spiders, including many pests, and are beneficial.

In the spring, adult females (queens) emerge from hibernation and search for prey and nectar from flowers and other sources. Each queen chooses a site and constructs a nest. Nests can be constructed in the ground, in a tree, or inside or outside a house, garage, or tool shed. Eggs are laid one to a cell in the nest; the queen hunts and brings back grubs or caterpillars to feed the larvae. The colonies grow over time and, in the case of hornets and yellowjackets, can eventually contain thousands of workers. Workers often take over the tasks of enlarging the nest and locating prey for the developing larvae while the queen concentrates on egg laying.

Later in the season, fertile males and females are produced; they leave the nest and mate; the males die and the fertilized females overwinter. After producing fertile males and females, the colony declines and eventually abandons the nest.

Warning Signs

Capture individual adults with a net or jar. If it's an isolated visitor, release it outdoors. If you have an infestation, freeze the captured individual and identify it yourself or take it to the county Cooperative Extension Service office or natural history museum. Vespid wasps are distinctive in having forewings that are folded in half at rest. If the wasp turns out to be a yellowjacket or hornet, take action

YELLOWJACKET

promptly, as their populations can grow to very large sizes.

Try to locate the nest. Look around porches or eaves and under air conditioners. If you suspect wasps are nesting inside walls or voids, tap the walls nearby and listen for rustling or buzzing.

When to Take Action

Act as soon as you see wasps inside your house. If you regularly see wasps, suspect a nest inside a wall or on the outside of the house.

Long-term Controls

Physical Controls. Find the wasps' entrance and seal it off. Look for holes in window screens or gaps around windows, in walls, and under doorways. Since adult wasps are beneficial, destroy the nest only if you have to. *Caution:* Wear protective clothing when working around wasps. Do not attempt removal yourself if you have a known allergy to bee- or wasp stings.

If you have a small nest hanging from tree or roof eaves, you can knock it to the ground. Use a 10-to-15-foot pole and do the procedure at

Observe the wasps to find out where they are entering the house. Then blow an insecticidal dust or pyrethrin spray through the entrance hole and onto the nest.

night when the wasps are not active. Wasps will abandon a downed nest. For large nests, hire a professional who is used to handling bees or wasps and has the appropriate protective clothing. Don't try to do it yourself! Vacuuming is particularly appropriate for underground nests.

Chemical Controls. For nests in the walls or voids of houses, removal is usually impractical; the chemical approach is more appropriate. Use a least toxic substance such as pyrethrum, pyrethrins, a pyrethroid, or an insecticidal dust such as silica gel or diatomaceous earth. Observe the wasps to find out where they are entering the house or wall void. Then blow an insecticidal dust or pyrethrin spray through the entrance hole and onto the nest if possible, and plug the hole with steel wool that has been dusted or sprayed. Wasps not immediately killed by the dust or spray will try to leave or enter the hole and will chew on the steel wool and die. Resmethrin repels wasps; spray it in an entrance hole to prevent the wasps from angrily pouring out after you.

Short-term Controls

If a single wasp flies into the house, capture it by covering it with a glass when it lands. Slide a card (or very stiff paper) under the rim of the glass to cover the opening. Take the glass outside and release the wasp. Another easy technique is to open a window; wasps fly toward daylight and it's often easy to coax them outside. If this approach fails, kill the wasp with a flyswatter.

Last Resorts

Chemical Controls. Many outdoor wasp sprays shoot a thin jet of pesticide up to 20 feet. Choose one that contains pyrethrum/pyrethrins or a pyrethroid. Revenge Deep Freeze Wasp and Hornet Killer contains highly volatile substances that have a "freezing" action, which should deter the wasps from coming after you.

If you must use a spray indoors, use one that contains only pyrethrum/pyrethrins or a pyrethroid.

SPIDERS

The thought or sight of spiders evokes fear in some people. Yet except for the tarantula, black widow, and brown recluse, which have painful or poisonous bites, spiders are relatively harmless. In fact, they are useful predators, feeding almost entirely on other arthropods, especially insects. Many species spin silken webs that they use to catch prey or to hide in, while other species actively hunt for their prey. Indoors, spiders and their webs may be found on floors, walls, ceilings, in corners, under furniture, and in other disturbed places.

> *Since spiders eat the insects we consider pests, resist the urge to kill a spider when you see it. It is probably in your house because its food source is there.*

Since spiders eat the insects we consider pests, resist the urge to kill a spider when you see it. It is probably in your house because its food source is there. However, if you cannot tolerate them, catch them and release them outside. *Caution:* Do not attempt to capture a spider with your bare hands. For a medium-size spider, put a small glass or cup over it, slide a piece of stiff paper underneath the cup, pick it up, and empty the contents outdoors. For removing a large spider (e.g., a tarantula), use a small coffee can. There is never any need to use a chemical pesticide against spiders.

BROWN RECLUSE SPIDER

Natural History

Because they have a very painful bite that causes the death of surrounding tissue and can leave a slow-to-heal sore, brown recluse spiders

BROWN RECLUSE OR VIOLIN SPIDER

Adults of this smallish spider (8.5 mm body length) are light to dark brownish with a distinctive dark violin-shaped pattern on their backs.

are considered potentially dangerous to humans. Most occur in the Southwest, although they are also found throughout the Midwest south to the Gulf of Mexico.

Adult females range in body size from ⅛ to ½ inch, while males are slightly smaller. They are oval-shaped, with long thin legs that are up to one and a half times as long as the body. They come in various shades of brown, but all have a distinctive violin-shaped marking on their backs (hence their other common name: violin spider), with the base of the violin near the eyes and the neck stretching toward the waist. As their name suggests, they are shy and like dark places. They spin a web on the ground. The spider hides in the web during the day and leaves it at night to hunt for prey. Wandering spiders often hide in boxes, shoes, clothing, and other items on the floor.

Warning Signs

Look for the spider or its web. If you see a spider, check for the characteristic violin shape on the back. Look for them in closets, storage areas, and bedrooms. Look carefully in boxes, piles of papers, shoes, blankets, or other items stored on the floor in dark areas.

When to Take Action

Take action whenever you see one of these spiders. Consider taking some preventive action if you live in an area where they are known to be abundant.

Long-term Controls

Physical Controls. Regularly vacuum floors, baseboards, and corners, especially in closets, storage areas, and bedrooms. Move boxes and papers off the floor; inspect them before moving them. Carefully shake out any clothing or blankets that are on the floor before using them. Store clothing, blankets, and shoes off the floor.

Short-term Controls

Physical Controls. Catch the spider in a glass and release it outdoors or squash it with your foot, a stick, or other tool. You never need to use pesticides.

CHART 6-1
Products to Control Cockroaches and Ants

The products listed below are a fraction of those pesticides marketed to control cockroaches and ants. These products were selected for listing based on CU's judgment that they can be effective when used in the context of an Integrated Pest Management strategy, and that they pose the least risk to humans, pets, or the environment, based on the active ingredients they contain.

Some products not listed here contain one or more of the same active ingredients as these products and may be substituted for them. But many widely available products are not listed because they contain active ingredients that, in CU's judgment, pose greater potential risks to health or the environment than the ingredients of products listed. In our view, effective pest control does not require use of more toxic pesticides, and we have chosen not to list products that contain them. Products are listed in alphabetical order.

Recommended products contain one or more of the least hazardous active ingredients, including one or more of the following: abamectin (AB), boric acid (BB), silica gel (SG), diatomaceous earth (DE), fatty acids (FA), hydroprene (HY), hydramethylnon (HM), sulfluramid (SL), pyrethrins (PY), or pyrethroids, such as allethrin (AL), fenvalerate (FV), permethrin (PM), phenothrin (PH), resmethrin (RS), tralomethrin (TR), or tetramethrin (TM).

Other products also contain synergists, such as MGK 264 (M2) or piperonyl butoxide (PB), and/or petroleum distillates (PD), whose chronic health effects remain unknown and thus may pose a greater risk to humans and the environment.

Pests controlled: ants (A), cockroaches (R).

BRAND NAME	ACTIVE INGREDIENT(S)	PEST(S) CONTROLLED
Recommended		
Ace Hardware House & Garden Bug Killer II	PH, TM	A, R
Black Flag Ant Control System	BB	A
Black Flag Roach Ender Roach Killing System	AB	R
Borid	BB	R
Combat Ant Control System	HM	A
Combat Roach Control System	HM	R
Copper Brite Roach Prufe	BB	R
Diatom Dust	DE	A, R
Drax Ant Kill Gel	BB	A
Drax Ant Kill-PF	BB	A
Dri Die Insecticide	SG	A, R
Enforcer Ant and Roach Home Pest Control	PM	A, R

Products to Control Cockroaches and Ants (continued)

Brand Name	Active Ingredient(s)	Pest(s) Controlled
Enforcer Home Pest Control	PM	A, R
Enforcer Roach Ridd	BB	R
Gro-Well Home Pest Control	PM	A, R
Harris Famous Roach Tablets	BB	R
Hot Shot House & Garden Bug Killer	PH, TM	A, R
Hot Shot Roach & Ant Killer 3	TR	A, R
Hot Shot Roach Powder	BB	R
K-mart House & Garden Bug Killer II	PH, TM	A, R
Ortho Ant-Stop Ant Killer Spray	PH, TM	A, R
Ortho Flea-B-Gon Flea Killer Formula 11	PH, TM	A, R
Ortho Home & Garden Insect Killer Formula II	PH, TM	A, R
Ortho Household Insect Killer Formula II	PH, TM	A, R
Ortho Insecticidal Soap	FA	A, R
Ortho Flying & Crawling Insect Killer	AL, PH	A, R
Ortho Roach-Kill Powder	BB	R
Perma-Guard Fossil Shell Flour	DE	A, R
Protexall Ant-Kill	BB	A
R-Value Roach Kill	BB	R
Raid Max Ant Bait	SL	A
Raid Max Roach Bait	SL	R
Raid Multi-Bug Killer Formula D39	AL, RS	A, R
Real-Kill House & Garden Bug Killer II	PH, TM	A, R
Revenge Ant Killer Liquid Bait	BB	A
Rid-a-Bug Home Insect Killer	TR	A, R
Ringer's Diatomaceous Earth	DE	A, R
Roxo Liquid Ant Bait	BB	A

Brand Name	Active Ingredient(s)	Pest(s) Controlled
Safer Insecticidal Soap Concentrate	FA	A, R
Safer Yard & Garden Insect Attack	FA, PY	A, R
Starbar Roach-Ban Aerosol	HY, PM	A, R
Other		
Black Flag Ant & Roach Killer Formula B	AL, PB, PM	A, R
Black Flag House & Garden Insect Killer II	AL, PB, RS	A, R
Black Flag Roach Powder	PB, PD, PY, SG	A, R
Black Flag Triple Active Bug Killer	PB, AL	A, R
Combat Ant and Roach Spray	AL, FV, M2, PB	A, R
Drione	PB, PY, SG	A, R
Enforcer Ant & Roach Killer III	M2, PM, PY	A, R
Hot Shot Fly & Mosquito Killer	AL, M2	A, R
Hot Shot Flying Insect Killer Formula 411	PD, PH, TM	A, R
Hot Shot House & Garden Bug Killer Formula 721	PD, PH, TM	A, R
Hot Shot Roach & Ant Killer Triple Strength	AL, M2, TR	A, R
Hot Shot Roach & Ant Killer 2	AL, M2, TR	A, R
Hot Shot Roach and Ant Killer Formula PRWB-1	M2, PB, PY	A, R
New Era Insect Spray	PB, PY	A, R
Perma-Guard Household Insecticide	DE, PB, PY	A, R
Raid Ant & Roach Killer 6 Spray	PB, PM, PY	A, R
Raid Flying Insect Killer 12	AL, PB, PH, TM	A, R
Raid House & Garden Formula 11	PB, PY, TM	A, R
Real-Kill Flying Insect Killer	PD, PH, TM	A, R
Real-Kill House & Garden Bug Killer II	PD, PH, TM	A, R
Revenge Home Exterminator	PB, PY, SG	A, R
Safer Crawling Insect Attack	PB, PY	A, R

CHART 6-2
Other Products to Control Cockroaches and Ants

These products are foggers, which, in general, we do not recommend. However, under exceptional circumstances (i.e., a household infestation of thousands of cockroaches that have not been controlled by the other recommended tactics, such as vacuuming, boric acid, baits, sprays, and so on), a fogger may be needed, BUT ONLY AS A LAST RESORT. We apply the same proviso to these products as to the products in Chart 6-1.

Recommended products listed below contain the least hazardous active ingredients, including one or more of the following: methoprene (MP), pyrethrins (PY), or pyrethroids such as allethrin (AL), cyfluthrin (CY), fenvalerate (FV), permethrin (PM), phenothrin (PH), resmethrin (RS), or tetramethrin (TM).

Other products listed below also contain synergists, such as MGK 264 (M2) or piperonyl butoxide (PB), and/or petroleum distillates (PD), whose chronic health effects remain unknown and thus may pose a greater risk to humans and the environment.

BRAND NAME	ACTIVE INGREDIENT(S)
Recommended	
Green Thumb Home Insect Fogger	PH, TM
Raid Fumigator Fumigating Fogger	PM
Spectracide Indoor Fogger	PM, TM
Zodiac House & Kennel Fogger	PM
Zodiac IGR Plus Fogger	MP, PM
Other	
Black Flag Adult Flea Killer Indoor Fogger Step 2	PD, PH, TM
Black Flag Automatic Room Fogger	PB, PD, PY
Black Flag Flea Killer System (two indoor foggers)	MP, PD, PH, TM
Black Flag Fogger Pine Scent	PD, PH, TM
Black Flag Large Area Automatic Room Fogger	PD, PH, TM
Black Flag Roach Fogger	PB, PD, PY
Black Jack Household Indoor Fogger Roach & Insect Killer X	PD, PH, TM
Four Paws Fast Killing Indoor Fogger	PD, TM
Holiday Household Insect Fogger	FV, M2, PB

BRAND NAME	ACTIVE INGREDIENT(S)
Holiday Household Insect Fogger New Pine Scent	PD, PH, TM
Ortho Hi-Power Indoor Insect Fogger Formula IV	FV, M2, PB, PY
Raid Fogger	M2, PB, PD, PY
Raid Fogger II	M2, PB, PH, PY, TM
Raid Max Fogger	CY, M2, PB, PY
Sergeant's Indoor Fogger	AL, FV, PD

CHART 6-3
Products to Control Flying Insect Pests

The products listed below are a fraction of those pesticides marketed to control flying insects. These products were selected for listing based on CU's judgment that they can be effective when used in the context of an Integrated Pest Management strategy, and that they pose the least risk to humans, pets, or the environment, based on the active ingredients they contain.

Some products not listed here contain one or more of the same active ingredients as these products and may be substituted for them. But many widely available products are not listed because they contain active ingredients that, in CU's judgment, pose greater potential risks to health or the environment that the ingredients of listed products. In our view, effective pest control does not require use of more toxic pesticides, and we have chosen not to list products that contain them. Products are listed in alphabetical order.

Recommended products listed below contain the least hazardous active ingredients, including one or more of the following: *Bacillus thuringiensis israelensis* (BTI), fatty acids (FA), sulfur (SU), pyrethrins (PY), or pyrethroids such as allethrin (AL), permethrin (PM), phenothrin (PH), resmethrin (RS), tetramethrin (TM) or tralomethrin (TR).

Other products listed below also contain synergists, such as M11 (M1), MGK 264 (or n-octyl bicycloheptene dicarboximide, M2) or piperonyl butoxide (PB), and/or petroleum distillates (PD), whose chronic health effects remain unknown and thus may pose a greater risk to health or the environment.

BRAND NAME	ACTIVE INGREDIENT(S)
Recommended	
Ace Hardware Flying Insect Killer II	PH, TM
Ace Hardware House and Garden Bug Killer II	PH, TM
Bactimos	BTI
Bactospeine	BTI
Combat Flying Insect Killer	PH, TM
d-Con Flying Insect Killer	TM
Enforcer Flying Insect Killer III	TM, PH
Enforcer Wasp & Hornet Killer	RS
Enforcer Wasp & Yellowjacket Foam	PM, TM
Green Thumb Flying Insect Killer	AL, RS
Green Thumb Home & Garden Insect Killer	RS
Hot Shot Flying Insect Killer	PH, TM

BRAND NAME	ACTIVE INGREDIENT(S)
Hot Shot Wasp & Hornet Killer II	RS
HWI Hardware Flying Insect Killer	PH, TM
K-mart Flying Insect Killer	PH, TM
K-mart House & Garden Bug Killer II	PH, TM
Mosquito Dunks	BTI
Ortho Home & Garden Insect Killer Formula II	PH, TM
Ortho Household Insect Killer Formula II	PH, TM
Ortho Professional Strength Flying & Crawling Insect Killer	AL, PH
Pic Sulfur Candle	SU
Raid Multi-Bug Killer Formula D39	AL, RS
Real-Kill House & Garden Bug Killer	PH, TM
Revenge Deep Freeze Wasp and Hornet Killer	RS
Rid-a-Bug Home Insect Killer	TR
Vectobac 12AS	BTI

Other

Black Flag House & Garden Insect Killer	AL, PB, RS
Black Flag Triple Active Bug Killer	AL, PB
Black Jack II Fly & Mosquito Spray	PD, PH, TM
Four Paws Super Fly Repellent	M1, M2, PB, PY
Hot Shot Fly & Mosquito Insect Killer	AL, M2, PD
Hot Shot Flying Insect Killer Formula 411	PH, TM
Hot Shot Flying Insect Killer Formula 611	AL, RS
New Era Insect Spray	PB, PY

Products to Control Flying Insect Pests *(continued)*

BRAND NAME	ACTIVE INGREDIENT(S)
Ortho Flying & Crawling Insect Killer	AL, PD, PH
Ortho Home & Garden Insect Killer Formula II	PD, PH, TM
Ortho Household Insect Killer Formula II	PD, PH, TM
Raid Flying Insect Killer 12	AL, PB, PH, TM
Raid House & Garden Formula 11	PB, PY, TM
Raid New Formula Flying Insect Killer	M2, PB, PD, PY, TM
Raid Professional Strength Flying Insect Killer	PB, TM
Real-Kill Flying Insect Killer	PH, PD, TM
Revenge Home Exterminator	PB, PY, SG
Revenge Farm & Home Fly Bomb (aerosol)	MS, PB, PY

Products to Control Houseplant Insect Pests

The products listed below are a fraction of those pesticides marketed to control houseplant insect pests. These products were selected for listing based on CU's judgment that they can be effective when used in the context of an Integrated Pest Management strategy, and that they pose the least risk to humans, pets, or the environment, based on the active ingredients they contain.

Some products not listed here contain one or more of the same active ingredients as these products and may be substituted for them. But many widely available products are not listed because they contain active ingredients that, in CU's judgment, pose greater potential risks to health or the environment than the ingredients of products listed. In our view, effective pest control does not require use of more toxic pesticides, and we have chosen not to list products that contain them. Products are listed in alphabetical order.

Recommended products listed below contain the least hazardous active ingredients, including one or more of the following: fatty acids (FA), disyston (DS), methoprene (MP), sulfur (SU), pyrethrins (PY), or pyrethroids such as allethrin (AL), phenothrin (PH), permethrin (PM), resmethrin (RS), or tetramethrin (TM).

Other products listed below contain ingredients that pose a somewhat greater risk to health or the environment, including the following: piperonyl butoxide (PB), rotenone (RO), and/or petroleum distillates (PD).

BRAND NAME	ACTIVE INGREDIENT(S)
Recommended	
Dexol Tender Leaf Spider Mite Killer	MP, RS
Dexol Tender Leaf Systemic Granules Insect Control	DS
Frank's Nursery & Crafts Insect Spray for House & Garden	AL, RS
Green Aid Insect Spray for Houseplants	PB, PY
New Plant Life Spider Mite and Mealybug Spray	MP, RS
Safer African Violets Insect Attack	FA
Safer Insecticidal Soap for House Plants	FA
Other	
Bachman's Insect Spray	PB, PY
Frank's N&C White Fly & Insect Spray	PB, PD, PY
Hyponex Bug Spray	PB, PY
Pergament Full Blast Insect Killer	PB, PY, RS
Plant Marvel I-Bomb Insecticide Spray	PB, PY, RO
Schultz Instant Houseplant & Garden Insecticide	PB, PY

7

CONTROLLING RODENT PESTS

Most people occasionally encounter a mouse or rat in their homes. Many also find rodents such as rats, mice, gophers, moles, and squirrels in their lawns and gardens. If you are having a problem with a rodent, remember the principles of IPM and don't overreact. These animals have a place in nature if not in your home or garden, and the best approach is to prevent them from entering a particular area, or to lure them away.

If you see evidence of rodent activity—gnawed packages of food or grain, gnaw marks or holes in various areas, droppings—or even the animals themselves in your home, don't panic. First find out what kind of rodent it is. For example, rustling, scratching, or squeaking in the walls or attic could be either squirrels or rats. In the vast majority of cases, however, your problem will be mice or rats.

SQUIRRELS

If the animal in the attic is a squirrel and you need to remove it, find out how and where it entered, then seal the entrance hole. Open any windows in the attic and encourage the squirrel to leave through one of them. If this doesn't work (or the attic has no windows), try capturing the squirrel in a live trap and releasing it outdoors. Once the squirrel is gone, find the entrance hole and seal it with metal screen and/or wood. If squirrels are a recurrent problem in your attic, consider cutting off any tree branches that overhang your roof.

RATS

Natural History

Although far less common a problem than mice, rats can eat and contaminate food; damage structures and machinery by gnawing; and harbor human diseases, including bubonic plague, salmonellosis, infectious jaundice, murine typhus, rickettsialpox, and rat-bite fever. (Most of these diseases are not problems in the United States.) In fact, rats bite more than 150,000 people every year.

Two species—the Norway rat (also called sewer rat or brown rat) and the roof (or black, tree, or European) rat—live in or around human habitations and cause virtually all the rat problems suffered by homeowners and apartment dwellers. (See Table 7-1 following.)

One or both species occurs in all 50 states. The Norway rat is far more common than the roof rat and predominates in the cooler parts of the country. Roof rats are most common in the South, especially Florida, the

Rabies has been increasing in wildlife and can be carried by raccoons and bats. If you suspect an animal is rabid, call your local game warden to request instructions.

Moreover, the bites of rats can cause infections and possible disease in humans. It is therefore extremely important that you do not corner or confront these animals or try to touch them with your bare hands. Use traps that work automatically and wear protective clothing and gloves to remove any living animals you plan to release in the wild.

TABLE 7-1

■ DIFFERENCES BETWEEN ADULTS OF THE MAJOR RAT SPECIES ■		
Characteristic	Norway, Sewer, or Brown Rat	Roof, Black, or Ship Rat
Color	brown, gray, white, black, pied	dark gray or brown above, gray or whitish below
Size (body + head)	about 10 inches	about 8 inches
Body	heavy, thick	light, slender
Weight	1 pound or greater	½–¾ pound
Tail length	shorter than head + body	longer than head + body
Fur	soft	somewhat stiff
Eyes	small	large
Ears	small, covered with short hair	large, almost naked
Nose	blunt	pointed
Preferred habitat		
outdoors	burrows in ground, sewers, under foundations	trees and dense vine growth
indoors	under floors, in walls	aboveground, in attics, walls
Preferred food	protein, fat	fruits and vegetables

Gulf Coast, and the coastal regions of California, although small colonies are found in virtually all seaport towns.

The two rat species have several features in common. Their front teeth grow continuously, and so they need hard things to gnaw on, such as wood, plastic, or metal. Both species can breed throughout the year, although spring and fall are peak seasons. Like many rodents, both have poor vision but excellent senses of smell, taste, hearing, and touch. When presented with foods they have not eaten before, they nibble a small bit of it. If they become sick within two to three days, they remember the food and will not consume it again. Mother rats have been known to take their young to poisonous food and make them taste it, so they will learn to avoid it in the future. This behavior makes it hard to poison them.

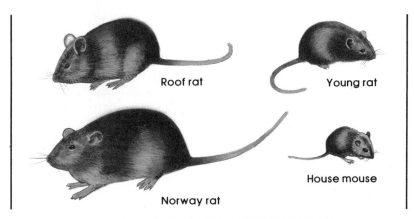

Roof rat

Young rat

Norway rat

House mouse

DISTINGUISHING AMONG THE ROOF RAT,
NORWAY RAT, AND HOUSE MOUSE

Their long whiskers are very sensitive to touch and are used to sense the environment. They tend to be creatures of habit and establish runways and passageways, frequently located next to a wall so that their whiskers will brush against it while they are running. They come out primarily at night, when fewer predators are active. They do not like to be in the open and prefer cover. Although aggressive, they actually have few defenses against predators and rely for protection on their ability to hide.

Warning Signs

The most important signs are gnaw marks on wood, plastic, metal, pipes, and foodstuffs; droppings; runways marked by grease stains made by the rat's fur; and tail and/or footprints. You may even see the rats themselves. In addition, your pets may show signs of agitation or excitement.

To be sure that an infestation is ac-

tive, look for fresh droppings. Place a thin layer of flour along suspected runways and look for tracks the following morning. If you find no tracks within three days, rats are probably not active in that area. Since rats run next to walls, you should find droppings there.

If droppings are out in the open, or if you find fresh droppings while the flour remains undisturbed, suspect roof rats. Look for exposed pipes, beams, or wires running close to the ceiling, particularly in attics and basements.

Finally, look for potential nest sites, such as holes in the wall and hidden spaces behind heavy appliances or piles of boxes or food, burrows, refuse or trash piles, or (in the case of roof rats) ivy-covered walls outdoors.

When to Take Action

Take action as soon as you have evidence of an active rat infestation.

Long-term Controls

Physical Controls. Long-term rat control requires reducing their access to food, water, and nesting places (harborage). Reducing nesting sites, combined with preventing the rats from coming indoors, plays a major component in a rat IPM program. Indoors, look for holes in the walls or baseboards and seal them with cement, heavy-gauge metal sheeting, or ¼-inch hardware cloth. Preventing rats from getting indoors can be a challenge. Rats can squeeze through holes you think are too small for them. As a rule of thumb, if a rat can get its skull through a hole, it can get its whole body through. Norway rats can swim extended distances and dive under water for up to 30 seconds.

❧ *As a rule of thumb, if a rat can get its skull through a hole, it can get its whole body through.*

They can come up a sewer pipe and enter a house through the water trap of a toilet, climb the inside of a vertical pipe from 1½ to 4 inches in diameter, and come up through floor drains. They can walk horizontally on pipes of any size and crawl up the outside of a pipe up to 3 inches in diameter. If the pipe is within 3 inches of the wall, they can climb the pipe no matter how big the diameter is by pressing their backs against the wall. Seal off potential entry areas around pipes. Norway rats are good burrowers and can come up through unpaved basement floors.

Outdoors, walk the perimeter of the house and seal holes in the house structure or burrows. If burrows are a severe problem, install underground fencing. Use a small-mesh hardware cloth, available at any hardware store. It should extend at least 1 foot aboveground and for 1½ to 2 feet underground. For best results, slant it away from the house in an L shape. Clear brush away from the edge of the house, because rats like cover. Put a grid over pipes.

Roof rats are good climbers and often use trees, poles, phone, cable TV, or utility wires to gain entrance. Put up metal cones, with the narrow end pointing toward the house, around wires, cables, and pipes. Cut tree limbs that either touch the roof or are within jumping distance of it. If you have wooden shingles on your roof, inspect them to make sure that a roof rat hasn't gnawed a hole through them to gain entry to the attic. Put a rat guard (a cylinder of metal) on poles or trees, making sure that it is at least 18 inches tall and that

---

❧ *To prevent roof rats from entering the house, put up metal cones, with the narrow end pointing toward the house, around wires, cables, and pipes.*

---

the nailheads are flush with the metal's surface. Remove outdoor trash, rubbish, and wood or stone piles that can serve as harborage sites. If you have problems with roof rats and have ivy-covered walls, get rid of the ivy.

If you find rat holes in your roof, repair them promptly. Other rodents, such as squirrels, sometimes get into attics through holes and then cannot find them again to leave.

Reduce food sources. Indoors, store bread, grains, cookies, etc., in glass jars or sealable plastic containers. Don't leave pet food and water out; remove it as soon as the pet is through eating or drinking. Put garbage in covered containers. Outdoors, put garbage in a can with a spring mechanism so that it won't open if knocked over. Remove pet feces, which also serve as rat food. Bird feeders are a major food source for rats in suburbs, so put a rat guard or a layer of sticky material (Tanglefoot, Tack Trap, or Stickem) on the pole so that rats can't climb it.

Short-term Controls

Physical Controls. If you don't also use long-term measures, you'll have to repeat the short-term techniques over and over again.

Traps are good for small infestations and allow easy removal of the rats, dead or alive. Traps come in live, snap, and glue types. *Snap* traps have the drawback that they can injure pets or children, while *glue* traps can be a nuisance if children or pets touch them. Place the snap traps in areas not accessible to pets and children, such as behind or underneath furniture. Place the traps along runways and passageways. Put snap traps perpendicular to the wall, with the trigger end closest to the wall. For roof rats, you may have to attach traps to the tops and/or sides of pipes being used as runways. Since rats have poor vision and a very good sense of smell, they are often suspicious of new things in their environment and may avoid them, a phenomenon known as "trap shyness." If this is a problem in your house, put an unsprung trap out for a few days with food so that the animal becomes used to it; then set and bait it. Put snap traps out in pairs separated by a few inches, so that if the rat attempts to jump over one, it's caught by the second one.

Bait is optional. Try peanut butter, cheese, fat, or dog food for Norway rats, and peanut butter or bits of fruit or vegetables for roof rats. If the food or traps remain untouched after four to five days, move them to a new area.

Glue traps, which often come prebaited, can be put out in pairs next to the wall along runways and passageways. Since rats are attracted to dark tunnels, put the traps under cover and your catch will probably increase. If you consider the snap or glue traps too inhumane, use a *live* trap. Put it parallel to, and within an inch of, the wall. Once you've caught the rat, release it far away from human habitation, such as in a nearby woods.

Finally, there are *ultrasound* devices. Intense high-frequency sound repels rodents and can be used as part of an IPM program. The sound pressure levels disrupt rat movement patterns by making them avoid areas with the ultrasonic devices. Since the rats simply move to other areas, you must use traps and bait stations in those areas to catch the diverted rodents.

Although some ultrasonic devices have been shown to work best in commercial facilities, barns, milking parlors, or other areas with high rat populations, in tests performed by the EPA the devices sold to consumers have been shown to be ineffective.

Last Resorts

Chemical Controls. Poison baits, the most commonly used technique, have a few drawbacks. First, there is the problem of accidental poisoning of pets, children, or other nontarget animals. Second, poisoned rats may die in an inaccessible place, such as a wall, or behind heavy appliances, filling the house with a foul odor. To minimize accidental poisoning, use covered bait stations. Poison baits come in two basic types: single-dose and multiple-dose. In general, the single-dose poisons, such as zinc phosphide, strychnine, or the botanical red squill, while very effective, are also acutely toxic and thus can represent a greater danger to children and nontarget animals.

The effectiveness of poison baits depends on the inert material used as well as the availability of alternate food sources. Indoors, put the baits along runways and passageways, preferably in areas where children and pets can't reach them. Outdoors, always use bait stations and put them along runways or in areas with a lot of brush cover. *Caution:* Wear gloves when handling these baits.

Put out poisonous tracking powders where the rats will walk through them. When they groom themselves, the rats ingest the poison and die. Do not use the powders around the house unless they are kept to areas where pets or children cannot reach them. *Caution:* Wear a dust mask and gloves when handling these powders.

Fumigants, such as cyanide or methyl bromide, are extremely toxic. If you have a huge infestation and all other techniques have failed, hire a professional fumigator.

---

*C*AUTION: *Traps, Baits, and Pesticide Powders*

1. To minimize accidental poisoning, use covered bait stations.
2. Do not use poisonous powders around the house unless they are kept to areas where pets and children cannot reach them.
3. Place snap traps and poison bait traps in areas not accessible to pets and children, such as behind or underneath furniture.

MICE

Natural History

Mice are far more common than rats in dwellings. Although a number of species can become a problem indoors, we focus on the house mouse. The house mouse has moderately large ears, short and broad feet, and a tail usually as long as the head and body. It is usually dusky gray toward the head, with a slate or black darkening along the middle of the back, which gradually pales to an ashy gray underside. The ears and tail are usually brownish.

Mice contaminate foodstuffs with droppings and urine and, in large infestations, destroy clothing, furniture, and books. Their gnawing can also create holes in floors, walls, and electrical equipment.

Mice can be responsible, either directly or indirectly, for a number of diseases. Food contaminated with their droppings may harbor *Salmonella* bacteria or tapeworm eggs. Mouse parasites carry rickettsialpox, bubonic plague, and epidemic or murine typhus. Mice have also been known to carry histoplasmosis, tularemia, and lymphocytic choriomeningitis.

Like other rodents, mice have front teeth that grow continually, which accounts for their gnawing behavior. Like rats, they have very poor eyesight, but keen senses of smell, touch, and taste. They tend to remain hidden and run close to walls, using their whiskers to sense their environment. They are creatures of habit, often visiting the same sites each day. Consequently, they are very wary of new objects (such as traps) in their environment.

Mice have much smaller home ranges than rats. If food is abundant, they may not wander more than 3 feet from their nest or hiding place. They are also adaptable and can live almost anywhere, in heating ducts in skyscrapers, in underground steam tunnels, and in ground-level burrows. They eat much the same food as people, particularly meats, grains, cereals, seeds, fruits, and vegetables. Unlike rats, they eat only a little bit at a time, feeding 15 to 20 times a day in different places, which makes them harder to poison than rats. They prefer sweet liquids to water for drinking.

Females build nests in hidden sites using cloth, cotton, and other materials. The need for nesting material explains the damage they can do to clothing, furniture, and wall insulation.

Warning Signs

The most obvious sign of mouse infestation is the presence of their tiny droppings, usually found next to walls along runways. Also look for nibbled foodstuffs, clothing, or tiny gnaw marks. In particularly large infestations, a pungent musty odor is

❧ *The most obvious sign of mouse infestation is the presence of their tiny droppings, usually found next to walls along runways.*

noticeable. Although they can cause problems year round, the peak season for mouse infestation is early fall, when they are busiest finding food and shelter for the winter.

When to Take Action
Take action when your nuisance level is exceeded.

Long-term Controls
Physical Controls. The key to mouse IPM is habitat management—preventing entry in the first place and reducing harborage sites by improved sanitation. Indoors, do not leave food around, and keep breads, cereals, grain, and pet foods in sealed containers. When feeding pets, remove the dish or bowl after feeding. Go through the house and look for holes and other hiding places. Plug up all the holes and clean any areas where you find droppings. Clean up piles of paper, clothing, or other items in the basement, closets, or attics. Outdoors, remove brush near your garden if the mice are damaging your vegetables.

If mice continue to be a problem, consider getting a cat. They readily catch and eat mice (rats tend to be too big for them).

Short-term Controls
Physical Controls. Put snap or glue traps out in pairs along runways or near the nest or harborage sites. Place the traps perpendicular to the wall or runway and within 3 feet of sites. If you have a heavy infestation, put traps out every 4 to 6 feet along walls in infested areas. If the mice become bait-shy, pre-bait unsprung traps for a few days. Glue traps are best put

❧ *Glue traps are best put under cover, because rodents are attracted to dark tunnels.*

under cover, because rodents are attracted to dark tunnels. (Wear gloves when handling sticky materials.) A number of live traps are also available that can catch more than one mouse. If you use live traps, release the mice away from the house. For more information on trapping, see the preceding section on rats.

Last Resorts
Chemical Controls. Poison baits, even with covered bait stations, should be a last resort. Indoors, a mouse may die and rot in a place where you can smell it but can't get to it. Outdoors, a mouse can die and potentially poison a pet that eats it.

Tracking powders and fumigants are not appropriate for mice. Ultrasonic and electromagnetic devices are a waste of money.

Tracking powders and fumigants are not appropriate for mice.

GOPHERS

Natural History

Gophers eat a variety of plants, particularly roots, bulbs, tubers, some grasses, and seeds. They can be a major cause of damage to lawns and gardens.

The main types of gophers in the United States are eastern pocket gophers, western pocket gophers, and the Mexican pocket gopher. In general, eastern pocket gophers are found in the East and Gulf Coast states, western pocket gophers west of the Rockies, and the Mexican pocket gopher in the Southwest. They are thickset, from 5 to 18 inches long, including a short, sparsely haired tail. They have small eyes and ears, short necks, long, chisel-like front teeth that grow continuously, long claws on their front feet, and large, fur-lined cheek pouches in which they store food. They live alone underground in extensive burrows, coming together only during breeding season, in spring and early summer. The burrows are usually from 6 to 18 inches below the soil surface. Gophers feed on the underground parts of plants they encounter while burrowing. They occasionally leave their burrows to feed on succulent herbaceous plants, which they cut up, stuff into their pouches, and take back to underground storage areas. They do not hibernate, and so must hoard food to tide them over during the winter.

Do not confuse a gopher mound with a molehill. Gophers make a small, fan-shaped mound of loose dirt with a closed burrow at one side of the mound. Molehills are usually constructed of softer soil and look like a small volcano, with the closed burrow exit at the center of a circle. Gophers are territorial, and you most likely have only one in your yard or

Do not confuse a gopher mound with a molehill. Gophers make a small, fan-shaped mound of loose dirt with a closed burrow at one side of the mound. Molehills are usually constructed of softer soil and look like small volcanoes, with the closed burrow exit at the center of a circle.

EASTERN POCKET GOPHER

Pocket gophers burrow, carrying foods such as roots, bulbs, seeds, and grasses in fur-lined pouches inside the mouth.

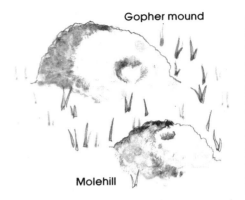

Gopher mound

Molehill

DISTINGUISHING BETWEEN GOPHER MOUNDS AND MOLEHILLS

Gophers make a small, fan-shaped mound of loose dirt with a closed burrow (called a plug) at one side of the mound. Molehills are usually constructed of softer soil and look like small volcanoes, with a closed burrow exit (or plug) at the center of the circle.

garden, even though you may see a number of mounds, unless you have an especially large yard. The mounds are produced when the gopher throws soil out of its burrow as it enlarges it. Besides the mound, you may notice small plugs of soil that are flush with the ground. These are exit holes.

Warning Signs

Look for mounds and exit holes.

When to Take Action

Unless the gopher is causing significant damage to your plants, ignore it. The burrows help aerate the soil and add organic matter to it.

Long-term Controls

Biological Controls. Encourage natural enemies (snakes, owls, weasels, badgers) by providing cover and appropriate habitat. Do not kill snakes, particularly in the West, where gopher snakes are effective control agents.

Physical Controls. You can also take advantage of the fact that gophers have distinct food preferences.

For example, western pocket gophers strongly dislike oleander. Plant a strip of oleander around the area you want to protect. Ask your county Cooperative Extension Service, land-grant university, natural history museum, or nursery for the identity of plants in your area that either repel gophers or are more attractive to gophers than the plants you want to protect.

To protect a small garden, either construct a small cement wall or, better yet, install a fence of small-mesh

 Plant a strip of oleander around the area you want to protect from gophers.

hardware cloth. The fence should extend about 9 inches aboveground and 2 feet underground. For best results, the fence should slant away from the garden and be L-shaped at the bottom. To protect seedling trees, sink an 18-inch-high cylinder of small-mesh hardware cloth 12 inches into the planting hole, leaving 6 inches aboveground.

Short-term Controls

Physical Controls. Short-term controls include flooding, fumigation, and trapping. First you need to find the main burrow, usually located 12 to 18 inches from the plug in the mound. The gopher mound is at the end of a lateral burrow. Probe the soil with a long, thin stick or coat hanger in a circle around the mound. When the probe suddenly drops a couple of inches, you've found the main burrow.

To flood a burrow, either insert a garden hose into the main burrow or dig into the mound until you expose a runway, and insert the hose there. Cover the hose with soil and turn on the water. Watch for the gopher trying to escape through one of the exit holes. Be prepared to kill it with a shovel. Flooding can be particularly effective in early summer to midsummer, when the female is raising young, because the flooding will drown them.

If you choose fumigation, use a commercial gas cartridge or smoke bomb, or make your own with a lawnmower. The gas cartridges are effec-tive, but the gas takes a while to diffuse throughout the entire length of the burrow. Cover the hole to prevent smoke from escaping.

You can modify a gas-powered riding lawnmower to fumigate burrows. You will need a 10-foot-long flexible metal exhaust pipe that fits over the mower's rigid exhaust pipe.

Dig up a mound until you expose the main runway. Squirt some motor oil into one end of the flexible exhaust pipe, put it over the mower's exhaust pipe, and stick the other end into the runway. Cover this end with soil, turn on the engine, and let it idle for 10 or 15 minutes. The carbon monoxide will kill the gopher(s), while the oil will smoke and allow you to see any open exit holes. Quickly cover these holes to suffocate the gopher, or kill it as it emerges from one of them. This method spreads the smoke throughout the burrow system quickly.

The lawnmower is the quickest, safest, and most effective means of killing gophers. If you don't have a riding lawnmower, use a car instead. However, it can be difficult to get the car close enough to a mound to use a flexible exhaust pipe. *Caution:* Be sure to leave the windows of the car open while using this procedure.

Traps are effective for most infestations that do not cover large areas. There are three types of lethal traps available, and all work well (see page 200). You can place them in the lateral burrow, but they are most effective when put in the main burrow.

Dig a hole to expose the main burrow and insert two traps, facing in opposite directions along the burrow. Set the traps and attach them to a stake in the ground with a cord or flexible wire (so the animal can't drag the trap away before it dies). Put a board over the hole and cover it with soil so no light gets through. Check the trap in a couple of days. If possible, set the traps in a burrow next to a recently created mound, as this is evidence that the burrow is being actively used.

Last Resorts

Chemical Controls. Poison baits are usually grain covered with strychnine

 Although effective, poison baits can accidentally poison nontarget organisms that might eat the dead animal.

alkaloid, zinc phosphide, or chlorphacinone. The baits are put into the tunnel system and then covered up. Although effective, poison baits can accidentally poison nontarget organisms that may eat the dead gopher. Use one of the other suppressive methods before resorting to a poison bait. *Caution:* Wear gloves when handling these baits.

MOLES

Natural History

Raised ridges running through the lawn means you have moles. Moles are solitary burrowing mammals exquisitely adapted to living in narrow underground tunnels. They are small, with bodies 5 to 7 inches long, fur that easily brushes in any direction, flat, pointed heads, no external ears, very small eyes covered with skin, and short legs and tails. The short front legs point outward and have broad claws used as shovels or as paddles for swimming. They are found in soft, moist, crumbly soil, through which they can often swim. They eat primarily insects (especially white grubs), earthworms, spiders, and other soil invertebrates.

Four species of moles cause homeowners problems by invading lawns in search of food. These are the star-nosed mole, the eastern mole, the hairy-tailed mole, and the California mole. The first three species occur east of the Rocky Mountains, and their ranges frequently overlap, while the fourth occurs in California and southern Oregon. None lives in the Great Basin, the Rocky Mountains, and the western Great Plains because the soil is too dry and stony.

The three eastern species can often be distinguished by their habits. The eastern mole prefers well-drained soils and does damage by tunneling 1 to 2 inches below the soil surface, which creates long, winding ridges.

MOLE

The star-nosed mole prefers poorly drained soils and raises numerous mounds rather than ridges because it usually tunnels 4 to 6 inches below the soil surface. During cool, moist weather when their food is close to the surface, star-nosed moles may also create long, winding ridges.

Moles construct deep permanent burrows, up to 3 feet down—used year-round for living and raising young—and shallow feeding burrows, just below the soil surface, some used just once. They also create exit holes, called molehills, that look like miniature volcanoes with plugged holes in the centers; molehills are often located close to the deep permanent burrows.

Moles are highly energetic creatures, active around the clock in short alternating cycles of work and rest. The bulk of their lives is spent expanding their burrow systems and

The near-constant activity and high metabolism of moles require them to be voracious feeders; some species consume more than their weight in food daily.

looking for food, although some species, such as the star-nosed mole, spend a significant amount of time aboveground. Their near-constant activity and high metabolism require them to be voracious feeders; some species consume more than their weight in food daily.

Warning Signs
Look for the raised ridges that characterize mole feeding burrows, or the volcano-shaped mound. Feeding-burrow ridges can be straight or twisted, and can be extensive. They usually appear in spring or summer.

When to Take Action
Take action when the number of ridges exceeds your personal threshold. However, the ridges and molehills are mainly an esthetic problem. The main damage caused is indirect; the feeding burrows lift turf or soil, exposing plant roots that then dry out. If this becomes a problem, tamp down the ridges and water them. Although feeding-burrow ridges complicate lawn mowing, tolerate moles as much as possible. They feed voraciously on many soil pests, especially grubs, which are their favorite food.

Long-term Controls
Biological Controls. Since white grubs, Japanese beetles, and billbugs are a major food source for moles, controlling the grubs will make your yard or garden less attractive to these animals. (See chapters 3 and 4.)

Physical Controls. Moles cannot dig well in highly compacted, stony, or heavy clay soils. For a small area, such as a garden, surround it with a barrier or border 6 inches to 1 foot wide and 2 feet deep, consisting of highly compacted soil, stones, and/or clay, and try to keep it dry. You may need a small fence 6 inches to 1 foot high to prevent the mole from walking over the barrier area. Or you can install a fence of small-mesh hardware cloth that extends 1½ to 2 feet underground and 6 inches aboveground. Slant the fence away from the protected area underground in an L shape.

Short-term Controls

Physical Controls. Use flooding, trapping, or fumigating to kill or remove the moles. Flooding works best for smaller burrowing systems, if you can locate the permanent burrows. Some West Coast species put up a number of molehills close to their deeper burrows. Probe around the molehill with a thin stick or coat hanger to find the deeper burrows. Once you've found them, dig them up, insert a garden hose, and turn it on. Watch the other molehills; if the mole exits, be prepared to kill it with a shovel. It may take a while to flush the mole out. Flooding during the spring kills the young in the nest.

❧ *Controlling grubs, their major food source, will make your garden less attractive to moles.*

If you don't want to kill the mole, use a live trap. First find an active surface tunnel. Look for a straight stretch of surface-burrow ridge, preferably one close to a molehill, and tamp it down with your foot. If it is an active feeding burrow, it will be repaired by the mole within a day. Expose the surface tunnel and bury a 3-pound coffee can with the open top level with the bottom of the runway. Cave in the runway on either side of the can and cover with a board to prevent any light from entering. The mole will push up the caved-in soil and fall into the can. Check the trap the following day and release the mole far away.

Lethal traps are spring-loaded and triggered when the mole tries to clear a caved-in burrow where the trap is set. There are three basic types of mole traps on the market: *harpoon* or *impaling* (Victor), *scissor-jawed* (Out O'Sight), and *choker* (Nash). Each needs to be set in a different fashion; follow the instructions. The harpoon traps are the easiest to use because they do not require that you dig up a surface burrow. To increase the effectiveness of the traps, set them along an active feeding burrow and do not disturb any other part of the burrow. If the trap is not sprung within two days, move it to a new location. *Caution:* Be sure to keep all these traps out of reach of children and pets.

You can use smoke bombs or a lawnmower to fumigate moles (see page 197). Fumigation is most effective when the smoke bomb or exhaust is put into the deeper burrows.

BATS

Natural History

Bats are the only mammals capable of true flight; their wings are actually modified hands. They also have received bad press, thanks to Hollywood, and are unfortunately associated with evil. Although bats can spread rabies and histoplasmosis to humans, the threat of the former has probably been overexaggerated, while there are many sources for the latter. The useful nature of bats has been generally overlooked. Virtually all 40 bat species found in the United States feed solely on insects. Indeed, these bats feed voraciously on insects. A little brown bat can easily consume 1,000 insects in an evening.

Bats forage almost exclusively at night, and spend their days in roosts. They occur singly or in colonies, some of which are enormous in size. Most colonies contain both sexes, although many species have special nursery colonies consisting solely of pregnant females. Pregnant female Mexican free-tailed bats migrate every summer to one of five caves in Texas where they form huge nursery colonies. Observers estimate that the five caves contain some 100 million female bats, and by the end of the season some 100 million offspring.

U.S. bats have the same basic life cycle. Adults mate in fall or winter, with a single young (a few species bear twins while the red bat bears litters of one to four) being born between May and July. By the end of the summer, the young forage with the adults at night.

Warning Signs

Detect bat presence via sight, noise, droppings, urine, or collective odor. If bats are roosting in or on your house or nearby, you should see them frequently, especially at dusk and during the early evening hours. High-pitched squeals or squeaks or rustling coming from walls or attics may awaken sleepers at night. Droppings and brown stains (usually a combination of urine, feces, and body secretions)—evidence of resting or sleeping roosts—may cause streaks on the outside of your house near entry and exit holes, or underneath resting roosts. If the bats have gotten into a false wall, a foul odor may permeate a part of the house.

LITTLE BROWN BAT

When to Take Action

Take action as soon as your personal threshold is exceeded. Some people tolerate bats living or roosting in their houses because they eat so many insects. Indeed, a growing number of people actively encourage bats to live in their area by erecting artificial bat houses.

Long-term Controls

Design your house to make it bat-proof. Be sure that all holes to the outside are sealed. Pay particular attention to eaves, awnings, wooden siding that is not quite snug, and areas around the attic. Bat-proof the house in early spring, before the bats arrive, or in the evening in late fall, while the adults and young are out hunting for insects.

If your house has bats, find the holes they use to enter your house and seal them. Locate holes either through the stains left on the house near the holes, or have a person stand on each side of the house for an hour around sundown to see where the bats are exiting. For small holes, use a high-grade caulking material. For larger holes, use an expandable foam, hardware cloth, or wood. Be sure to seal the holes in the evening after all the bats have left.

If there is a colony of bats in the walls, attic, or siding of the house, do not seal the hole immediately as you may trap bats inside where they will die and then start to stink as they decay. In these cases, staple (or use duct tape) flexible bird netting about 1 foot above the entry/exit hole and let the netting fall loosely to about 3 feet beneath, and 2 feet to the sides of, the entry/exit hole. Attach (use staples or duct tape) the sides of the netting to the house but leave the bottom of the netting open. The bats will be able to leave from the entry hole, fall to the bottom of the netting and fly away, but they will be unable to get past the netting when they return the following morning. Leave the netting up for 7 to 10 days, then take it down and seal up the hole(s).

To encourage bats to stay in the area, especially if you are ejecting them from your house, put up artificial bat houses for them (see References and Resources).

Short-term Controls

If a stray bat flies into your house, don't panic. Confine it to one room by closing all the doors and then open up the windows in the room and allow the bat to leave.

Since bats are such beneficial creatures, there should be no need to use poison baits or fumigants against them.

PART FOUR

PESTS

O·N

PETS

8

CONTROLLING FLEAS, TICKS, AND MITES

More than half of the estimated 200 million pets in the United States are dogs and cats. Fifty-two percent of households own at least one cat or dog. Given the prevalence of these pets, and that most are permitted outdoors, many owners have to cope with pet pests.

We shouldn't use anything toxic on our furry friends—especially since we pet them, hug them, and often sleep with them. Moreover, we should guard against overreacting to their pests. The presence of a couple of fleas on your dog probably worries you more than it bothers your pet.

 FLEAS

Natural History

Every dog or cat that spends time outdoors has probably had fleas at some point in its life. Fleas are small, wingless, dark-colored insects, with flat bodies and strongly developed hind legs especially adapted for jumping. They can jump distances 200 times their body length. Adults have sucking mouthparts. Toxins in fleas' saliva often cause intense itching and severe inflammation of the skin on humans and pets.

The life cycle of a flea has four stages: egg, larva, pupa, and adult. Eggs may be laid on the host (and fall off) or in the environment (usually the host's nest or sleeping areas). Most eggs are laid when temperatures exceed 65 degrees and humidity is high. Flea larvae resemble whitish maggots. They do not live on the host, but feed on dried blood and other organic material in the pet's environment. For most species, the dried blood is a necessary part of the larval diet; without it they cannot develop into adults. Larvae avoid light and require a moist environment; they have a thin skin and can dry out easily. At the end of the larval stage, the larva spins a silken cocoon and enters the pupal, or resting, stage. Pupae are very sensitive to mechanical disturbances. Vibrations that signal the presence of an animal trigger emergence of the adult flea from the cocoon, maximizing its chance of finding a host. This explains why great numbers of fleas suddenly appear in a house when the residents come back from an extended absence.

Most fleas lay their eggs when temperatures exceed 65 degrees and humidity is high.

Females are bigger than males, but adults of both sexes consume blood of mammals and birds. Many species only visit the host briefly to feed, spending a good deal of time nearby. Other species, such as the sticktight flea, remain on a single host for the bulk of their adult lives. In the absence of hosts, some species slow their metabolic activity and pass many months in a state of virtual suspended animation similar to that of the pupa. For other species, the absence of a host in hot weather may lead to death in just two to five days. In general, adult fleas shun light and seek out warmth and moisture.

Fleas are generally most abundant in the summer and fall. Smaller populations develop in hot, dry summers owing to the sensitivity of the larvae to drying out. Humid, rainy summers favor flea development and often lead to population outbreaks. Fleas usually overwinter in immature stages in the North and as adults on a host in the South.

Most problems on pets are caused by one of four species: cat, dog, human, and sticktight fleas. Despite the names, all four species attack cats, dogs, and other animals. Cat and dog fleas are very closely related. The cat flea is found throughout the United States, causing the biggest problem in the East and far West. The dog flea is found throughout the United States except for the Rocky Mountains.

The human flea is primarily a pest of buildings, causing the most problems in the Midwest, South, and Pacific Coast. It lives indoors and can breed in areas where dust and organic debris accumulate.

The sticktight flea primarily attacks poultry, but can infest both pets and humans. It is particularly prominent in the South and Southwest. After mating, the female remains on a single host for the rest of her life, often embedding herself in a short burrow, which may ulcerate. Larvae live either in the material used for bedding or in the burrow or ulcer on the host. Adults mainly infest the ears, particularly the ridges, in dogs and cats. Ulceration and subsequent infection can make large infestations potentially life-threatening to an animal.

Warning Signs

Starting in spring or early summer, check your pet every week or two for adult fleas or evidence of their presence. Monitor more frequently during summer and fall, especially if the summer has been wet.

Suspect fleas if your pet scratches a lot. Look for little white or black specks (eggs and fecal pellets) known as "salt and pepper"; bite marks, which are small red puncture sites surrounded by red haloes; or the adult fleas themselves. Concentrate on the pet's head and neck area, which fleas prefer; in larger infestations, adults can be found on all parts of the body. Check the edges of the ears. Since they shun light, adults stay close to the skin, so inspect the fur carefully. Comb the pet with a flea comb, and count the number of fleas you remove. If you find only a couple, don't

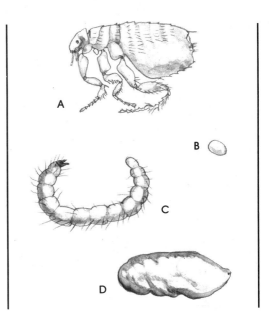

LIFE CYCLE OF THE CAT FLEA

The adult cat flea (A) lays its eggs (B) on the host (eggs then fall off) or in the environment. The eggs usually hatch in 1 to 12 days, and larvae (C) seek out a dark, moist area in the pet's environment, where they feed on dried blood or other organic matter. Next, larvae spin silken cocoons and enter the pupal stage (D), which can last for months if a host is not present.

get upset; they could be transients. If the numbers steadily rise over the course of a few weeks, take action. You may wish to take a few fleas to your county Cooperative Extension Service or land-grant university for identification.

When to Take Action

Take action when the number of fleas steadily increases.

Long-term Controls

To control fleas successfully, you need to act against the adult, both on the animal and in the environment, and against the immature stages, located in the environment.

Controlling fleas on your pet is half the battle. To prevent immediate reinfestation, you also need to control the fleas in the pet's bedding, furniture, carpets, wherever fleas may lurk. Steps to control adult and immature stages in the environment are complicated. They aim to kill the fleas and remove food sources, particularly the dried blood that the larvae require.

Physical Controls. Determine where the fleas are breeding; it can be indoors or outdoors, usually where the animal sleeps, or spends most of its time. Indoors, vacuum pet bedding and nearby carpeting thoroughly and vigorously to remove the adult fleas and eggs. To remove more larvae (they coil up and hang on to the fibers when disturbed), use a vacuum with a beater-bar attachment. Vacuuming also removes dried blood and other organic debris that flea larvae eat. If the infestation is particularly bad, rent a steam-cleaning vacuum cleaner or hire a professional to do the job. Wash the pet's bedding and clothes. Outdoors, fleas can infest lawns and other environments as well as the pet's sleeping area.

Chemical Controls. Getting rid of the larval food source can be ex-

tremely difficult because there is extensive organic debris in the environment. You must therefore kill the fleas or prevent them from reproducing. Any of the substances used to kill adults on your pets (fatty acids, citrus oil extracts [*d*-limonene, linalool], diatomaceous earth, silica gel, pyrethrum/pyrethrins, and pyrethroids) kill adults in the environment, either indoors or outdoors. Many of these substances also kill larvae. The citrus oil linalool kills all flea stages, including the eggs, and so is particularly useful. Apply the product to the animals' bedding and the surrounding environment. Do not use pyrethroids near water; they are highly toxic to fish and some aquatic invertebrates. *Caution:* Some people are allergic to the pulverized flowers that are the source of pyrethrum. Wear a dust mask or respirator if you are allergic to pollen and are spraying pyrethrum, or choose a different chemical. Note that synthetic pyrethroids, if overused, can stimulate pesticide resistance in flea populations; they are also toxic to natural enemies. Use them outdoors only when you absolutely have to, and minimize the amount of area over which you spray them. Silica gel and diatomaceous earth lose much of their effectiveness when wet, and thus aren't very useful outdoors.

The insect growth regulators methoprene (Precor) and fenoxycarb are useful for environmental flea control. They prevent larvae from turning into adults, breaking the reproduc-

CAUTION: *All Pesticides Have Some
Degree of Toxicity*

1. Store all pesticides in their
   original containers in areas
   where children and pets cannot
   get at them.
2. Read the label thoroughly before
   using the product.
3. Minimize your exposure to the
   pesticide.
4. Wear protective clothing,
   including goggles, hat, long
   pants, long-sleeved shirt, rubber
   boots, and unlined rubber gloves
   when necessary.
5. Always wear either a dust mask
   or a respirator mask when
   spraying pesticides.
6. Thoroughly wash application
   equipment, hands, and clothing
   after using pesticides.

tive cycle, and are relatively nontoxic to pets and people. Indoors, spray methoprene or fenoxycarb on the pet's bedding, its surroundings, and anywhere else the pet rests. Outdoors, spray bedding and the surrounding environment. You can mix

*The insect growth regulators
methoprene and fenoxycarb are
useful for environmental flea control.*

the methoprene or fenoxycarb with an insecticide for double protection. *Caution:* Wear a dust mask and protective clothing when applying these substances.

Short-term Controls

Physical Controls. Physically remove adult fleas from your pet with a flea comb. Put petroleum jelly at the base of the tines so fleas will stick to it. Flick the captured fleas into a can of sudsy water.

Chemical Controls. Some chemical controls are also relatively safe and effective. Try a product containing fatty acids or an insecticidal soap, in either shampoo or spray form. You can also sprinkle an insecticidal dust, composed of either diatomaceous earth or silica gel, on the animal. Silica gel absorbs the wax from the flea's exoskeleton, whereas diatomaceous earth cuts up the exoskeleton; in both cases, the flea dries out and dies. Products containing citrus oils (particularly linalool) also kill adults. *Caution:* Wear a dust mask and protective clothing when handling these products. Products containing pyrethrum or pyrethrins, synthetic pyrethroids, or rotenone can also be used relatively safely. *Caution:* Some people are allergic to the pulverized flowers that are the source of pyrethrum. Wear a dust mask or respirator if you are allergic to pollen and are spraying pyrethrum, or choose a different chemical.

If the infestation is severe, shampoo or dip once a week or every other week, and use the sprays or dusts in between washings or dippings. If your animal is particularly sensitive to flea bites, look for a shampoo or dip that contains lanolin or aloe vera or one such as Dermapet, which is hypoallergenic, to soothe the skin.

Cats receive greater exposure to pesticides applied to their fur than do dogs, because cats lick their fur. Cats are often more sensitive to pesticide toxicity. There have been reports of cats being poisoned by as little as 0.4 percent pyrethrins, or by citrus oil sprays. To be absolutely safe, use only insecticidal soap, diatomaceous earth, or silica gel to kill fleas on cats.

Once fleas are off your animal, you need to keep them off—especially if there is a bad infestation in your house—to prevent the adults from continuing to excrete dried blood, a necessary larval food. If the flea season is bad, monitor your pet every two or three days and remove any newly arrived fleas with a flea comb.

Flea collars or sprays that contain oils such as pennyroyal oil, citronella,

Cats receive greater exposure to pesticides applied to their fur than do dogs, because cats lick their fur. To be absolutely safe, use only insecticidal soap, diatomaceous earth, or silica gel to kill fleas on cats.

or citrus oils repel fleas in the lab, though there is little scientific data in the field. Diet supplements that contain, among other things, brewer's yeast (for vitamin B_1) and garlic are supposed to make the animals' blood less palatable to fleas as well as to make the animal more tolerant of flea bites, but there is limited evidence for this. The compounds benzyl benzoate, diethyltoluamide (DEET), permethrin, and resmethrin also make effective repellents.

Diet supplements that contain, among other things, brewer's yeast and garlic may make animals' blood less palatable to fleas as well as make the animal more tolerant of flea bites.

You may be tempted to try one of the widely advertised (and expensive) "ultrasonic" or "electronic" flea collars. Our advice: Save your money. These devices do not kill fleas, cause changes in flea behavior, or cause fleas to leave animals.

Last Resorts

Chemical Controls. Except for extreme infestations, we do not recommend standard flea collars. These emit neurotoxic organophosphorus or carbamate pesticide vapors or dusts at levels high enough to kill the fleas, but theoretically low enough not to harm the pet. But questions re-

Caution: Standard flea collars emit neurotoxins that may have deleterious long-term effects on cats and dogs.

main about the long-term effects of exposure to such neurotoxins. These pesticides interfere with nerve impulse transmission, the mechanism of which is the same in animals and insects, so some deleterious effect on animals is possible. The chemicals in the collars can also irritate, or even ulcerate, the necks of sensitive animals. However, if there is a severe flea infestation outdoors, you may feel forced to use one. To minimize your pet's exposure to the chemicals, put the collar on before the pet goes outdoors and take it off as soon as it comes back inside. Between uses,

store the collar in a sealed bag or jar in the freezer or refrigerator.

If you have a very severe flea infestation indoors, and none of the long- or short-term controls has worked, you may have to resort to using an indoor fogger. In general, we recommend against them because they leave a film of pesticide over everything in a room, not just the areas where the fleas are located. However, desperate problems require desperate measures. Choose a flea fogger that contains methoprene (Gencor) plus a pyrethroid. *Follow the label directions carefully.* Some contain flammable propellants or solvents, so extinguish all pilot lights on the kitchen range to prevent the possibility of an explosion. Better yet, look for foggers that have a water-based propellant. Read chapter 9, Safe Use, carefully before using a fogger.

 # TICKS

Natural History

In addition to being a nuisance, ticks can spread Lyme disease, Rocky Mountain spotted fever, Q fever, and tularemia—as well as tick paralysis. Indeed, ticks can pass on more diseases to humans than any other arthropod except mosquitoes. Ticks are arachnids, not insects (they have eight legs, not six); they are closely related to mites and spiders. They can

feed on the blood of many different vertebrates. They have one or more hosts and are slow-feeding and long-lived.

Common pest ticks belong to two families: hard ticks and soft ticks. Both families have the same generalized four-stage life cycle: egg, larva, nymph, and adult. Hard ticks lay eggs in one massive clutch, usually containing several thousand eggs; soft

🌸 *Ticks can pass on more diseases to humans than any other arthropod except mosquitoes.*

ticks lay a number of smaller clutches. The larvae, which have only six legs, find a suitable host. After feeding once, the larvae drop off the host and molt into eight-legged nymphs. The nymphs feed once, drop off their host, and then molt, either into adults or into another nymphal stage. Adults feed on a host, drop off, and then reproduce, starting the whole cycle over again.

Some ticks, such as the spinose ear tick, feed on a single host during their lifetime. Many other species feed on more than one host, and over 90 percent of hard tick species have three hosts—one each for the larval, nymphal, and adult stages. Some ticks have many potential hosts, while others prefer specific hosts. Between blood meals, the ticks usually hide in cracks and crevices in the vicinity of the host they just left.

To find a host, most hard ticks engage in a form of behavior called *questing*. The larvae crawl up vegetation, anchor themselves with their hind legs, and wave their other two pairs of legs in an attempt to grasp a passing animal. The height of the ticks' perch determines the size of the host they catch; the host can be a small animal such as a mouse, or a larger one such as a dog, deer, or human. Though questing behavior is strongest in larvae, nymphs and adults also exhibit it. Some ticks follow chemical signals—such as carbon dioxide or butyric acid, given off by the animal—to track down a host. Some taste host tissue or blood before feeding, to be sure they've made a good choice.

Most soft ticks feed for a short period, often at night, and then drop off; they are rarely found on a host. Most hard ticks feed for a few days or even weeks, and may wander around on a host for hours before settling down to feed. Many species also exude a cementlike chemical in their saliva, which anchors them to the host, making them extremely difficult to remove.

The tick species most likely to infest pets and transmit diseases are the spinose ear tick, brown dog tick, American dog tick, Rocky Mountain spotted fever tick, northern deer tick, and western black-legged tick. Only the spinose ear tick is a soft tick; the others are hard ticks.

Spinose Ear Tick

The spinose ear tick infects primarily birds and is found in the West and Southwest. The larva finds a host, burrows deeply into the ear, feeds, molts to a nymph, feeds again, molts to a second nymphal stage, feeds once more, drops off the host, molts to an adult, then mates and lays eggs, usually in cracks and crevices on wooden

A. Brown dog tick

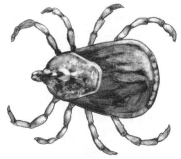

B. American dog tick

C. Female northern deer tick

COMMON TICK PESTS

(A) **Brown dog tick.** *The most widely distributed tick species, these are 3 mm (before feeding) with a solid, reddish brown, elongated body. (B)* **American dog tick.** *These are abundant, particularly east of the Rocky Mountains. This large tick (6 mm before feeding) has an oval-shaped, dark brown body with white mottling on its shield. (C)* **Northern deer tick.** *This small tick (3 mm before feeding) is dark brown and is particularly abundant on the East Coast, where it has spread Lyme disease.*

objects. The second nymphal stage has a number of spines, giving rise to the common name.

Brown Dog Tick

The brown dog tick is probably the most widely distributed tick pest of dogs and cats. It is found throughout the United States. In the North, the brown dog tick is frequently found indoors. In the South, it is also found in kennels and lawns. This tick rarely bothers people, but in dogs it can cause irritation, anemia, paralysis, and canine piroplasmosis. The adult females are about ⅛ inch long, reddish brown, and have a longer body than other ticks. Dogs are the main host for all three stages, although larvae and nymphs feed on a variety of small mammals while the adults may feed on other large mammals. On dogs, adults usually feed in the ears or between the toes, while the larvae and nymphs are most often found on the back. Between blood meals, or when searching for nesting sites, these ticks climb upward looking for cracks and crevices in which to hide. They may be found indoors around window moldings, behind pictures, furniture, and curtains, near the ceiling, or on vertical surfaces.

American Dog Tick

The American dog tick is common east of the Rocky Mountains, although it has now spread along the Pacific Coast. In the East, it is the species most likely to be found on humans. It can transmit Rocky Mountain spotted fever and tularemia, as well as cause tick paralysis in pets and humans. It is a fairly large tick; adult females are ¼ inch long, oval-shaped, and dark brown with white mottling on the shield. The larvae and nymphs prefer field mice or voles as hosts, while the adults prefer dogs, opossums, raccoons, and squirrels. It is most abundant in wooded areas and fields.

Rocky Mountain Spotted Fever Tick

The Rocky Mountain spotted fever tick is common in many parts of the western United States. Adults feed primarily on large mammals, both wild and domestic, while the larvae and nymphs feed primarily on small mammals. It is particularly abundant in bushy vegetation, a good cover for small mammals, and forage grasses or other vegetation suitable for large mammals. It transmits Rocky Mountain spotted fever, Q fever, tularemia, and Colorado tick fever, and causes tick paralysis. Feeding adults, the stage most likely to bite humans, are most abundant in summer. It looks much like the related American dog tick, but is paler in color.

Northern Deer Tick

The northern deer tick lives in the Midwest and Northeast and is the main carrier of Lyme disease on the East Coast. Although it has been found on many mammal and bird species, the adults primarily use white-tailed deer as hosts, while the larvae and nymphs use white-footed mice. Eggs are laid in the soil in early spring. The larvae hatch in early summer, find a host in late summer, feed, drop off, molt to a nymph the following spring, feed in early summer, drop off, molt to an adult in early fall, find a host, feed again and mate on a host, then drop off. Only the fertilized female overwinters. It is a very small tick; adults are the size of a sesame seed, while nymphs are almost too small to see. The nymphs transmit Lyme disease more readily than the adults, so the greatest danger of infection comes in early summer to mid-summer.

Western Black-legged Tick

The western black-legged tick lives in the Pacific coastal states, Idaho, and Nevada. It is found primarily in the humid coastal areas, especially where grasses and bushy vegetation are abundant. It is the main carrier of Lyme disease in the western United States and has a painful bite that takes days to heal. It feeds on more than 80 species of vertebrates. The adults prefer large mammals as hosts, while the larvae and nymphs prefer small mammals, reptiles, or birds, with the nymphs occasionally attacking large mammals. Adults are most active from November to May, during which time they find a large mammalian host, feed, and drop off to lay eggs. Larvae and nymphs are most abundant from March to June. This tick is larger than its East Coast relative; females are $\frac{1}{10}$ inch long while the males are smaller. Females have a red-brown body and males a brown-black one; both sexes have black legs.

Warning Signs

When, where, and how frequently to monitor for ticks depends on the tick species and its biology, the season, and what habitat you or your pets visit. For example, the peak season for the brown dog tick inside homes is fall and winter, while deer tick nymphs are most abundant in late spring and early summer.

Monitor ticks both on the host and in the environment. Carefully search your body after visiting a habitat that may have ticks—usually areas of high vegetation, fields, woods, and woodlots. Brush dogs and cats with a flea comb twice a week to remove unattached ticks. For very small ticks, such as the northern deer tick, use a lint roller to remove unattached ones. For different tick species, concentrate on different parts of the animal: ears and between the toes for adult brown dog ticks, the back and neck for larvae and nymphs of brown dog ticks, ears for the spinose ear tick (see page 212).

Monitor ticks visually in the environment using *flags* and *drags*, or traps. Indoors, particularly in areas where pets sleep, check cracks and crevices near baseboards, around windows, along the walls or near the ceiling, and on curtains or furniture. Outdoors, check around the doghouse, kennel, or run area; inspect cracks and crevices near the roof of the animal housing and those on nearby wood poles, such as fencing. Any ticks you find can be removed by hand and put in sudsy water.

The tick flag or drag takes advantage of the questing behavior of hard ticks. Flags and drags are rectangular pieces of light-colored (preferably white) cloth with a heavy nap, such as flannel. A flag is a smaller rectangle (less than 10 square feet) attached to a stick or pole; a drag is a large rectangle (approximately 4 feet wide and 6 feet long), weighted down at one end and attached to a bar at the other end. Sweep the flag or drag over vegetation that you think is tick-infested. Wear long pants, a long-sleeved shirt, and socks. Use the flag for vegetation of medium height and the drag for low vegetation such as grasses or lawns.

Questing ticks mistake the moving cloth for a host; they cling to it and show up well against the white background. Sweep the vegetation near the doghouse, kennel, or run to collect ticks for later identification. You can use a flag when picnicking or hiking in fields or wooded or semi-wooded areas, to clear ticks from your path or picnic. In areas where Lyme disease is prevalent, sweep the lawn with a drag before letting children play, to test for presence of northern deer ticks. Also, listen to

In areas where Lyme disease is prevalent, sweep the lawn with a drag before letting children play, to test for the presence of northern deer ticks. Also, listen to what health officials are saying about the risk in your area.

what health officials are saying about the risk in the area. Remove the ticks by hand and dispose of them, or soak the flag or drag in sudsy water.

Carbon dioxide traps are used by tick researchers to monitor ticks in the wild. To make a trap, punch four holes in the sides near the bottom of a covered Styrofoam ice bucket. On a 1-foot-square piece of plywood, apply masking tape, sticky side up, or paper covered with a sticky material (Tanglefoot, Tack Trap, or Stickem) around the edges to create a sticky barrier. Fill the bucket with dry ice (2 pounds is good for about 3 hours) and place it in the center of the plywood square. Ticks in the vicinity are attracted to the carbon dioxide, thinking that it is the exhaled breath of a host, and get caught on the sticky material. When the dry ice is gone, check the ticks and sweep the surrounding ground with a drag. Many nymphs will be within a few feet of the trap. Collect some for later identification.

When to Take Action

Take action when you find ticks during monitoring.

Long-term Controls

Long-term control of tick problems relies on avoidance and monitoring. Learn about the biology and ecology of disease-transmitting ticks and their hosts in your area. Find out in which habitats and at what times of year the various life stages are the biggest problems, then avoid those habitats during peak tick periods. Stay away from trash, brush, piles of wood or rocks, and other areas that harbor rodents, the major hosts for the larval and nymphal stages.

Physical Controls. If you or your pet can't avoid such areas, physically remove ticks with a flag or drag in the area that you are using. Inspect yourself and your pet thoroughly after spending time in potentially tick-rich habitats. When hiking or traveling in known tick country, wear light-colored socks and pants, and tuck the pants into the socks. Use a flag to sweep the path in front of you. Check your clothing, particularly the pants, every hour or two and remove any crawling ticks. When you return home, take a shower immediately to help remove unattached ticks.

Modify the habitat to make it less appealing to ticks and their hosts. Indoors, seal small cracks and crevices around baseboards, window molding, and ceilings, and fill large cracks or holes with plaster or wood in all the rooms to which your pet has access. Outdoors, do the same for doghouses or kennels. Cut vegetation and grass to reduce humidity, as ticks are susceptible to drying out. Clear woodpiles, brush, and other materials that serve as homes to rodents. In the East, where Lyme disease is present, prevent or discourage deer from coming on your property. If they are feeding on your vegetable garden, erect a fence, preferably an electric one, or use a repellent such as blood meal or Deer Away. Blood meal sprayed on plants or hung in cheesecloth bags on stakes every 3 feet around your garden repels deer and rabbits for a couple of weeks in dry weather. Deer

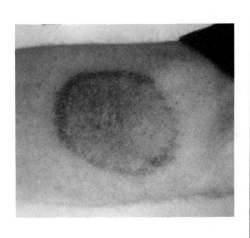

SKIN RASH FROM LYME DISEASE

A typical early rash—a circle, 2 inches in diameter, on the leg—may have a clear area in the center. (Photo courtesy of Central Research Division, Pfizer Inc.)

Blood meal sprayed on plants or hung in cheesecloth bags on stakes every 3 feet around your garden repels deer and rabbits for a couple of weeks in dry weather.

Away can also be sprayed on plants to protect them from deer. *Caution:* Wear a dust mask when spraying this substance. Either repellent must be reapplied after a rain. Blood meal has the extra benefit of being an effective fertilizer.

Biological Controls. Since white-footed mice are a major host for northern deer tick larvae and nymphs, eliminate the mice or control the ticks on them. Modify the habitat to reduce living spaces for the mice (see chapter 7). Damminix consists of tubes with small cotton balls treated with permethrin. Put the tubes in woodpiles or rock piles, dense vegetation, tall grass, or around the perimeter of your property. The mice use the cotton as nesting material and the permethrin kills ticks, fleas, and mites on the female mouse and her babies. Pesticide-coated baits dramatically reduce tick infestation on the mice.

Short-term Controls

Physical Controls. Although you can remove unattached ticks from pets or people with your bare hands, do not try this with attached ticks. Embedded ticks must be carefully removed so that the head and mouthparts do not remain behind in the skin; if they do, they may cause infection. Cover the tick with petroleum jelly, which suffocates it and usually makes it detach itself within half an

To dislodge embedded ticks:
1. Cover the tick with petroleum jelly, which suffocates it and usually makes it detach itself within half an hour.
   Or:
2. Use tweezers or your hand, protected by gloves or a tissue, and gently but steadily pull until the tick is detached.
3. Do *not* twist or crush the tick.
4. Do *not* use alcohol, gasoline, turpentine, or heat to try to dislodge ticks. These methods will kill the tick but leave it embedded and infectious.

hour. Don't use alcohol, gasoline, turpentine, or heat to try to dislodge ticks; these remedies kill the tick but leave it attached and infectious. Use tweezers or your hand, protected by gloves or a tissue, and gently but steadily pull until either it detaches or you pull it out. Do not twist or crush the tick; that can leave the head and mouthparts behind or release body fluids that contain disease organisms. Put the tick in alcohol to preserve it for later identification, or kill it by dropping it in alcohol or sudsy water.

Monitor your pet frequently during peak tick season. If the animal gets severely infested, keep it indoors during the worst part of the tick season or restrict its movements to prevent it from entering prime tick habitat. Use tweezers to remove attached ticks and, depending on species, use a flea comb or lint roller to remove any unattached ticks, or pick them off by hand.

For a large indoor infestation of brown dog ticks, use a vacuum or carbon dioxide traps. Vacuum the area thoroughly, particularly cracks and crevices where ticks hide. Construct a carbon dioxide trap and put it in the infested room. Close the doors to the room and keep pets and people out. At the end of the day, remove the trap and vacuum the immediate vicinity.

Chemical Controls. A variety of relatively safe chemicals kills ticks on animals or in the environment. These compounds should be used as part of an IPM program and not just by themselves (i.e., monitor pets and environment regularly). On pets, use a product containing diatomaceous earth, silica gel, or insecticidal soap. Citrus oil products (those with linalool and/or *d*-limonene), pyrethrins or pyrethrum, or permethrin can also be used safely on dogs. *Caution:* Some cats are sensitive to these compounds and may be poisoned by them. Some people are allergic to the pulverized flowers that are the source of pyrethrum. Wear a dust mask or respirator if you are allergic to pollen and are spraying pyrethrum, or choose a different chemical. To treat infested cats, restrict products to those that cats can usually tolerate, such as insecticidal soap, diatomaceous earth, or silica gel.

Last Resorts

Chemical Controls. If tick problems are especially bad, or if you cannot avoid tick habitats, spray an insect repellent or insecticide on your clothing, particularly your pantlegs, socks, and shirt sleeves. Of the insecticides tested, only permethrin, when applied to clothing, is 100 percent effective against many ticks, including the major species that transmit Lyme disease. Use it, or a repellent such as DEET, benzyl benzoate, resmethrin, or indalone. Do not apply the insecticide or repellent to your skin, only to clothing.

For indoor infestations, if you still have a problem after vacuuming and using a carbon dioxide trap, spray an insecticidal dust containing either diatomaceous earth or silica gel into wall voids. Wear a respirator mask

> *INDOOR CAUTIONS: Pesticides*
>
> 1. Extinguish the pilot light on your gas stove before setting off a fogger or insect bomb.
> 2. Air out the room before reentering.
> 3. Cover any food or dishes before using pesticide sprays.
> 4. Wear a dust mask or respirator mask.

when dusting. If necessary, also apply one of the safer compounds (diatomaceous earth, silica gel, insecticidal soap, *d*-limonene, linalool, pyrethrins, or permethrin) to cracks and crevices where the ticks hide.

MITES

Several different mites attack cats and dogs, among which the ear mite and the sarcoptic mange mite are the most prominent.

Ear Mites

Natural History

Ear mites can infest the ears of cats and dogs; the tip of the tail and the feet may be attacked in heavy infestations. Mites occur more commonly in cats than dogs; an estimated half of all cats have them at one time or another during their lives. Ear mites burrow into the ear and feed off the lymph from pierced cells. Their activity can lead to inflammation, crusting, and scabbing of tissue. Severe infestations can permanently damage the middle ear, which may cause the animal to hold its head to one side and wander around in circles. Heavy scratching of the ear can also lead to ear canker, which, if left untreated, can cause loss of hearing. Some people also have allergic reactions to ear mites—either to their bites or to small particles (such as feces) given off by the mites or to both.

Warning Signs

Routinely check your pet's ears every week or two to catch any mite infestation early. If your pet scratches its ears frequently or shakes its head vigorously, the culprit may be ear mites. (It might also be fleas or ticks.) Ear mites produce a black tarry discharge, a mixture of their excreta, dried blood, and wax, which accumulates in the ear. Ordinarily a pet's ear canal is clean; black waxy material is usually a sign of pest infestation.

Ear mites produce a black tarry discharge, a mixture of their excreta, dried blood, and wax, which accumulates in the ear of the pet.

Examine the ears. If you see something moving, it's probably a flea or a tick; ear mites are too small to see readily with the unaided eye. If no fleas or ticks are present, check for ear mites by removing some of the wax from the ear with a cotton swab and examining it with a magnifying glass or hand lens. If you see any small moving specks in or on the wax, your pet has ear mites. If not, take your pet to a veterinarian.

When to Take Action

Take action as soon as you find evidence of an ear mite infestation.

Long-term Controls

Since your pet picks up ear mites from other animals or, more rarely, from the environment, the only sure control for ear mites is to keep your pet indoors.

Short-term Controls

All commercial products for ear mite control contain an insecticide and an antibiotic. You do not need them; plain mineral oil works fine for light or moderate infestations. The oil suffocates the mites and loosens the wax and other material in the ear canal. (It also makes ticks and fleas leave the ear.) Apply the mineral oil with an eyedropper and massage the outside of the ear to work the oil down into the ear canal. You can remove excess ear wax or mite excreta with a cotton swab, but don't probe too deeply, or you may damage the pet's eardrum. Repeat this process every four or five days for at least three weeks so that you treat any new generations that hatch.

All commercial products for ear mite control contain an insecticide and an antibiotic. You do not need them; plain mineral oil works fine for light or moderate infestations.

Last Resorts

If the symptoms do not improve after two weeks of mineral oil use, your pet has a severe infestation and perhaps a secondary infection. Take the animal to a veterinarian for more intensive treatment. Ask the vet about prescribing abamectin.

Mange Mites

Natural History

Mange is a persistent contagious skin disease caused by parasitic mites and characterized by loss of hair, inflammation, thickening and wrinkling of the skin, and oozing lesions that form crusts or scabs. The animal experiences intense itching and often scratches constantly. If left untreated, the mite infestation eventually can cover the entire body, and the animal can lose all its hair. Mange affects both dogs and cats, although dogs are more susceptible.

There are two types of mange: *sarcoptic* and *deomodectic* (or red) mange. Sarcoptic mange is more common than red mange and is covered here. Red mange mites burrow into hair follicles, are extremely difficult to control (they can't be removed), and require treatment by a veterinarian.

The sarcoptic mange mite burrows into the animal's skin, especially the softer skin on the head and legs, and then the skin on the body. Once the female enters the burrow, she never leaves it. She feeds on the liquid that oozes from pierced cells, continues to enlarge the burrow throughout her life, and reproduces there, usually laying two to three eggs a day. The intense itching that accompanies mange usually occurs a month to six weeks after the mites have invaded, because it takes the body's immune system that long to react. The itching is usually accompanied by a reddening or swelling at the site of the burrow, and the pet often scratches constantly. The scratching kills many mites, but also can cause secondary infections that obscure a proper diagnosis. The pet should be treated for sarcoptic mange as soon as possible.

The mite is the same species that causes scabies in humans, and some cross-infection is possible, although the mites die in humans within a couple of weeks. Thus, it is a good idea to keep children away from mangy dogs. If either you or your children begin to itch around the hands, you may need to be treated yourself for scabies.

Warning Signs

Monitor your pet for mange mites every week or two. However, monitoring can be difficult. If your pet scratches incessantly, look for the

The mite that infects dogs is the same species that causes scabies in humans. It is therefore a good idea to keep children away from mangy dogs.

source of the itch. If no fleas or ticks are readily apparent, but the pet is losing hair or you see raised reddish burrows, a generalized rash, or a number of tiny scabs or raised bumps on your pet's skin, suspect mange mites. (In highly infected cases, you may even see the female mites crawling over the surface of the skin.) Look carefully for new burrows that have not started itching. The burrows are less than 1 inch long, and you can see the mite at the end as a small raised whitish oval less than a millimeter in size, with a dark dot at one end.

When to Take Action

Take action as soon as you find evidence of mange mites.

Long-term Controls

Infected animals transmit mange to uninfected ones. The only way to prevent infection is to keep your pet from contact with other animals.

Short-term Controls

Chemical Controls. The only feasible controls for mange are chemical. Try using sulfur; mites are very sensitive to it. You can make a simple sulfur preparation yourself, or buy a preformulated product from a drug or pet store. To make your own for-

---

The only feasible controls for mange are chemical. The least toxic chemicals are sulfur and permethrin.

---

mulation, buy inorganic sulfur from a plant nursery, or technical-grade sulfur from a chemical supply house. Dissolve enough sulfur in a petroleum jelly base to make a 6 percent solution by weight (e.g., dissolve .6 ounce of sulfur in 9.4 ounces of jelly). Sulfur-based products normally used to treat humans, such as Eucerin or Aqua-aquphor, can be used on your pet and are available at the drugstore. So is Eurax, which contains crotamiton. Eurax has anti-itching properties, but can be irritating with prolonged use or on highly irritated skin. Be sure to wear gloves when applying these substances.

Apply either the sulfur-based product or crotamiton nightly for three days. Before the first application, gently wash and dry the pet. Then vigorously rub the product into the skin over all the infected areas, avoiding the eyes, mouth, and genitals. Since the itching can persist for a while after the mites have been killed, you may need to give the pet an anti-itch treatment. Ointments that contain cortisone are available from your veterinarian. Do not apply an anti-itch preparation prior to treatment for mites because the steroid may suppress your pet's immune response to the mites, and prolong or even worsen the infection.

Permethrin is faster acting than sulfur or crotamiton, requiring only one application. Elimate, a product formulated for use in humans and sold in pharmacies, can be used on your pet.

Last Resorts

Chemical Controls. If your pet has a severe case of mange that does not respond to the suggested controls, or you do not want to undertake these controls yourself, take it to your vet. Ask the vet to prescribe, if possible, abamectin. This chemical, which is derived from a soil bacteria, is highly effective against a range of pests, including mange mites, at *very* low application levels. The animal is given the abamectin orally.

CHART 8-1

Products to Control Pet Pests in the Environment

The products listed below are a fraction of those pesticides marketed to control pests on pets. These products were selected for listing based on CU's judgment that they can be effective when used in the context of an Integrated Pest Management strategy, and that they pose the least risk to humans, pets, or the environment, based on the active ingredients they contain. Products listed for use on the pet may also be used in the pet's environment (see Chart 8-3).

Some products not listed here contain one or more of the same active ingredients as these products and may be substituted for them. But many widely available products are not listed because they contain active ingredients that, in CU's judgment, pose greater potential risks to health or the environment than the ingredients of listed products. In our view, effective pest control does not require use of more toxic pesticides, and we have chosen not to list products that contain them. Products are listed in alphabetical order.

Recommended products listed below contain the least hazardous active ingredients, including one or more of the following: fatty acids (FA), diatomaceous earth (DE), fenoxycarb (FB), hydroprene (HY), methoprene (MP), silica gel (SG), pyrethrins (PY), or pyrethroids, such as allethrin (AL), cyfluthrin (CY), fenvalerate (FV), permethrin (PM), phenothrin (PH), resmethrin (RS), tetramethrin (TM), or tralomethrin.

Other products listed below also contain synergists, such as MGK 264 (M2) or piperonyl butoxide (PB), and/or petroleum distillates (PD), that pose a somewhat greater risk to health or the environment.

BRAND NAME	ACTIVE INGREDIENT(S)
Recommended	
Black Flag Flea Ender Spray	HY, PM
Dexol Flea Free Carpet Treatment	PY, SG
Diacide	DE, PY
Enforcer Flea Killer for Carpets V	PH
Enforcer Precor Concentrate	MP
Enforcer 7-Month Flea Spray for Homes	MP, PM
Flea Stop Concentrate and Yard Spray	FV
Gro-Well Home Pest Control	PM
Ortho Flea-B-Gon Flea Killer Formula II	PH, TM
Ortho Household Insect Killer— Formula II	PH, TM
Ortho Insecticidal Soap	FA

Brand Name	Active Ingredient(s)
Ortho Professional Strength Flying & Crawling Insect Killer	AL, PH
Ortho Total Flea Killer	MP, PM
Raid Multi-Bug Killer Formula D39	AL, RS
Rid-a-Bug Flea & Tick Killer	TR
Safer Flea & Tick Attack Premise Spray	FA, PY
Siphotrol + Area Treatment	MP, PM
Spectracide Flea & Tick Killer	TR
Spectracide Flea & Tick Killer 3	PM
Spectracide Home Insect Control 3	TR
Sulfodene Scratchex Flea & Tick Spray	PM, PY
Sulfodene Scratchex Power Guard	PM, PY
Zodiac Fleatrol Premise Spray	MP, PM
Other	
Ace Hardware Flea & Tick Killer	AL, M2, PH
Blue Lustre Flea Killer for Carpets	PB, PY
Cardinal Clean-Scent Carpet & Rug Pest Control Room Deodorizer	PB, PY
Cardinal Rid Flea & Tick Spray	PB, PM, PY
Daltek Flea & Tick Carpet Powder	PB, PY
Diacide Pet Powder	DE, PB, PY
Drione	SG, PB, PY
Hartz 2 in 1 Time Release Flea & Tick Killer	M2, PB, PY
Hartz 2 in 1 Time Release Household Flea & Tick Killer	M2, PD, PY
Hot Shot Flea & Tick Killer	AL, M2, PH
K-mart Pet & Home Flea Killer	AL, M2, PH
Natra Pet House & Carpet Spray	PB, RS
Ortho Pet Flea & Tick Spray— Formula III	PB, PY

Products to Control Pet Pests in the Environment *(continued)*

Brand Name	Active Ingredient(s)
Ortho Total Flea Killer Spray	M2, MP, PB, PY
Perma-Guard Household Insecticide	DE, PB, PY
Raid Flea Killer	M2, PB, PY, TM

Raid Flea Killer Plus Egg
    Stop Formula
	M2, MP, PB,
	    PY, TM

Raid House & Garden Formula 11	PB, PY, TM
Real-Kill Pet and Home Flea Killer	AL, M2, PD, PH
Revenge Home Exterminator	SG, PB, PY
Sergeant's Rug Patrol	PB, PH
Vetchem Siphotrol + Premise Spray	M2, MP, PB, PY
Vetchem Siphotrol + II House Treatment Spray	MP, PB, PM
Zodiac Fleatrol Carpet Spray	M2, MP, PB, PY
Zodiac Fleatrol Indoor Spray	MP, PB, PM

CHART 8-2
Other Products to Control Pet Pests in the Environment

The products listed below are foggers, which, in general, we do not recommend. However, under exceptional circumstances (i.e., a household infestation of thousands of fleas that have not been controlled by the other recommended control tactics, e.g., vacuuming, use of sprays, etc.), a fogger may be needed, BUT ONLY AS A LAST RESORT. We apply the same proviso to these products as to the products in Chart 8-1.

Recommended products listed below contain the least hazardous active ingredients, including one or more of the following: methoprene (MP), pyrethrins (PY), or one of the pyrethroids, such as allethrin (AL), cyfluthrin (CY), fenvalerate (FV), permethrin (PM), phenothrin (PH), resmethrin (RS), tetramethrin (TM), or tralomethrin.

Other products listed below contain synergists, which pose a somewhat greater risk to health or the environment, including the following: MGK 264 (M2) or piperonyl butoxide (PB), and/or petroleum distillates (PD), whose chronic health effects remain unknown.

BRAND NAME	ACTIVE INGREDIENT(S)
Recommended	
Enforcer Flea Fogger	MP, PM
Enforcer Four Hour Fogger	PM, PY
F Ketchem VetFog	PM
Flea Stop Fogger	PH
Green Thumb Home Insect Fogger	PH, TM
Hot Shot Fogger 3	PM, TM
Hot Shot Fogger	PH, TM
Raid Fumigator Fumigating Fogger	PM
Rid-a-Bug Flea Fogger	TR
Rid-a-Bug Indoor Fogger	PM, TM
Siphotrol + Fogger	MP, PM
Spectracide Professional Flea Control Fogger	MP
Zodiac Fleatrol Fogger	MP, PM
Zodiac House & Kennel Fogger	PM
Other	
Black Jack Household Indoor Fogger Roach & Insect Killer X	PD, PH, TM

Other Products to Control Pet Pests in the Environment *(continued)*

BRAND NAME	ACTIVE INGREDIENT(S)
Flea Stop Fogger	PH, PB
Four Paws Fast Killing Indoor Fogger	PD, TM
Holiday Household Insect Fogger	FV, M2, PB
Holiday Household Insect Fogger New Pine Scent	PD, PH, TM
Ortho Hi-Power Indoor Insect Fogger Formula IV	FV, M2, PB, PY
Raid Flea Killer Plus Egg Stop Formula	MP, M2, PB, PY, TM
Raid Fogger	M2, PB, PY
Raid Fogger II	M2, PB, PH, TM
Raid Max Fogger	CY, M2, PB, PY
Sergeant's Indoor Fogger	AL, FV

CHART 8-3
Products to Control Pests on Pets

The products listed below are a fraction of those pesticides marketed to control pests on pets. These products were selected for listing based on CU's judgment that they can be effective when used in the context of an Integrated Pest Management strategy, and that they pose the least risk to humans, pets, or the environment, based on the active ingredients they contain.

Some products not listed here contain one or more of the same active ingredients as these products and may be substituted for them. But many widely available products are not listed because they contain active ingredients that, in CU's judgment, pose greater potential risks to health or the environment than the ingredients of products listed. In our view, effective pest control does not require use of more toxic pesticides and we have chosen not to list products that contain them. Products are listed in alphabetical order.

Recommended products listed below contain the least hazardous active ingredients, including one or more of the following: fatty acids (FA), diatomaceous earth (DE), citrus oils (*d*-limonene [DL] or linalool [LL]), hydroprene (HY), methoprene (MP), pennyroyal oil (PO), pyrethrins (PY), or pyrethroids (allethrin [AL], permethrin [PM], phenothrin [PH], resmethrin [RS], tetramethrin [TM], or tralomethrin [TR]).

Other products listed below contain synergists, such as MGK 11 (M1), MGK 264 (M2), or piperonyl butoxide (PB), rotenone (RO), and/or petroleum distillates (PD), whose ingredients pose a somewhat greater risk to health or the environment.

BRAND NAME	ACTIVE INGREDIENT(S)
Recommended	
Balance Australian Eucalyptus Flea and Tick Repellent Shampoo	EU
Balance Pennyroyal Flea and Tick Repellent Shampoo	PO
Daltek 14 Day Flea & Tick Spray for Cats	PM, PY
Daltek Organic Dip for Dogs with d-Limonene	DL
Daltek Organic Shampoo for Dogs and Cats with d-Limonene	DL
Daltek Organic Spray for Dogs and Cats with d-Limonene	DL
Dermapet Shampoo	FA
Diatom Dust	DE

Products to Control Pests on Pets *(continued)*

BRAND NAME	ACTIVE INGREDIENT(S)
Enforcer Flea & Tick Spray for Dogs II	PM, PY
Enforcer Flea Spray for Cats	PM, PY
Flea Stop Concentrated Shampoo for Cats	LL
Flea Stop Dip	DL
Flea Stop Flea Repel for Cats	PM, PY
Flea Stop Flea Repel for Dogs	PM, PY
Flea Stop Pet Spray Aerosol	DL
Flea Stop Pet Spray	DL, LL
Flea Stop Shampoo	DL
Four Paws Ear Mite Remedy	PY
Four Paws Flea & Tick Spray	PM, PY
Perma-Guard Fossil Shell Flour	DE
Rid-a-Flea Flea & Tick Killer for Dogs and Cats	RS
Safer Flea and Tick Attack	FA, PY
Safer Flea Soap for Cats	FA
Safer Flea Soap for Dogs	FA
Sulfodene Scratchex Flea & Tick Spray	PM, PY
Sulfodene Scratchex Power Guard	PM, PY
Victory Veterinary Formula Flea & Tick Spray for Dogs	PM, PY

Other

BRAND NAME	ACTIVE INGREDIENT(S)
Ace Hardware Flea & Tick Killer	AL, M2, PH
Adams 14 Day Flea Dip	M1, M2, PB, PM, PY
Adams Flea and Tick Mist	M1, M2, PB, PY
Adams Flea-Off Pyrethrin Dip	M1, M2, PB, PY
Adams Flea-Off Shampoo	M1, M2, PB, PD, PY

Brand Name	Active Ingredient(s)
Adams Residual 14 Day Flea-Off Mist	M1, M2, PB, PY
Adams Tick Killer	M1, M2, PB, PM, PY
Cardinal Flea & Tick Powder for Cats	PB, PY
Cardial Flea & Tick Powder for Dogs	PB, PY
Cardinal Flea & Tick Shampoo	PB, PY
Cardinal Flea & Tick Shampoo for Cats and Kittens	PB, PY
Daltek Concentrated 3:1 Flea & Tick Shampoo	PB, PY
Daltek Ear Mite Lotion	RO
Daltek Flea & Tick Shampoo with Pyrethrin	PB, PY
Daltek Flea & Tick Shampoo for Cats	PB, PY
Daltek Flea & Tick Shampoo for Puppies	M2, PB, PY
Daltek Timed-Release 8-Day Flea Foam for Cats	PB, PD, PY
Enforcer Flea & Tick Powder for Pets	M2, PB, PD, PY
Enforcer Flea & Tick Shampoo for Pets	M2, PB, PD, PY
Flea Stop Dermatological Quick Kill Flea & Tick Spray for Cats	PB, PY
Flea Stop Dermatological Quick Kill Flea & Tick Spray for Dogs	PB, PY
Flea Stop Mist	PB, PY
Flea Stop Pet Spray for Cats	DL, LL, PB
Flea Stop Pyrethrin Flea & Tick Dip	PB, PY

Products to Control Pests on Pets *(continued)*

BRAND NAME	ACTIVE INGREDIENT(S)
Flea Stop Pyrethrin Dip for Cats	PB, PY
Flea Stop .26% Pyrethrin Shampoo	PB, PY
Four Paws Extra Strength Flea & Tick Shampoo	M2, PB, PD, PY
Four Paws Flea Foam Dry Bath for Cats	M2, PB, PD, PY
Four Paws Flea Foam Dry Bath for Dogs	M2, PB, PD, PY
Four Paws Flea & Tick Soap	M2, PB, PY
Four Paws Magic Coat Flea & Tick Shampoo for Dogs	M2, PB, PD, PY
Four Paws Super Fly Repellent	M1, M2, PB, PY
Hartz Cat Flea Powder	RO
Hartz 2-in-1 Dog Flea Soap	M2, PB, PD, PY
Hartz 2-in-1 Flea & Tick Dip for Dogs/Cats	M2, PB, PD, PY
Hartz 2-in-1 F&T Killer for Cats Fine Mist Spray	M2, PB, PD, PY
Hartz 2-in-1 F&T Killer for Dogs Fine Mist Spray	M2, PB, PD, PY
Hartz 2-in-1 Luster Bath for Cats	M2, PB, PD, PY
Hartz 2-in-1 Luster Bath for Dogs	M2, PB, PD, PY
Hartz 2-in-1 Luster Bath Mousse for Cats & Dogs	M2, PB, PD, PY
Hartz 2-in-1 Rid Flea Shampoo for Dogs	M2, PB, PD, PY
Hot Shot Flea & Tick Killer	AL, M2, PH
K-mart Pet & Home Flea Killer	AL, M2, PH
Mycodex Pet Shampoo with 3x Pyrethrin	PB, PY
Mycodex Aqua Spray with Pyrethrins	PB, PD, PY

Brand Name	Active Ingredient(s)
Mycodex 14 Pet Spray	PB, PM, PY
Natra Aloepet Shampoo for Dogs & Cats	PB, PY
Natra D-Flea Shampoo	PB, PY
Natra Flea & Tick Killer for Cats	PB, PY
Natra Flea and Tick Killer for Dogs	PB, PY
Natra Flea & Tick Killer for Puppies and Kittens	PB, PY
Natra Flea Shampoo	PB, RS
Natra Pet Flea & Tick Killer for Cats	PB, RS
Natra Pet Flea & Tick Killer for Dogs	PB, RS
Natra Pet Flea & Tick Killer for Puppies and Kittens	PB, RS
Ortho Pet Flea & Tick Spray Formula III	PB, PY
Ovitrol + for Puppies & Kittens, Dogs & Cats	M2, MP, PB, PY
Petland Flea/Tick Shampoo	PB, PD, PY
Raid Flea Killer	M2, PB, PY, TM
Real-Kill Pet & Home Flea Killer	AL, M2, PD, PH
Rid-a-Bug Flea & Tick Killer	TR
Rid-a-Bug Shampoo	AL, M2, PH
Sergeant's Ear-Mite Preparation for Cats	RO
Sergeant's Skip-Flea Shampoo for Dogs	PB, PM
Sulfodene Scratchex Flea & Tick Shampoo	M2, PB, PY
Sulfodene Scratchex Power Dip	PB, PY

Products to Control Pests on Pets *(continued)*

BRAND NAME	ACTIVE INGREDIENT(S)
Vetchem Flea & Tick Shampoo for Dogs & Cats	PB, PD, PY
Vetchem Flea & Tick Shampoo Plus	M1, M2, PB, PY
Vetchem Flea & Tick Pump Spray for Dogs and Cats	M2, PB, PY
Vetchem Fleatrol Spray for Kittens and Cats	M2, MP, PB, PY
Vetchem Fleatrol Spray for Puppies and Dogs	M2, MP, PB, PY
Vetchem Pyrethrin Dip for Dogs & Cats	PB, PY
Vetchem Super Concentrated Flea & Tick Shampoo for Dogs & Cats	M2, PB, PY
Vetchem Water-Based Flea & Tick Pump Spray for Dogs and Cats	M2, PB, PY
Victory Flea Soap for Dogs	M2, PB, PD, PY
Zenox Flea & Tick Shampoo for Dogs	PB, PY
Zodiac F&T Pump Spray for Dogs and Cats	M2, PB, PY
Zodiac Flea & Tick Shampoo for Dogs & Cats	PB, PD, PY
Zodiac Fleatrol Carpet Spray for Dogs	M2, MP, PB, PY
Zodiac Fleatrol Spray for Cats and Kittens	M2, MP, PB, PY
Zodiac Fleatrol Spray for Dogs and Puppies	M2, MP, PB, PY
Zodiac Pyrethrin Dip for Dogs & Cats	PB, PY
Zodiac Super Concentrated Flea & Tick Shampoo for Dogs & Cats	PB, PY

Brand Name	Active Ingredient(s)
Zodiac Triple Action Conditioning Flea & Tick Shampoo with Aloe Vera for Dogs & Cats	M1, M2, PB, PY
Zodiac Water-Based Flea & Tick Pump Spray for Dogs & Cats	M2, PB, PY

PART FIVE

SAFER PESTICIDES

9

SAFE USE

Although it is true that a pest is any organism that people consider a nuisance because it carries disease, kills crops or flowers, or simply annoys people, it is important to remember that *pest* is a relative term, not an absolute one. For example, most Americans view dandelions as a weed while many Europeans view them as beautiful wildflowers. Most so-called pests have an important role in the balance of nature. Instead of viewing pest control as all-out war, it is more logical and productive to think in terms of management and containment.

The IPM approach to home and garden pest control relies on nonchemical techniques as much as possible. But now and then you may need to resort to chemical controls. When

---

❧   *If you need to use a pesticide, choose one that's effective against the problem pest, is relatively safe to apply, and has minimal adverse impact on nontarget species and the environment.*

---

that need arises, which product, with which pesticide ingredients, should you choose? Ideally, you'll pick one that's effective against the problem pest, is relatively safe to use, and has minimal adverse impacts on nontarget species and the environment. To choose well, you need to know something about the chemicals used in the products. Read chapter 10 and the appendixes carefully and consult the toxicity scores in chapter 10 to help you choose the safest and most effective pesticide for the problem at hand.

PESTICIDE FORMS AND APPLICATION EQUIPMENT

Learning how to handle, use, and dispose of pesticides safely requires knowing something about the various forms of pesticides as well as the ways they are applied. Pesticides are sold in solid, liquid, and gaseous forms. Solid pesticides consist primarily of dusts, granules, and baits. Dusts include sulfur, boric acid, and diatomaceous earth. Many of the insecticides and herbicides used on lawns and gardens come in granular form. Most rodenticides come in bait form, as do a number of insecticides used against cockroaches and ants. Liquid pesticides consist primarily of sprays and concentrates. Most of the pesticides you will encounter for indoor or outdoor use come in liquid form. Gaseous pesticides are primarily fumigants. The smoke bombs or cartridges used to kill gophers and moles release gases. Some pesticides are sold in forms that combine properties of solids and liquids—for example, powders that are mixed with water, before spraying, such as the bacteria *B. thuringiensis*, and flowables, such as many copper and sulfur compounds. Foggers or bombs that are used primarily against insects indoors and occasionally outdoors combine properties of liquids and gases; they are fine mists of liquid droplets.

To apply pesticides, special equipment is needed:

- *Dusts.* Use a bulb duster to blow dust into cracks and crevices or to dust pets. Use a compressed-air duster for larger jobs, such as blowing dust into wall voids or other confined spaces.
- *Granules.* Use a push-type granule applicator when applying granules to lawns or turf.
- *Liquids.* Many household insect sprays come in aerosol cans or trigger-pump sprays. Use aerosol sprays to treat cracks and crevices, indoor plants, and pets. Use trigger-pump sprays to treat houseplants or pets. Use a compressed-air sprayer, which resem-

bles a large can with a handle and a long tube with a nozzle at the end, to apply insecticides indoors as well as insecticides, herbicides, and fungicides outdoors. Outdoors, use trigger-pump sprays to kill insects, weeds, or diseases in small areas. For large lawns and gardens, you may need to use a sprayer that connects to your garden hose to kill certain insects, weeds, or diseases. You can use backpack sprayers to apply insecticides, herbicides, and some fungicides in your garden. A wick applicator, which resembles a sponge mop with a hollow handle, lets you apply herbicides directly to many weeds.

 # HANDLING AND USE

To use pesticides safely, follow these precautions:

- Store *all* pesticides in their original containers in areas where children and pets cannot reach them. Indoors, keep them locked up or in cabinets or on shelves that are too high for children to touch. Do not leave them in a cabinet under the kitchen sink. Outdoors, put lawn and garden pesticides in a shed that can be locked or on shelving that is too high for children to reach.
- Read the label thoroughly *before* using the product. The label contains important information on how to use the pesticide, what precautions to take, and what to do in case of accidental poisoning.
- A major goal of safe use is to minimize your exposure to the pesticide. Use pesticides only as a last resort, when genetic, cultural, physical, or biological controls have failed. If you must use pesticides, use spot application only where the pests are a problem. Do not spray the entire garden to control aphids; spray only infested plants, and spray only those parts of the plant where the aphids have congregated.
- Follow commonsense guidelines.

Outdoors

Mix only the amount of pesticide you can use at one time.

- Do not apply liquid pesticides during windy weather. If possible, apply them on perfectly calm days. If this is not possible, and there is a slight breeze, be sure the wind is at your back; never spray pesticides into the wind.

- Do not apply large quantities of liquid pesticides to a lawn or garden if you have a creek, river, lake, or pond near your house; they can pollute the groundwater.
- Do not smoke while applying liquid pesticides.
- Refrain from using granular pesticides, such as diazinon granules, on the lawn. The granules are often consumed by birds and can kill them.
- If you use poison baits for rodents or slugs, use a covered bait station to ensure that nontarget animals, such as dogs, do not eat the bait.

Indoors

Use aerosol sprays judiciously and wear a respirator mask if there is any chance of your inhaling the pesticide. Don't use a continuous spray to kill just a couple of cockroaches or a few flies.

- If you use a fogger or insect bomb in the kitchen, extinguish the pilot light if you have a gas range. Many foggers have flammable materials in them.
- Follow all the label directions, such as covering dishes and any food.
- When using a fogger, air out the room before reentering.
- When using an insecticidal dust such as boric acid or diatoma-

ceous earth, apply the dust in areas where pets and children cannot easily get at them.
- Always wear a dust mask to prevent inhalation of dust particles.

Protective Clothing

Most important, use protective clothing and equipment. Pesticides can enter the body via exposed skin or by inhalation. The lungs and respiratory passages absorb most pesticides the quickest of all. The eyes are also highly susceptible to pesticide absorption, as are the scalp and ears.

Read the pesticide label, which should contain cautionary statements such as "Take precautions to prevent exposure," "Avoid contact with skin," "Avoid contact with eyes," or "Avoid inhalation." Follow this advice by wearing appropriate protective clothing. For example, if the label reads "Take precautions to prevent exposure," wear long pants, a long-sleeved shirt, a hat, and a respirator. If the label reads "Avoid contact with skin," wear long pants, a long-sleeved shirt, a respirator, rubber gloves, goggles, and a hat. Wear rubber boots if the pesticide is in toxic category I or II. (We recommend against using any pesticides in these categories; see chapter 10.)

Clothing and equipment exist to protect all parts of the body. To protect the body and arms, wear long pants and a long-sleeved shirt or jacket. Waterproof pants and jackets work best, but they can be extremely uncomfortable during hot weather. Pants and shirts made of closely woven cotton are a good second choice.

To protect your hands, use unlined (the lining may absorb pesticide) rubber or latex gloves that extend at least halfway up the forearm. Wear a long-sleeved shirt outside of the gloves to prevent pesticides from reaching your skin. If you are spraying trees or anything overhead, tuck the shirt sleeves into the gloves, so that any liquid running down the glove will fall on the shirt sleeve and not your skin.

To protect your feet, wear rubber boots; leather or cloth shoes readily absorb liquids. Put the pant legs on the *outside* of the boots.

To protect your scalp and ears, wear a hat—preferably one with ear flaps—or a jacket with a hood.

Use goggles to protect your eyes and a respirator or dust mask to protect your mouth and prevent inhalation of pesticides. A respirator that has a cartridge containing activated charcoal removes both dust and chemical vapors. A dust mask, frequently made of paper and disposable, removes only dust particles.

When using liquid pesticides outdoors, it is a good idea to wear virtually all items of protective clothing mentioned above. If you are using a backpack sprayer or a sprayer that you connect to a hose, wear long pants and a long-sleeved shirt, a hat, gloves, and a respirator. If the backpack sprayer leaks at all, or if there is

any chance of becoming wet from the spray, wear a waterproof jacket and pants. When using a pump sprayer or aerosol spray, wear long pants and a long-sleeved shirt at a minimum. If there is any possibility of breathing the spray—this is especially true if the spray comes out as a mist—wear a respirator. Use goggles if the label recommends avoiding eye exposure. If you apply granular pesticides by hand, wear gloves.

Some people are allergic to the pulverized flowers that are the source of pyrethrum. If you are allergic to pollen, be sure to wear protective clothing and a mask.

After you have finished using pesticides, carefully clean up. Thoroughly wash application equipment as well as your hands and clothing. If no pesticide has spilled on your clothing, wash it in the regular cycle, but separate from other items.

If pesticide spills on your clothing, either discard the clothing or wash it. If you discard the clothing, bundle it in a plastic bag and dispose of it in the trash. If you wash the clothing, wash it separately in hot water and use the longest wash cycle. Run the clothing

Remember that all pesticides have some degree of toxicity or they wouldn't work. Use them carefully, sparingly, and only after other controls have failed.

through two cycles. Remove the clothes and then run the washer through a third wash cycle, using hot water to clean it of any residual pesticide.

If pesticide spills on your skin, wash the area thoroughly with soap and lots of warm water as soon as possible. For pesticide spills on the ground, either remove the soil and discard it, or if the spill is light allow the microbes in the soil to degrade the pesticide. If you take the latter approach, fence off the area so that children and pets cannot reach it.

To discard pesticide products, such as empty spray cans, follow the label directions, if there are any. You can also wrap them in newspaper and call your local poison hotline and ask where you can dispose of the item. If the pesticide container is partially full, dispose of it by an approved system in your county. Call the local poison hotline. Never put either partially full or empty pesticide containers in the trash; they may pollute the local landfill. Never reuse containers.

10

THE
CHEMICALS

The word *pesticide* is a broad term, referring to a substance that kills pests. Rather than classifying pesticides by the chemical families they belong to, most scientists categorize them by the type of organism they are used against. Thus, substances used against insect pests are called *insecticides*; against weeds, *herbicides*; against fungal diseases, *fungicides*; against rodents, *rodenticides*; against mollusks (slugs and snails), *molluscicides*.

At the beginning of this project, Consumers Union sent shoppers to hardware and garden stores in the Northeast (Westchester County, New York), the Southeast (Port Richey, Florida), the Midwest (Newton, Iowa), and the West Coast (San Francisco and Berkeley). Their instructions were to buy one sample of each household and home lawn or garden pesticide product they found. The samples were all shipped to our headquarters, so that we could examine the products and determine the ingredients each contained. By law, pesticide products' labels must include the names and amounts of active ingredients in the products.

Our shoppers collected 299 differ-

By law, pesticide labels must include the names and amounts of active ingredients in the products. Read the labels carefully and then check our toxicity chart before buying a product.

ent products. In addition, we copied ingredient information from other products in local stores. In all, we obtained data on the ingredients of 513 different products, including the concentrations of active ingredients for 397. The products contained 107 different active ingredients in all. After some study, we chose for detailed research 89 active ingredients found in a significant number of products.

This research served as the basis for the products listed in the charts at the ends of the earlier chapters. For a complete listing and discussion of all substances tested and their comparative toxicity sorces, see the Appendix.

PESTICIDE CLASSIFICATION

According to the EPA, a pesticide's active ingredient is any chemical that "will prevent, destroy, repel or mitigate any pest or will alter the growth or maturation or other behavior of a plant, or cause the leaves or foliage to drop from the plant, or accelerate the drying of plant tissue." About 600 active ingredients are used in this country. They are combined in many ways to create an estimated 45,000 differently formulated pesticide products. A product contains one or more active ingredients plus one or more of some 1,200 so-called inert ingredients. The EPA defines an inert ingredient as "any intentionally added ingredient in a pesticide product which is not pesticidally active" against the pest(s) targeted by the particular product. These ingredients are added to give the product a desired attribute, such as to dissolve, dilute, propel, stabilize, or enhance the action of the active ingredient(s). Products come in many forms, including dusts, granules, liquids, emulsions, and aerosol sprays.

Given such a variety of products, classification is important. Pesticides can be classified by their use, or by their chemical structure. The former scheme groups active ingredients according to the kind of pest they are used against, such as insecticides and herbicides.

Classifying active ingredients by their chemical nature separates them according to chemistry and how they work. However, this scheme ignores the kind of pest an ingredient controls.

The scheme we use is a hybrid of biological and chemical classifications.

We group the active ingredients by biological use, then by chemical structure within types. For example, insecticides are considered a group, then separated into chemical categories such as organochlorines, organophosphates, and carbamates.

This book does not cover the whole range of pesticides. We have omitted wood preservatives and bactericides (also called disinfectants). Instead, we concentrate on the more traditional categories of pesticides—insecticides, herbicides, fungicides, rodenticides, and molluscicides.

How many of the more than 600 active ingredients, 1,200 inert ingredients, and 45,000 products are used around the home and garden? The EPA estimates that 210 active ingredients are used solely for nonagricultural purposes and the remaining 390 have both agricultural and nonagricultural uses. Nonagricultural uses include use in homes, gardens, restaurants, hospitals, airplanes, offices and other public buildings, supermarkets, highway rights-of-way, and so forth. The EPA has estimated that 25 percent of the 1.1 billion pounds of pesticide active ingredients used in the United States in 1991—about

273 million pounds—were used for nonagricultural purposes. Of this, 69 million pounds were applied by consumers in and around their homes and gardens, in the form of insecticides (30 million pounds), herbicides (25 million pounds), fungicides (11 million pounds), and other (3 million pounds). This represents roughly 7 percent of all pesticide active ingredients used in the United States in 1991. It does not include the pesticides applied in and around homes and gardens by exterminators or lawn-care companies.

Good statistics are not readily available on the number and relative amounts of individual chemicals used in and around households and gardens. But a comparatively small number account for the majority of the poundage used. One recent analysis of the market for home pesticides indicated that approximately 12 compounds account for the lion's share of poundage of active ingredients used.

Our survey of 89 different active ingredients comprises the vast bulk of the poundage of active ingredients used in and around the house and garden.

 ## RATING INGREDIENTS FOR SAFETY

To help you choose a safe, effective pesticide, we have reviewed available knowledge of the possible adverse health and environmental effects of

the 89 ingredients on our list. Each chapter summarizes the data for ingredients of appropriate pesticides, and at the end of each chapter we rated the toxicity of the ingredients. Ingredients in the *Recommended* products sections pose minimal threats to health or the environment; those listed as *Other* products or not listed at all are the most toxic, the most hazardous to the environment, or both.

Though the information in the charts comes from many sources, the bulk of it, particularly on toxicity, comes from the EPA's Office of Pesticide Programs and the California Department of Food and Agriculture.

Toxicity

All pesticides are inherently toxic; after all, they are designed to kill pests. But some ingredients are much more toxic to humans than others. For example, many insecticides are nerve poisons. Since the nerves of insects and humans work in much the same way, what poisons one can poison the other.

The health risk to humans of pesticide exposure has two components: the inherent toxicity of the chemical and the degree of exposure. Information on both inherent toxicity and degree of exposure is needed to estimate risk.

The toxicity charts summarize available information about the toxicity of the active ingredients. Most data come from laboratory tests involving animals, microbes, or cell cultures. Under law, many of the tests

are paid for and furnished to the EPA by the manufacturer of the chemical. The EPA then reviews the test data and decides to register or reregister the chemical for specific pesticide uses. Where data from humans (either case studies or epidemiological data) are available, we have noted that in the comments section.

The toxicity charts summarize what is known about ten different measures of toxicity. Three measures are for acute effects (acute oral toxicity [i.e., the LD_{50}], skin irritation, and eye irritation); two are for subacute effects (acetylcholinesterase inhibition and tissue and organ effects); four are for chronic effects (carcinogenicity, mutagenicity, reproductive toxicity, and teratogenicity [birth defects]); one measure—neurotoxicity—is broad and includes both subacute and chronic effects.

The chronic effects of pesticides are generally the most serious and most controversial. They are serious because such effects—especially cancer, birth defects, and mutations—are progressive, self-propagating, and essentially nonreversible. They are controversial both because of their dread nature and because of the large scientific uncertainties inherent in efforts to assess the risks of such effects that a chemical may cause.

The criteria used to rate the active ingredients differ according to the distinctive nature of each type of toxicity. The specific criteria and toxicity scale are spelled out below for each type of effect. These are guidelines for interpreting the toxicity charts. Chronic effects are dealt with first, followed by acute effects, then by subacute effects.

Chronic Effects

CARCINOGENICITY

Carcinogenicity is a chemical's ability to cause cancer or malignant tumors. The ability to cause tumors, whether benign or malignant, is called *oncogenicity.* There is a debate among cancer researchers about whether a compound that induces benign tumors will also cause malignant tumors, and whether all benign tumors will eventually become malignant. The EPA has taken the stance that evidence of oncogenicity is evidence of potential carcinogenicity. We agree with this approach, which is weighted on the side of protecting public health.

There is also debate about whether the results of laboratory animal studies can be extrapolated to humans. According to both the International Agency for Research on Cancer (IARC) and the U.S. National Toxicology Program, all chemicals known to cause cancer in humans that have been adequately tested in animals also cause cancer in laboratory animals. While the reverse may not always be true, we judge it appropriate to assume that a substance that causes cancer in animals poses some risk of causing cancer in humans.

It is appropriate to assume that a substance that causes cancer in animals poses some risk of causing cancer in humans.

In the charts, a score of +2 means there is positive evidence of oncogenicity either in human studies or in two or more valid animal studies involving different species, or that there is unequivocal evidence in a single species (i.e., positive evidence from both sexes, and dosage dependency). Similarly, a score of −2 means there is negative evidence in two or more valid animal studies involving different species. A score of +1 means there is some positive evidence of oncogenicity, but not enough to warrant +2—for instance, data for a single species, or methodologically limited studies showing a positive result. A score of −1 means that there are a number of slightly flawed studies that do not show a positive result. A score of 0 means either that no data exists; that data has not yet been reviewed by the EPA and so is unavailable to the public; that the data is inconclusive; or that the data is from seriously flawed experiments.

MUTAGENICITY

Mutagenicity is a chemical's ability to cause mutations or permanent

Mutation is believed to be the first step in the process of carcinogenesis.

changes in the genetic material that may be passed on to offspring. Mutation is also believed to be the first step in the process of carcinogenesis, and mutagens may also be potential carcinogens. Mutations in and of themselves may cause potentially serious effects. There are many ways that a compound can cause mutations and many different tests used to screen for mutagenicity. The EPA routinely gathers data on at least seven different tests for this effect. Most tests use bacteria or mammalian (including human) cell cultures, although some use live animals. A compound that causes mutations in one or more bacterial tests may prove not to cause mutations in tests of mammalian cells. Thus, our chart uses a weight-of-evidence approach. A score of +2 refers to definite positive evidence from three or more valid bacterial tests (in the absence of a mammalian cell test), or from one valid test involving mammalian cells or whole animals. A score of +1 refers to definite positive evidence from one or two valid bacterial tests. A score of 0 means no data at all or no data from valid tests or negative data from only one or two bacterial tests. A score of −1 means negative evidence from

several bacterial tests or from at least one mammalian cell culture test. A score of −2 means negative evidence from a broad array of bacterial tests *plus* evidence from at least one test involving mammalian cell cultures or live animals.

REPRODUCTIVE TOXICITY

Reproductive toxicity is a chemical's ability to cause adverse effects on the reproductive process, such as sterility or decreased fertility, reduced litter size, lower birth weight, poor viability of offspring, or increased spontaneous abortion or fetal resorption rates. The studies are done on a number of mammalian species, with rats being the most common, and usually consist of exposing three successive generations to the compound and observing the success of reproduction by the first two generations. A score of +2 or −2 means there is unequivocal positive or negative evidence, respectively, in a single species. A score of +1 or −1 means there is some positive or negative evidence of reproductive toxicity, but in fewer studies or methodologically weaker studies than would warrant +2 or −2. A score of 0 means no data, data that has not been reviewed yet by EPA, totally inconclusive data, or data from seriously flawed experiments. As is the case with teratogenicity, negative results mean no observed effects on reproductive processes at dosages lower than those that have other clear toxic effects on the females.

TERATOGENICITY

Teratogenicity is a chemical's ability to cause birth defects or malformations in the offspring of pregnant animals exposed to the compound. As in the case of carcinogenicity, a substance that causes birth defects in animals may be assumed to pose a similar risk in humans. In our charts, a score of $+2$ means positive evidence of birth defects in either two or more valid animal studies involving different species or unequivocal evidence in a single species (i.e., positive evidence at a range of doses in multiple studies). A score of -2 means negative evidence in two or more valid animal studies involving different species—either no birth defects at all, or no effects at dosages lower than those that have obvious harmful effects in the mothers as well. A score of $+1$ means there is some positive evidence of teratogenicity, but fewer studies or less methodologically limited ones, or that there are only one or two acceptable negative studies. A score of 0 is reserved for no data, data

---

A substance that causes birth defects in animals may be assumed to pose a similar risk in humans.

---

that has not yet been reviewed by EPA, inconclusive data, or data from seriously flawed experiments.

Acute Effects

ACUTE ORAL TOXICITY

Acute oral toxicity refers to the oral LD_{50}, the dose fed to the animals that kills half of the exposed population. (Acute toxicity data is also available for different routes of exposure, such as absorption through the skin or by inhalation, and for various species; we limit ourselves here to oral exposure in rats.) Acute toxicity is not necessarily predictive of chronic toxicity; a low oral LD_{50} does not imply low chronic toxicity. However, the oral LD_{50} is itself a key indicator of acute hazards in some contexts, such as accidental ingestion. All poisons are classed into four standard toxicological categories based on their acute toxicity: I (highly toxic), II (moderately toxic), III (slightly toxic), and IV (practically nontoxic). Using the oral LD_{50} values in rats we have scored categories I, II, III, and IV as $+2$, $+1$, -1, and -2, respectively. If no oral LD_{50} data from rats is available for a compound, it was scored 0.

SKIN IRRITATION

Skin irritation is a compound's ability to irritate the skin. The effect in humans can range from no effect, through mild irritation, to severe skin lesions. The standard test consists of shaving a patch of hair on an animal's body (usually a rabbit), applying the chemical to the skin, then observing the patch of skin for a set period of time and scoring it for the degree of irritation. The result is an absolute number. However, the EPA groups these scores into four standard toxicological categories—I, II, III, and IV—with category I meaning severe irritation and category IV meaning mild irritation or no sign of irritation. Categories I, II, III, and IV are scored as $+2$, $+1$, -1, and -2, respectively. Lack of skin irritation data is scored as 0.

EYE IRRITATION

A compound's ability to irritate the eyes can range, in humans, from mild discomfort to severe eye damage that, if not treated promptly, could result in blindness and perhaps loss of the eye. The usual test, called a Draize test, involves applying the chemical directly to a rabbit's eye, then observing the eye for a set period of time and scoring it for the degree of irritation. The resulting scores are absolute numbers. However, the EPA groups the scores into four standard toxicological categories—I, II, III, and IV— with category I referring to severe irritation (such as ulceration of the cornea) and category IV referring to mild irritation or no sign of irritation. Categories I, II, III, and IV are scored

as $+2$, $+1$, -1, and -2, respectively. Lack of eye irritation data is scored as 0.

Subacute Effects

ACETYLCHOLINESTERASE INHIBITION

Acetylcholinesterase inhibition is one measure of potential neurotoxicity. The neurotransmitter known as *acetylcholine* transmits some impulses from nerve to nerve and nerve to muscle in all animals and insects, and is particularly important in the brain. An enzyme, acetylcholinesterase, normally breaks down the acetylcholine as soon as a nerve signal has been transmitted. But some chemicals, including two classes of insecticides, *organophosphates* and *carbamates*, destroy or block the action of acetylcholinesterase. In the absence of the enzyme, the nerve remains "on," transmitting impulses longer than normal and proving unable to respond to external stimuli. In humans, this effect can lead to a wide range of symptoms, depending on degree of enzyme blockage, such as sweaty hands, cramps, convulsions, twitching, seizures, paralysis, and numerous mental problems.

Some degree of acetylcholinesterase inhibition occurs even at very low levels of exposure to acetylcholinesterase-inhibiting pesticides in both humans and animal species. At present there is no consensus among experts about the degree of depression of acetylcholinesterase levels needed for the effect to be considered serious. Given this uncertainty, we have only two scores for this effect: 0 and +2. A score of +2 means the substance causes acetylcholinesterase inhibition; a score of 0 means it does not. The scores are based either on tests in animals or on the pesticide's chemical structure; all organophosphates or carbamates cause this effect.

TISSUE/ORGAN DAMAGE

Tissue/organ damage is a chemical's ability to damage various internal tissues or organ systems. This category is something of a catchall and covers effects on the kidney, spleen, liver, circulatory system, and other systems not specifically covered by other categories. The data is usually gathered at the same time as carcinogenicity data. The tests generally involve feeding animals (rats, mice, hamsters, dogs, or monkeys) the compound daily for 90 days to one year, then sacrificing the animal and examining it for damage. A score of +2 means there is evidence of adverse effects on at least two different internal systems in one species, or a strong adverse effect on the same system in two different species. A score of +1 means there is evidence of adverse effects on at least one internal system or evidence, either from a number of different systems or different species, from tests that are methodologically

somewhat weak. A score of 0 means no data, data that has not been reviewed yet by EPA, totally inconclusive data, or data from seriously flawed experiments. A score of −1 means there are some valid tests showing no evidence of negative effects on any internal system for at least two different species. Because of the near impossibility of ruling out all forms of possible toxicity in this broad category, there is no −2 score here.

NEUROTOXICITY

Neurotoxicity is a compound's ability to poison components of the nervous system. Neurotoxicity can produce sensory, motor, cognitive, or behavioral effects. Most classes of insecticides work by poisoning nerves in one way or another, and acute exposure to some herbicides, fungicides, and molluscicides may also be neurotoxic in humans. In the past, the EPA did not routinely require pesticides to be tested for potential neurotoxic and neurobehavioral effects,

Most classes of insecticides work by poisoning nerves in one way or another, and acute exposure to some herbicides, fungicides, and molluscicides may also be neurotoxic in humans.

except for some very narrow and minimal testing of organophosphates. The EPA recently proposed requiring testing for some additional subtle neurotoxic and neurobehavioral effects, but the vast majority of pesticides have not been tested for such effects. In the ratings, there are only two scores for this effect: 0 and +2. A score of +2 means that at least one methodologically sound study has shown neurotoxic or neurobehavioral effects in humans or animals. (Acetylcholinesterase inhibition is not included here; see above.) For example, clinical case reports show that the insect repellent DEET caused seizures in a total of six girls. A score of 0 means no data or insufficient data. In our judgment the tests required by EPA in the past were too narrow to assess potential neurobehavioral effects.

Comments

Miscellaneous or additional facts about a pesticide's toxicity that do not fit into the preceding headings are contained in comments in the Appendixes. Examples include discussion of potential contaminants or breakdown products that may contribute to a compound's toxicity and comments about exposures via routes other than home pesticide use, such as residues in the diet or potential contamination of drinking water.

 # ENVIRONMENTAL EFFECTS

What kind of adverse effects do the pesticides that we use around our homes and gardens have on the environment? As with health effects, some ingredients are more hazardous to the environment than others. The hazards can range from a potential for contamination of soil, surface water, and groundwater to the disruption of ecosystems through the mechanisms of pest resistance and pest resurgence. For our purposes, we restricted ourselves to a fairly limited set of environmental effects: the toxicity of the ingredients to various nontarget organisms, including fish, aquatic invertebrates, birds and bees, and pests' natural enemies.

The charts summarize available information on these potential environmental effects. The data comes from laboratory tests involving the nontarget organisms. In general, the data is in two forms, either LD_{50} or LC_{50}; LC_{50} is the lethal *concentration* (rather than *dose*), either in the water (in the case of fish and aquatic invertebrates) or in the diet (in the case of birds) needed to kill half of the exposed population of organisms. Both

LD_{50} and the LC_{50} yield an absolute number (usually expressed as mg/kg of body weight, mg/liter of water, or ppm [parts per million] in the diet, respectively), with a low number meaning high toxicity, and vice versa. For the LD_{50} data for birds, we use the four standard toxicological categories: I (highly toxic), II (moderately toxic), III (slightly toxic), and IV (practically nontoxic). There are no equivalent standard toxicological categories for the LC_{50} data for aquatic organisms, or the LD_{50} data for bees and natural enemies, but we designed our own four categories, using a logarithmic scale. As with the LD_{50} data in the toxicity section, we scored categories I, II, III, and IV as $+2$, $+1$, -1, and -2, respectively. If no data is available for a compound, it scored 0. The scoring scheme is the same for all five effects, and the category descriptions below provide information about the test species used and how toxicity was measured.

Fish

Toxicity to fish is usually tested on rainbow trout (a cold-water species) and the bluegill or sunfish (warm-water species). The fish are put in water containing different concentrations of a pesticide active ingredient and observed for a standard time, generally 96 hours. The LC_{50} is the concentration that kills half of the fish by the time 96 hours have elapsed. In cases where toxicity

---

TOXICITY RATINGS FOR ENVIRONMENTAL EFFECTS OF PESTICIDES

Ratings were attained for the following target organisms:
1. Fish
2. Aquatic invertebrates
3. Birds
4. Bees
5. Natural enemies of pests

---

scores for trout and bluegill are widely different, we have used the more toxic score.

Aquatic Invertebrates

The usual test species is the water flea, *Daphnia* spp., a common small crustacean. As are the fish, the *Daphnia* are put in water containing different concentrations of a pesticide active ingredient and observed for 96 hours, and the LC_{50} is the concentration that kills half of the *Daphnia* by 96 hours.

Birds

Three test species are commonly used: mallard duck, bobwhite quail, and ring-necked pheasant. (These are the most important avian game species.) The pesticide is fed to the birds in all cases, but sometimes the data is presented as LC_{50}, other times as LD_{50}. (Sometimes the concentration of the pesticides in the birds' diets is both more relevant and easier to measure than the amount or dose ingested. But for a pesticide used in granular form, where birds can eat the granules themselves, it makes more sense to look at the actual doses ingested and derive the LD_{50}.) Birds are fed or exposed to the pesticide for eight days. As with fish, if the different bird species have widely differing toxicities, we have used the most toxic rating for our score. Variability is noted in the Appendixes.

Bees

The test species is the honeybee. The test data generally takes the form of a topical LD_{50}, which is the lethal dose when applied to the bee's body, and is usually expressed in terms of micrograms/bee.

CHART 10-1
Common and Chemical Names for Pesticide Active Ingredients

COMMON NAME	CHEMICAL NAME
Insecticides	
Acephate	*O,S*-dimethyl acetylphosphoramidothioate
Allethrin	(*RS*)-3-allyl-2-methyl-4-oxocyclopent-2-enyl (1*RS*)-*cis,trans*-chrysanthemate *or* 2-methyl-r-oxo-3-(2-propenyl)-2-cyclopent-1-yl 2,2-dimethyl-3-(2-methyl-1-propenyl) cyclo-propanecarboxylate
Bacillus thuringiensis	*Bacillus thuringiensis*
Bendiocarb	2,2-dimethyl-1,3-benzodioxol-4-yl methylcarbamate *or* 2,3-isopropylidenedioxyphenyl methylcarbamate
Boric acid	B
Carbaryl	1-naphthyl methylcarbamate
Chlordane	1,2,4,5,6,7,8,8-octachloro-2,3,3a,4,7,7a-hexahydro-4,7-methanoindene
Chlorpyrifos	*O,O*-diethyl *O*-3,5,6-trichloro-2-pyridyl phosphorothioate
Cyfluthrin	(*RS*)-α-cyano-4-fluoro-3-phenoxybenzyl (1*RS*,3*RS*:1*RS*,3*SR*)-3-(2,2dichlorovinyl)-2,2-dimethylcyclopropanecarboxylate *or* cyano(4-fluoro-3-phenoxyphenyl)methyl 3-(2,2-dichloroethenyl)-2,2-dimethylcyclopropanecar-boxylate
DEET	*N,N*-diethyl-*m*-toluamide
Demeton, or demeton-0 + demeton-S	mixture of *O,O*-diethyl *O*-[2-(ethylthio)ethyl]phos-phorothioate and *O,O*-diethyl *S*-[2-(ethylthio)ethyl] phosphorothioate
Diazinon	*O,O*-diethyl *O*-2-isopropyl-6-methylpyrimidin-4-yl phosphorothioate
Dichlorvos, DDVP	2,2-dichlorovinyl dimethyl phosphate
Dicofol	2,2,2-trichloro-1,1-bis(4-chlorophenyl)ethanol
Dimethoate	*O,O*-dimethyl *S*-methylcarbamoylmethyl phosphorodithioate
Disulfoton	*O,O*-diethyl *S*-2-ethylthioethyl phosphorodithioate

Common and Chemical Names for Pesticide Active Ingredients (continued)

COMMON NAME	CHEMICAL NAME
Fatty acids	potassium salts of fatty acids
Fenoxycarb	ethyl 2-(4-phenoxyphenoxy)ethylcarbamate *or* ethyl [2-(*p*-phenoxy)ethyl]carbamate
Fenvalerate	(*RS*)-α-cyano-3-phenoxybenzyl (*RS*)-2-(4-chlorophenyl)-3-methylbutyrate *or* cyano(3-phenoxyphenyl)methyl 4-chloro-α-(1-methylethyl)benzeneacetate
Heliothis NPV	*Heliothis zea* NPV
Hydramethylnon	5,5-dimethylperhydropyrimidin-2-one 4-trifluoromethyl-α-(4-trifluoromethylstyryl)-cinnamylidenehydrazone
Hydroprene	ethyl (*E,E*)-3,7,11-trimethyldodeca-2,4-dienoate *or* (*E,E*)-ethyl 3,7,11-trimethyl-2,4-dodecadienoate
Lead arsenate	acid orthoarsenate
Lindane	1-α,2-α,3-β,4-α,5-α,6-β-hexachlorocyclohexane *or* γ-1,2,3,4,5,6-hexachlorocyclohexane
Malathion	S-1,2-bis(ethoxycarbonyl)ethyl O,O-dimethyl phosphorodithioate
Methomyl	S-methyl N-(methylcarbamoyloxy)thioacetimidate
Methoprene	isopropyl(E,E)-(RS)-11-methoxy-3,7,11-trimethyldodeca-2,4-dienoate
Methoxychlor	1,1,1-trichloro-2,2-bis(4-methoxyphenyl)ethane
MGK 11	2,3:4,5-bis(2-butylene) tetrahydro-2-furaldehyde
MGK 264	N-octyl bicycloheptene dicarboximide
MGK 874	2-hydroxyethyl-n-octylsulfide
Mineral oil	mineral oil
Naled	1,2-dibromo-2,2-dichloroethyl dimethyl phosphate
Oftanol	1-methylethyl 2[[ethoxy[(1-methylethyl)amino]phosphinothioyl]oxy]benzoate
Paradichlorobenzene	paradichlorobenzene
Permethrin	3-phenoxybenzyl (RS)-cis,trans-3-(2,2-dichlorovinyl)-2,2-dimethylcyclopropanecarboxylate
Petroleum distillates	same

COMMON NAME	CHEMICAL NAME
Phenothrin	3-phenoxybenzyl 2,2-dimethyl-3-(2-methylprop-1-enyl)cyclopropanecarboxylate *or* 3-phenoxybenzyl (±)-*cis,trans*-chrysanthemate
Phosmet, PMP	*S-O,O*-dimethyl phosphorodithioate with *N*-(mercaptomethyl)phthalimide *or* *O,O*-dimethyl *S*-phthalimidomethyl phosphorodithioate
Piperonyl butoxide	5-[2-(2-butoxyethoxy)ethoxymethyl]-6-propyl-1,3-benzodioxole
Propoxur	2-isopropoxyphenyl methylcarbamate *or* *O*-isopropoxyphenyl methylcarbamate
Pyrethrins	mixture of six chemicals: pyrethrin I and II, jasmolin I and II, and cinerin I and II
Resmethrin	5-benzyl-3-furylmethyl (1*RS*)-*cis,trans*-chrysanthemate
Rotenone, Derris cubé	(2*R*,6a*S*,12a*S*)-1,2,6,6a,12,12a-hexahydro-2-isopropenyl-8,9-dimethoxychromeno[3,4-*b*]furo[2,3-*h*]chromene-6-one *or* 1,2,12,12a-tetrahydro-8,9-dimethoxy-2-(1-methylethenyl-[1]benzopyrano-[3,4-*b*]furo[2,3-*h*][1]benzopyran-6[6*H*]-1
Silica gel	silica gel
Sodium arsenate	sodium orthoarsenate
Sulfluramid	N-ethyl Perflourooctanesulfonamide
Tetrachlorvinphos	(*Z*)-2-chloro-1-(2,4,5-trichlorophenyl)vinyl dimethyl phosphate
Tetramethrin	3,4,5,6-tetrahydrophthalimidomethyl (±)-*cis,trans*-chrysanthemate *or* cyclohex-1-ene-1,2-dicarboximidomethyl (1*RS*)-*cis,trans*-2,2-dimethyl-3-(2-methylprop-1-enyl)cyclopropanecarboxylate
Tralomethrin	(*S*)-α-cyano-e-phenoxybenzyl (1*R*,3*S*)-2,2-dimethyl-3-[(*RS*)-1,2,2,2-tetrabromoethyl]-cyclopropanecarboxylate *or* cyano(3-phenoxypenyl)methyl 2,2-dimethyl-3-(1,2,2,2-tetrabromoethyl)cyclopropanecarboxylate
Xylene	xylene

Common and Chemical Names for Pesticide Active Ingredients *(continued)*

COMMON NAME	CHEMICAL NAME
Herbicides	
Ammonium sulphamate	ammonium sulphamidate *or* monoammonium sulfamate
Atrazine	2-chloro-4-ethylamino-1,3,5-triazine *or* 2-chloro-4-ethylamino-6-isopropylamino-*s*-triazine
Benfluralin, benefin	*N*-butyl-*N*-ethyl-α,α,α-trifluoro-2,6-dinitro-*p*-toluidine
Chlorthal dimethyl, DCPA	dimethyl tetrachloroterephthalate
Dicamba	3,6-dichloro-*o*-anisic acid (formulated as an amine or sodium salt) or 3,6-dichloro-2-methoxybenzoic acid
Dichlobenil, DBN	2,6-dichlorobenzonitrile
Diquat dibromide	1,1′-ethylene-2,2′-dipyridylium dibromide
Fluazifop-butyl	butyl (*RS*)-2-[4-(5-trifluoromethyl-2-pyridyloxy)phenoxy]propionate
Glyphosate	*N*-(phosphonomethyl)glycine
Mecoprop	(±)-2-(4-chloro-2-methylphenoxy)propionic acid *or* (*RS*)-2-(4-chloro-*o*-tolyloxy)propionic acid
Pendimethalin	*N*-(1-ethylpropyl)-2,6-dinitro-3,4-xylidine *or* *N*-(1-ethylpropyl)-3,4-dimethyl-2,6-dinitrobenzenamine
Prometon	2,4-bis(isopropylamino)-6-methoxy-1,3,5-triazine *or* 2,4-bis(isopropylamino)-6-methoxy-*s*-triazine
Siduron	1-(2-methylcyclohexyl)-3-phenylurea *or* *N*-(2-methylcyclohexyl)-*N*′-phenylurea
Simazine	2-chloro-4,6-bis(ethylamino)-1,3,5-triazine *or* 2-chloro-4,6-bis(ethylamino)-*s*-triazine
Triclopyr	3,5,6-trichloro-2-pyridyloxyacetic acid
Trifluralin	α,α,α-trifluoro-2,6-dinitro-*N*,*N*-dipropyl-*p*-toluidine
2,4-D	(2,4-dichlorophenoxy)acetic acid [35 different esters and salts]
Fungicides	
Benomyl	methyl 1-(butylcarbamoyl)benzimidazol-2-ylcarbamate

Common Name	Chemical Name
Bordeaux mixture	mixture of calcium hydroxide and copper sulfate *or* tribasic copper sulfate
Captan	*N*-trichloromethylthio-4-cyclohexene-1,2-dicarboximide *or* 1,2,3,6-tetrahydro-*N*-(trichloromethylthio)-phthalimide
Chlorothalonil	2,4,5,6-tetrachloroisophthalonitrile
Copper sulfate	copper sulfate
Dinocap	2(or 4)-(1-methylheptyl)-4,6(or 2,6)-dinitrophenyl crotonate
Folpet	*N*-(trichloromethylthio)phthalimide
Manzoceb	manganese ethylenebis(dithiocarbamate) (polymeric) complex with zinc salt
Maneb	manganese ethylenebis(dithiocarbamate) (polymeric)
Sulfur	sulphur
Triadimefon	1-(4-chlorophenoxy)-3,3-dimethyl-1-(1*H*-1,2,4-triazol-1-yl)butanone
Triforine	1,1'-piperazine-1,4-diyldi-[*N*-(2,2,2-trichloroethyl)-formamide]
Zineb	zinc ethylenebis(dithiocarbamate) (polymeric)
Rodenticides	
Brodifacoum	3-[3-(4'-bromo[1,1-biphenyl]-4-yl)-1,2,3,4-tetrahydro-1-naphthalenyl]-4-hydroxy-2*H*-1-benzopyran-2-one *or* 3-[3-(4'-bromobiphenyl-4-yl)-1,2,3,4-tetrahydro-1-naphthyl]-4-hydroxycoumarin
Bromadiolone	3-[3-(4'-bromobiphenyl-4-yl)-3-hydroxy-1-phenylpropyl]-4-hydroxycoumarin *or* 3-[3,(4'-bromo[1,1'-biphenyl]-4-yl)-3-hydroxy-1-phenylpropyl]-4-hydroxy-2*H*-1-benzopyran-2-one
Chlorophacinone	2-[(*p*-chlorophenyl)phenylacetyl]-1,3-indandione *or* 2-[(4-chlorophenyl)phenylacetyl]-1*H*-indene-1,3(2*H*)-dione
Cholecalciferol	activated 7-dehydrocholesterol (vitamin D_3)

Common and Chemical Names for Pesticide Active Ingredients *(continued)*

COMMON NAME	CHEMICAL NAME
Diphacinone	2-(diphenylacetyl)indan-1,3-dione *or* 2-(diphenylacetyl)-1*H*-indene-1,3(2*H*)-dione
Fumarin	3-(α-acetonylfurfuryl)-4-hydroxycoumarin
Pindone	2-pivalyl-1,3-indandione
Strychnine sulfate	strychnine sulfate *or* strychnidin-10-one
Warfarin	3-(α-acetonylbenzyl)-4-hydroxycoumarin *or* 4-hydroxy-3-(3-oxo-1-phenylbutyl)-2*H*-1-benzo-pyran-2-one
Zinc phosphide	zinc phosphide *or* trizinc diphosphide
Molluscicides	
Metaldehyde, metacetaldehyde	r-2,c-4,c-6,c-8-tetramethyl-1,3,5,7-tetraoxocane acetaldehyde homopolymer
Methiocarb	4-methylthio-3,5-xylyl methylcarbamate

CHART 10-2
Insecticide Toxicity Scores

Key	−2	0	+2
	Safe	Toxicity Unknown	Hazardous

Ingredient Common Name/Brand Name	Carcinogenicity	Mutagenicity	Reproductive Toxicity	Teratogenicity	Acute Toxicity	Skin Irritation	Eye Irritation	Acetylcholinesterase Inhibition	Tissue/Organ Damage, Misc. Effects	Neurotoxicity
	Chronic Effects				Acute Effects				Subacute Effects	
Acephate Orthene, Ortran	1	2	−1	−1	−1	−2	−2	2	0	0
Allethrin Pyrocide, Pynamin	0	−1	0	1	−2	−1	−1	−2	0	0
Bacillus thuringiensis Dipel, Vectobac, Thuricide, Bactimos	−2	−2	−2	−2	−2	−1	−1	−2	0	0
Bendiocarb Ficam, Dycarb, Tatoo, Turcam, others	−1	−1	1	1	1	−1	−1	2	1	0
Boric acid Boric	0	1	1	0	−1	−1	−1	2	0	0
Carbaryl Sevin, Denapon, Dicarbam, Murvin	−1	0	−1	−2	1	−2	−1	2	1	0
Chlordane Octachlor, Chlor Kil, Velsicol 1068	2	1	2	0	1	−1	−1	−2	2	0
Chlorpyrifos Dursban, Lorsban, Pyrinex	0	1	1	−2	1	−1	−1	2	0	0
DEET Metadelphene, OFF, MGK Diethyltoluamide	1	−1	1	0	−1	−2	2	−2	1	2
Demeton, or demeton-0 + demeton-S Systox	0	0	0	1	2	−1	−1	2	0	0

Insecticide Toxicity Scores *(continued)*

Ingredient Common Name/Brand Name	Chronic Effects				Acute Effects				Subacute Effects	
	Carcinogenicity	Mutagenicity	Reproductive Toxicity	Teratogenicity	Acute Toxicity	Skin Irritation	Eye Irritation	Acetylcholinesterase Inhibition	Tissue/Organ Damage, Misc. Effects	Neurotoxicity
Diazinon Spectracide	0	0	0	0	1	−2	−1	2	0	0
Dichlorvos, DDVP Vapona, No Pest Strip, Ciavap, Atgard	2	2	0	1	1	−2	−1	2	2	2
Dicofol Kelthane, Acarin	1	−2	1	−2	1	2	−1	−2	1	0
Dimethoate Cygon, Rogor	1	1	−1	−1	1	−2	1	2	0	0
Disulfoton Di-Syston, Dithisystox, Thiodemeton	0	2	1	1	2	−2	−1	2	0	0
Fatty acids none	0	0	0	0	−2	−1	−1	−2	0	0
Fenvalerate Sumicidin, Pydrin	1	−1	0	0	−1	−1	−1	−2	1	0
***Heliothis* NPV** Elcar, Biotrol VHZ	−1	−1	−1	−1	−2	−1	−1	−2	0	0
Hydramethylnon Amdro, Combat	1	−2	1	0	−1	−2	−1	−2	1	0
Hydroprene Gencor	0	−1	0	0	−2	−2	−1	−2	0	0
Lead arsenate Gypsine, Security, Talbot	2	2	2	2	2	−1	−1	−2	2	0
Lindane Lindex, Gammex, Lintox, Isotox	1	−1	0	−2	1	2	−1	−2	1	2
Malathion Cython	−2	−1	1	−1	−1	−2	−1	2	0	0

Ingredient Common Name/Brand Name	Chronic Effects				Acute Effects				Subacute Effects	
	Carcinogenicity	Mutagenicity	Reproductive Toxicity	Teratogenicity	Acute Toxicity	Skin Irritation	Eye Irritation	Acetylcholinesterase Inhibition	Tissue/Organ Damage, Misc. Effects	Neurotoxicity
Methomyl Lannate, Nudrin	-2	-2	1	-1	2	-2	2	2	1	0
Methoprene Altocid, Precor, Kabat, Apex, Diacon, Pharorid	-2	-2	-1	-2	-2	-2	-2	-2	0	0
Methoxychlor Pyrocide, Marlate	-1	0	1	1	-2	-1	-1	-2	0	0
MGK 11 MGK 11	0	0	0	0	-2	-1	-1	-2	0	0
MGK 264 MGK 264	0	0	0	0	-1	-1	2	-2	0	0
MGK 874 MGK 874	0	0	1	0	-1	-1	-1	2	0	0
Mineral oil none	0	0	0	0	-2	-2	-1	-2	0	0
Naled Dibrom	1	1	1	1	1	1	2	2	0	0
Oftanol none	-1	0	1	-1	2	-2	-2	2	0	0
Paradichlorobenzene Paradichlorobenzene	1	-1	0	0	-1	-1	-1	-2	2	0
Permethrin Ambush, Pounce, Atroban, Outflank, Talcord	1	-1	1	0	-2	-1	-1	-2	1	0
Petroleum distillates none	1	0	0	0	-1	-1	-1	-2	0	0
Phenothrin Sumithrin	0	-2	1	-1	-2	-1	-1	-2	1	0
Phosmet, PMP Imidan, Prolate	1	2	1	0	1	-1	2	2	1	2

Insecticide Toxicity Scores (continued)

Ingredient Common Name/Brand Name	Chronic Effects				Acute Effects				Subacute Effects	
	Carcinogenicity	Mutagenicity	Reproductive Toxicity	Teratogenicity	Acute Toxicity	Skin Irritation	Eye Irritation	Acetylcholinesterase Inhibition	Tissue/Organ Damage, Misc. Effects	Neurotoxicity
Piperonyl butoxide Butacide, Pybuthrin	0	−1	0	0	−1	−1	−1	−2	0	0
Propoxur Baygon	2	−1	1	−1	1	−2	−2	2	2	0
Pyrethrins Pyrethrum	0	0	0	−1	−1	0	0	−2	0	0
Resmethrin Chryson, Synthrin	1	0	2	−1	−2	−1	−1	−2	1	0
Rotenone, Derris cubé Prentox, Noxfire, Chem-Fish	−1	1	1	0	1	−1	−1	−2	1	0
Silica gel none	0	0	0	0	−1	−1	−1	−2	0	0
Sodium arsenate none	0	2	0	2	0	−1	−1	−2	0	0
Tetrachlorvinphos Gardona	1	−1	1	−1	−1	−1	−1	2	2	0
Tetramethrin Neo-Pynamin	1	0	1	−1	−2	−1	−1	−2	1	0
Xylene none	0	0	0	0	−1	1	1	−2	0	0

CHART 10-3
Herbicide Toxicity Scores

Key	−2	0	+2
	Safe	Toxicity Unknown	Hazardous

INGREDIENT COMMON NAME/BRAND NAME	CHRONIC EFFECTS				ACUTE EFFECTS				SUBACUTE EFFECTS	
	CARCINOGENICITY	MUTAGENICITY	REPRODUCTIVE TOXICITY	TERATOGENICITY	ACUTE TOXICITY	SKIN IRRITATION	EYE IRRITATION	ACETYLCHOLINESTERASE INHIBITION	TISSUE/ORGAN DAMAGE, MISC. EFFECTS	NEUROTOXICITY
Ammonium sulphamate Ammate	0	0	0	0	−1	−1	−1	−2	0	0
Atrazine Gesaprim, AAtrex	2	−1	0	−1	−1	−2	1	−2	0	0
Benfluralin, benefin Balan	0	0	0	0	−2	−2	1	−2	0	0
Chlorthaldimethyl, DCPA Dacthal, Wegro	−1	1	1	−1	−2	−2	−1	−2	0	0
Dicamba Banvel, Banex, Mediben	0	0	0	0	−1	−1	2	−2	1	0
Diquat dibromide Reglone, Midstream	1	1	1	1	1	−1	−1	−2	1	0
Fluazifop-butyl Fusilade	−1	−1	2	2	−1	−1	−1	−2	0	0
Glyphosate Roundup, Rodeo, Shackle	1	−2	1	−2	−2	−2	−1	−2	0	0
Mecoprop MCPP, Duplosan	0	0	0	0	−1	−1	−1	−2	0	0
Pendimethalin Prowl, Herbadox, Stomp	0	0	−1	1	−1	−1	−1	−2	1	0
Prometon Pramitol	0	0	0	−1	−1	−1	−1	−2	0	0

Herbicide Toxicity Scores *(continued)*

Ingredient Common Name/Brand Name	Chronic Effects				Acute Effects				Subacute Effects	
	Carcinogenicity	Mutagenicity	Reproductive Toxicity	Teratogenicity	Acute Toxicity	Skin Irritation	Eye Irritation	Acetylcholinesterase Inhibition	Tissue/Organ Damage, Misc. Effects	Neurotoxicity
Siduron Tupersan	0	0	−1	0	−2	−1	−1	−2	0	0
Simazine Algae-A-Way, Algicide, Aquazine, Princep	1	−1	0	−1	−2	−1	−1	−2	0	0
Triclopyr Garlon, Dowco 233, Turflon	1	1	1	−1	−1	−1	−1	−2	1	0
Trifluralin Treflan, Crisalin, Trim, TR-10	−1	−2	−1	−1	−1	−1	2	−2	0	0
2,4-D Hedonal, Weedone, Weed-B-Gon	2	1	2	−1	−1	−1	−1	−2	0	0

CHART 10-4
Fungicide Toxicity Scores

Key	−2	0	+2
	Safe	Toxicity Unknown	Hazardous

Ingredient Common Name/Brand Name	Chronic Effects				Acute Effects				Subacute Effects	
	Carcinogenicity	Mutagenicity	Reproductive Toxicity	Teratogenicity	Acute Toxicity	Skin Irritation	Eye Irritation	Acetylcholinesterase Inhibition	Tissue/Organ Damage, Misc. Effects	Neurotoxicity
Benomyl Benlate, Tersan 1991, Arbotrine	1	2	2	2	−2	−2	1	−2	2	0
Captan Merpan, Orthocide, SR-406	2	1	0	1	−2	−2	2	−2	0	0
Chlorothalonil Daconil, Bravo, Brovomil, Repulse, Mold-Ex	1	−2	1	−2	−1	−2	2	−2	0	0
Copper sulfate Bluestone, Copper sulfate	0	1	0	0	1	1	2	−2	0	0
Dichlobenil, DBN Decabane	1	−1	1	0	−1	−2	−1	−2	1	0
Dinocap, DPC Karathane	1	1	1	1	−1	−1	−1	−2	2	0
Folpet Folpan, Folpex, Phatan, Thiophal	2	2	1	2	−1	−2	1	−2	1	0
Mancozeb Dithane ultra, Dithane M-45 or 945	2	−1	0	1	−2	−1	−1	−2	0	0
Maneb Manzate, Tersan LSR, Dithane	1	1	0	0	−2	−1	−1	−2	2	0

Fungicide Toxicity Scores *(continued)*

| Ingredient Common Name/Brand Name | Chronic Effects | | | | Acute Effects | | | | Subacute Effec |
	Carcinogenicity	Mutagenicity	Reproductive Toxicity	Teratogenicity	Acute Toxicity	Skin Irritation	Eye Irritation	Acetylcholinesterase Inhibition	Tissue/Organ Damage, Misc. Effects	Neurotoxicity
Sulfur Sulfur	0	0	0	0	−2	−2	−1	−2	1	0
Triadimefon Bayleton	1	−1	1	2	1	−1	−1	−2	1	0
Triforine Funginex, Saprol	−1	−1	1	−2	−1	−2	2	−2	0	0
Zineb Dithane Z 78	2	−2	0	1	−2	0	1	−2	0	0

CHART 10-5
Rodenticide Toxicity Scores

Key	−2	0	+2
	Safe	Toxicity Unknown	Hazardous

Ingredient Common Name/Brand Name	Chronic Effects				Acute Effects				Subacute Effects	
	Carcinogenicity	Mutagenicity	Reproductive Toxicity	Teratogenicity	Acute Toxicity	Skin Irritation	Eye Irritation	Acetylcholinesterase Inhibition	Tissue/Organ Damage, Misc. Effects	Neurotoxicity
Brodifacoum Talon, Klerat, Mouser, Brodifacoum	0	−1	0	−1	2	−1	−1	−2	2	0
Bromadialone Maki, Super-Caid, Super-Rozol	0	0	0	0	2	−1	−1	−2	2	0
Chlorophacinone Rozol	0	0	0	0	−1	−2	−1	−2	2	0
Cholecalciferol none	0	0	0	0	−1	−1	−1	−2	2	0
Diphacinone Diphacins	0	0	0	0	2	−1	−1	−2	2	0
Fumarin F	0	0	0	0	−1	−1	−1	−2	0	0
Pindone Parakakes, Pival	0	0	0	0	−1	−1	−1	−2	0	0
Strychnine sulfate none	0	0	0	0	−1	−1	−1	−2	0	0
Warfarin Warfarin	0	0	0	2	1	−1	−1	−2	2	0
Zinc phosphide Gopha-Rid	0	0	0	0	2	−2	−1	−2	2	0

CHART 10-6
Molluscicide Toxicity Scores

	Key	−2	0	+2
		Safe	Toxicity Unknown	Hazardous

Ingredient Common Name/Brand Name	Chronic Effects				Acute Effects				Subacute Effects	
	Carcinogenicity	Mutagenicity	Reproductive Toxicity	Teratogenicity	Acute Toxicity	Skin Irritation	Eye Irritation	Acetylcholinesterase Inhibition	Tissue/Organ Damage, Misc. Effects	Neurotoxicity
Metaldehyde, Metacetaldehyde Ariotox	0	−1	1	0	1	−2	−1	−2	0	0
Methiocarb Mercaptodimethur, Mesurol	−1	0	0	−1	2	−1	−1	2	1	0

APPENDIXES

APPENDIX A

INSECTICIDES

Chemicals used to fight insects are called *insecticides.* This general heading also includes the *acaricides*, which are used to fight mites and ticks (which are *not* insects; they belong to the order Acarina, which also contains spiders). Given the prime importance and popular conception of insects as pests, we were not surprised to find that 47 of the 89 active ingredients included in our project were insecticides. EPA statistics show that of the more than 70 million pounds of active pesticide ingredients used around the home and garden each year, approximately half are insecticides.

This appendix groups the 47 insecticides into seven different categories based for the most part on similarities in chemistry or mode of action. After a brief description of the general traits of each category, its drawbacks and advantages, we provide a capsule summary of what is known about each individual active ingredient in the category.

Among the facts presented is the standard chemical name (as approved by the International Union of Pure and Applied Chemistry) of the active ingredient; this is the name that is generally on the label of pesticide products. We describe its mode of entry (how it enters the pest's body) and its mode of action (how it works).

There are basically three different

modes of entry: through the *mouth* during feeding, through *respiration*, or by *absorption* directly through the exterior covering of the body. Chemicals of the first type have to be ingested before they can work and are said to have "stomach action." Chemicals of the second type are called *fumigants*. Chemicals that the insects absorb directly are said to have "contact action."

Insecticides kill in a number of specific ways, although all have one characteristic in common: they poison nerves in one way or another. Examples of different modes of action include interfering with the process of breathing, interfering with nerve impulse transmission, paralyzing muscles, inhibiting metabolic pathways, and disrupting hormone biosynthesis. Some chemicals that kill insects also have a repellent effect in that they cause insects to try to avoid contact with them.

Also indicated is whether an insecticide is *systemic* or *nonsystemic*. A systemic pesticide is taken up by a plant and incorporated into its various tissues, where it is not washed away by the rain or degraded by sunlight. A nonsystemic pesticide stays on the surface of the plant and can potentially be washed off.

EARLY INSECTICIDES: BOTANICALS AND INORGANICS

The earliest insecticides generally came from plants, although some were no doubt inorganic chemicals. Among the early botanicals that are still used today are rotenone and pyrethrum. Two early inorganic insecticides still used today are both based on arsenic: lead arsenate and sodium arsenate. We provide specifics on the botanicals and then on the two inorganic insecticides.

ROTENONE

Rotenone, also known by the unwieldy chemical name (2R,6aS,12aS)-1,2,6,6a,12,12a-hexahydro-2-isopropenyl-8,9-dimethoxychromeno[3,4-b]furo[2,3-h]chromene-6-one, is derived from the roots of 68 species of legumes, but most commercial sources come from species in the genera *Derris* and *Lonchocarpus*. Its toxic properties have long been known. It remains a popular "organic" insecticide, used against a number of home and garden pests including chiggers, scabies, fire ants, and mosquitoes.

Mode of Action

Rotenone is a somewhat selective insecticide that works as both a contact and stomach poison. It is a slow-acting nerve poison that inhibits cell respiration and energy metabolism.

Toxicity—Acute Effects

Rotenone has a moderate acute oral toxicity and causes some irritation of the skin and eyes.

Toxicity—
Subacute and Chronic Effects

Rotenone has never been tested properly for subacute and chronic effects. For a long time it was considered safe and was exempted from the pesticide tolerance-setting process in 1955. The EPA is now establishing tolerances for rotenone, and has issued a Data Call-in for residue data. There are some reports that it has produced tumors in rats and hamsters, reproductive problems in hamsters, and liver changes in dogs and rats. All these studies are considered significantly flawed by both the EPA and the California Department of Food and Agriculture (CDFA). The questions raised by these studies, as well as the mutagenicity studies, finally stimulated the EPA in May 1986 to require manufacturers to provide better data on chronic effects.

Mutagenicity

There is some evidence that rotenone may pose a risk of genetic abnormalities in frog eggs and chick embryos.

Ecological Effects

Rotenone poses some environmental or ecological hazards. It is highly toxic to fish, aquatic invertebrates, and birds. Indeed, many indigenous people use root shavings from *Derris* and *Lonchocarpus* to catch fish; they just throw the shavings into the water and wait for the stunned fish to float to the surface. Rotenone is relatively nontoxic to honeybees.

PYRETHRUM

This natural botanical insecticide and the chemically similar synthetic pyrethroids are discussed later in this chapter.

LEAD ARSENATE

Lead arsenate is known chemically as acid orthoarsenate. Lead arsenate was an early chemical pesticide. Until the end of World War II, it was one of the two top-selling insecticides in the United States. However, the inherent toxicity of both of its primary components—lead and arsenic—has drastically curtailed its use. It is no longer used for any agricultural purposes, but we found it in one home product: Gator Roach Hives.

Mode of Action

Lead arsenate is a systemic insecticide and a plant growth regulator.

Toxicity—Acute Effects

Lead arsenate is highly acutely toxic, but only mildly irritating to the skin and eyes.

Toxicity— Subacute and Chronic Effects

Lead arsenate causes cancer, mutations, birth defects, reproductive problems, and damage to internal organs in laboratory animals.

Mutagenicity

Lead arsenate is considered a mutagen.

Ecological Effects

Lead arsenate is slightly toxic to fish and moderately toxic to aquatic invertebrates and birds.

SODIUM ARSENATE

Sodium arsenate has not been very well characterized toxicologically. It is used to kill a variety of insects in the garden and home.

Mode of Action

Sodium arsenate is a stomach poison.

Toxicity—Acute Effects

Sodium arsenate is moderately acutely toxic and causes some irritation of the skin and eyes.

Toxicity—
Subacute and Chronic Effects

Data is lacking concerning carcinogenicity, reproductive toxicity, and effects on internal organs and tissues. Laboratory studies show that sodium arsenate causes birth defects.

Mutagenicity

Sodium arsenate causes mutations.

Ecological Effects

Insufficient information is available, although it is probably hazardous.

 ORGANOCHLORINES

The first widely used synthetic organic pesticides were the organochlorines. Chemically they contain carbon, hydrogen, chlorine, and occasionally oxygen atoms. There are two major groups of organochlorines: the *chlorinated hydrocarbons* (also called *bridged diphenyls* because of their chemical structure) and the *cyclodienes* (so called because of their cyclical chemical structure). Examples of the former include DDT, dicofol, and methoxychlor; of the latter, chlordane, heptachlor, dieldrin, and aldrin.

The overwhelming success of DDT during and immediately after World War II stimulated tremendous market growth in insecticides, which were perceived as "magic bullets." A number of characteristics made DDT and the other organochlorines almost ideal insecticides: they were cheap and easy to make; they were highly acutely toxic to insects, while only moderately acutely toxic to mammals; and they were resistant to degradation, both in the environment and in biological systems. This persistence was initially seen as a positive attribute, since the chemicals did not have to be applied repeatedly at short intervals.

We now realize that the persistence of the organochlorines (with the exception of methoxychlor) in the environment and their resistance to biological degradation resulted in contamination of the world's ecosystem. The widespread use of these insecticides produced high levels of genetic resistance in many pests once controlled by them. Furthermore, because they readily dissolve in fat, organochlorines tend to become increasingly concentrated as organisms are eaten by other organisms and the pesticide is passed up the food chain. For example, within five years of the first use of DDD (a relative of DDT) to control gnats in Clear Lake, California, western grebes began dying around the lake. DDD levels in the birds' body fat were 100,000 times higher than the level in the lake water. The DDD had been transferred and concentrated, from water, to plankton in the water, to plankton-eating fish, to fish-eating fish, and to the grebes.

Organochlorines have also bioaccumulated in human fat and breast milk—entering the human food chain through meat, dairy products, fish, eggs, fruits, and vegetables. Indeed, organochlorines have been found in the fat tissue and breast milk of people all over the world including those, such as Eskimos in the Arctic, who live where these insecticides have never been used.

Such food-chain exposure raises health concerns because many organochlorines cause a wide range of chronic toxic effects in animals, including cancer, neurotoxicity, birth defects, reproductive effects, and damage to various organs and tissues (such as the liver, spleen, and kidneys). Indeed, as of 1972, 19 of 25 organochlorines tested had been found to be carcinogenic in animal tests, while data on the other 6 was incomplete. All the cyclodiene organochlorines (chlordane, heptachlor, dieldrin, and aldrin) have been linked to cancer, genetic damage, adverse reproductive effects, birth defects, and neurotoxicity. The dramatic decline in reproduction of larger fish-eating birds (such as bald eagles, peregrine falcons, brown pelicans, and ospreys) in the 1960s and 1970s was linked to organochlorine pesticide residues, which, among other effects, weakened their eggshells.

The concern over potential chronic health effects has been heightened by what little we know about how the organochlorines function. In general, they are nerve poisons, although the exact mechanism for this action is still not known for most compounds. The organochlorines also stimulate the activity of enzymes called *mixed function oxidases*, which detoxify numerous poisons (including insecticides) and play a role in steroid metabolism. The reproductive and other toxic effects of organochlorines may occur because of the effects of organochlorines on steroid metabolism.

Because of the combination of chronic toxicity and persistence in the environment, most organochlorine insecticides have been banned or severely restricted in most of the developed countries, including the United States. However, our shoppers did find a few home pesticide products that contained members of this class.

CHLORDANE

Chlordane is known chemically as 1,2,4,5,6,7,8,8 - octachloro - 2,3,3a,7, 7a-hexahydro - 4, 7 - methanoindene. Like most organochlorines, it has long residual activity. Chlordane was once used against a wide variety of insect pests, but most agricultural and home and garden uses of this insecticide were canceled or phased out by the EPA in 1978. By then, enough animal data had accumulated to show that chlordane was a carcinogen, reproductive toxin, and neurotoxin, and was suspected of being a teratogen and mutagen. The EPA allowed continued use of chlordane for control of subterranean termites, reasoning that subsurface application of chlordane should not result in either human or environmental exposure. Until 1987, about a million houses a year were treated with chlordane for this purpose. However, as data became available in the 1980s, it became clear that exposure did occur. Chlordane is quite volatile; vapors do penetrate into the home and are present in treated houses. Given the chronic toxicity and the persistent exposure in a house, various negative health effects may occur. A number of homeowners claim that exposure to chlordane has caused health problems in their families. They have banded together and many groups have gone to court to sue the manufacturer or applicator. In February 1987, the National Coalition Against the Misuse of Pesticides

(NCAMP) petitioned the EPA to cancel chlordane's termiticide uses. In July 1987, NCAMP and others sued the EPA to cancel chlordane's termiticide uses. Also that year, Velsicol (the manufacturer of chlordane) admitted that its own data demonstrated persistent residues of chlordane in more than 90 percent of monitored homes. In August 1987, the EPA reached a consent agreement with the manufacturer to halt all sales (to avert a ban) until it had developed safe application methods. In return, the EPA would permit all existing stocks in the market to be sold. In February 1988, a federal court ruled in NCAMP et al.'s favor, and ordered the EPA to ensure that all commercial application of existing stocks of chlordane halt by April 15, 1988. The EPA appealed the ruling, and in February 1989, a federal circuit court of appeals reversed the lower court and ruled that the EPA could make settlements with companies that would permit the sales of existing stocks. One issue that the EPA regulations did not deal with was the stocks of the insecticide in the possession of homeowners, some of whom may still have chlordane supplies purchased before sales were suspended.

Mode of Action

Chlordane is both a contact poison and a stomach poison, and also interferes with respiration.

Toxicity—Acute Effects

Chlordane has a moderate acute oral toxicity, and is a mild skin irritant and a severe eye irritant.

Toxicity—
Subacute and Chronic Effects

Chlordane causes liver tumors in mice, rats, and hamsters, and chronic damage to the liver and kidneys. A number of laboratory studies have also found adverse effects on reproduction, although the studies are not definitive. There is no solid evidence, one way or the other, on whether chlordane causes birth defects.

Some data on human health effects exists for chlordane. An epidemiological study of 1,403 workers involved in the manufacture of chlordane and heptachlor found no statistically significant increase in cancer rates. However, the study was too limited to prove that no cancer risk exists. Another study reported five cases of neuroblastomas in children, all in the same pediatric hospital and all of whom had prenatal and/or extensive environmental exposure to chlordane; one case of aplastic anemia in a boy exposed to chlordane and lindane; and a young girl with leukemia who had been exposed to chlordane. There have been numerous other reports of illness among people whose houses were treated with chlordane, but most such reports are anecdotal rather than clinically documented case histories.

Mutagenicity

There is some evidence from microbial systems that chlordane causes genetic damage, but not enough to class it definitely as a mutagen.

Ecological Effects

Chlordane is very highly toxic to fish and aquatic invertebrates, moderately to highly toxic to birds, and toxic to bees and other beneficial insects. It is also very persistent in the soil, and it bioaccumulates. Although it was thought to be immobile in the soil, chlordane has shown up in groundwater and wells in some states.

Given the ecological, environmental,

and potential health hazards of chlordane, consumers should choose safer alternatives for termite control (see chapter 6).

DICOFOL

Dicofol, known chemically as 2,2,2-trichloro - 1,1 - bis(4-chlorophenyl)ethanol, is used to control a wide variety of mite species on both indoor and outdoor plants. Dicofol is very similar to DDT in chemical structure, and products containing dicofol were often contaminated with DDT and related compounds (e.g., DDT and its breakdown products DDD, DDE, etc.; hereafter just DDT) in amounts as great as 10 percent or higher. But formulations of dicofol containing more than 2.5 percent DDT and related compounds were banned in 1986, and products with contaminants exceeding 0.1 percent were banned in 1989.

Mode of Action
Dicofol is a nonsystemic acaricide that acts as a contact poison.

Toxicity—Acute Effects
Dicofol has a moderate acute oral toxicity and causes slight skin irritation but severe eye irritation.

Toxicity—
Subacute and Chronic Effects
Dicofol has been inadequately tested for most subacute and chronic effects. Several studies in mice and rats found liver damage, tumors in the liver and lungs, and reproductive effects such as decreased litter size and weight of pups at weaning. These studies, however, although suggestive, all contained methodological flaws. In addition, the potential contamination of dicofol with DDT makes it difficult to tell whether negative effects, if real, resulted from exposure to dicofol or a contaminant. Acceptable studies, done with both rats and rabbits, have found no evidence that dicofol causes birth defects.

Mutagenicity
Tests using both microbial and mammalian cell culture systems indicate that dicofol does not cause mutations.

Ecological Effects
Dicofol appears to be very highly toxic to fish and aquatic invertebrates, only moderately toxic to birds, and basically nontoxic to bees. Good ecological studies are lacking, however; existing ones have the same problem of potential contamination with DDT present in toxicity studies.

Given the lack of good data on environmental and chronic health effects and dicofol's similarity to DDT, it is prudent to avoid using dicofol. There are several safer compounds that control mites (see chapter 8).

LINDANE

Lindane, known chemically as 1-α,2-α,3-β,4-α,5-α,6-β-hexachlorocyclohexane, is used to control a variety of insects—including termites and moth caterpillars—around the home and garden, to kill fleas on pets, and even, in some prescription shampoos, for controlling head and body lice and scabies, a mite infestation. Indeed, in 1988, more than 3 million children in the United States required treatment for head and body lice, and the lindane-based shampoo Kwell was a common remedy prescribed by pediatricians.

Mode of Action

Lindane acts as both a contact poison and a stomach poison, as well as interfering with respiration.

Toxicity—Acute Effects

Lindane has a moderate to high acute oral toxicity and is a slight skin irritant and a severe eye irritant.

Toxicity—
Subacute and Chronic Effects

Few methodologically solid studies exist on subacute and chronic effects, but some data suggests liver and kidney toxicity, and one study linked aplastic anemia in a boy with exposure to chlordane and lindane. There is some evidence that it causes liver tumors in mice and rats, while mouse, rat, and rabbit studies have found no evidence of birth defects. These studies were methodologically flawed and need to be redone, although the EPA is satisfied with the results of the teratogenicity studies. Valid tests on reproductive effects are also lacking.

Several clinical case reports involving use of Kwell shampoo or lotion suggest that lindane is neurotoxic in humans. The cases involved young children for the most part, who suffered seizures and other neurotoxic effects when lindane was absorbed through the skin. The risk seems to be higher with treatment for scabies, which may require application of the insecticide to large areas of the body.

Ecological Effects

Lindane is very highly toxic to fish, aquatic invertebrates, bees, and other beneficial insects, and moderately toxic to birds. Use of lindane to control garden pests may be counterproductive if it kills off the beneficial insects that feed on the pests one is trying to control. Unlike most other organochlorines, lindane is broken down fairly readily in the environment and poses fewer of the severe environmental problems associated with the more persistent organochlorines.

The EPA proposed canceling about 80 percent of lindane's uses in 1983, but decided in 1985 against the cancelation on risk/benefit grounds, imposing label changes instead. Use against termites is restricted to unoccupied buildings. At present, lindane is in the regulatory status known as Special Review, triggered because data suggest a significant hazard to human health or the environment. Stay away from lindane; there are many safer insecticides, both for anti-lice and for home and garden uses. (Several effective over-the-counter anti-lice preparations use pyrethroids or sulfur, which are much less toxic to people; see later section of this appendix.)

METHOXYCHLOR

Methoxychlor, known chemically as 1,1,1-trichloro-2,2-bis(4-methoxyphenyl) ethane, has a structure similar to that of DDT and is used against a wide array of insect pests, both indoors and outdoors.

Mode of Action

Methoxychlor is both a contact poison and a stomach poison.

Toxicity—Acute Effects

Methoxychlor is an eye and skin irritant, but is practically nontoxic when ingested.

Toxicity—
Subacute and Chronic Effects

There is a dearth of good studies on subacute and chronic effects, with most dating from the 1950s and 1960s, when studies were not up to current toxicological standards. A flawed study found evidence of oncogenicity, but more studies are needed. Two rat studies found some evidence of birth defects and reproductive effects, but both studies have flaws. No evidence of damage to tissues and organs has been found, but no valid studies exist.

Mutagenicity

The EPA and CDFA have no studies on file.

Ecological Effects

Methoxychlor is practically nontoxic to birds and mammals, extremely toxic to fish and aquatic invertebrates, and toxic to beneficial insects. It is biodegradable and relatively nonpersistent in the environment compared to most organochlorines; however, its half-life in water is about 46 days. This, combined with its very high toxicity to aquatic organisms, means that methoxychlor should not be used near any body of water.

  # ORGANOPHOSPHATES

The organophosphates were the second wave of synthetic organic insecticides, and are currently the most widely used pesticides in the United States. They were developed as by-products of nerve-gas research in Germany in the 1930s, then came to prominence as replacements for organochlorines that had lost their effectiveness as major insect pests became resistant to them.

Chemically, the organophosphates are esters of one of the phosphoric acids. Like other esters, they are hydrolyzed easily, which means that as a group they tend to be broken down more readily than the organochlorines and so are less persistent.

The organophosphates vary greatly in both chemical and biological properties. Toxicity to insects and mammals ranges from very low to very high, with little correlation between the two. While many degrade quickly in the environment, others are relatively stable. On the whole, however, organophosphates are less persistent in the environment, more likely to biodegrade, less stable in sunlight, and less likely to bioaccumulate than the organochlorines; consequently they cause much less long-term environmental contamination. On the other hand, most are more acutely toxic to humans in the short term. Indeed, the vast bulk of acute pesticide poisoning cases are attributable to organophosphates.

Mode of Action

Organophosphates kill insects by inhibiting acetylcholinesterase, an enzyme that breaks down acetylcholine, a neurotransmitter of impulses between nerves and between nerve and muscle in most animals. Death from organophosphate poisonings occurs from asphyxiation resulting in large part from paralysis of the respiratory muscles. Because the exact mechanism of organophosphate poisoning is known, it is possible to develop antidotes. Atropine sulfate, one such antidote, blocks certain poisoning symptoms and is the primary treatment.

Organophosphates inhibit acetylcholinesterase by physically binding to it (or attaching), thereby destroying its activity. The poisoned enzyme does not recover and must be replaced by newly produced enzyme, a process that takes from 30 to 60 days in humans. Therefore, effects of organophosphate exposure over such intervals are cumulative; repeated exposure to doses that bind to a certain percentage of acetylcholinesterase can build up in effect until an additional small dose can trigger symptoms of severe poisoning. This hazard is particularly severe for farmworkers and pesticide applicators, but should not be a problem for most consumers.

Toxicity— Subacute and Chronic Effects

Many organophosphates also cause delayed neurological effects that may produce changes in personality, memory, behavior, and nerve function. Besides the neurological and neurobehavioral effects, some organophosphates cause chronic effects such as cancer, mutations, birth defects, reproductive problems, and tissue and organ damage.

Ecological Effects

Organophosphates tend to cause fewer environmental effects than the organochlorines because they degrade more quickly. Although development of insect resistance is a problem, resistance levels are lower than with persistent organochlorines.

One particularly negative ecological drawback is that organophosphates generally are not selectively toxic to target pests; in fact, they are frequently far more toxic to beneficial insects than to pests. Thus they must be used with great care, if they are used at all.

ACEPHATE

Acephate, known chemically as O,S-dimethyl acetylphosphoramidothioate, is used to control a wide range of chewing and sucking insect pests in the home and on lawns, turfs, and ornamental plants, such as aphids, scale, caterpillars, and flea beetles.

Mode of Action

Acephate is a systemic insecticide that works as both a contact poison and a stomach poison.

Toxicity—Acute Effects

Acephate has a low acute oral toxicity to mammals, and is a severe eye irritant but not a skin irritant.

Toxicity— Subacute and Chronic Effects

There are no methodologically solid studies on subacute and chronic effects, but some data raises disturbing possibilities. One study on mice found evidence of liver tumors in females at the highest dose level and evidence of dose-related damage to the lungs and liver in both males and females. The EPA considered that study of minimal quality, and the evidence for carcinogenicity is still limited. A metabolite, methamidiphos, is a suspected human carcinogen. A rat reproduction study found reduced fertility in males and decreased pup survival. A subsequent study, which had a better design, found no such effects except where there were also significant effects on the parents. A rat study and a rabbit study failed to find evidence of birth defects, but the former study is considered somewhat flawed. More and better studies are needed on most of these effects.

Mutagenicity

Sufficient evidence from both microbial and mammalian cell culture shows acephate to be a weak mutagen.

Ecological Effects

Acephate is highly toxic to bees, moderately toxic to birds, and only slightly toxic to fish and aquatic invertebrates. Methamidophos, which is formed in the avian's gut when acephate is digested, is highly toxic to birds.

CHLORPYRIFOS

Chlorpyrifos, known chemically as O,O-diethyl O-3,5,6-trichloro-2-pyridyl phosphorothioate, is sold widely as Dursban or Lorsban. It is used to control a wide range of insect pests both indoors and outdoors, including ants, termites, fleas, ticks, cockroaches, spider mites, mealybugs, and wasps.

Mode of Action

Chlorpyrifos acts as both a contact poison and a stomach poison, and interferes with respiration.

Toxicity—Acute Effects

Chlorpyrifos has a moderate acute oral toxicity and is a slight eye and skin irritant.

Toxicity—
Subacute and Chronic Effects

There are few methodologically solid studies. Some studies in mice and rats found no evidence of carcinogenicity or reproductive toxicity; both studies are flawed and new ones are needed. A teratogenicity study on mice found evidence of skeletal defects at high doses. A tera-togenicity study on rats found no evidence of birth defects at any level tested. A recent Italian report of an acute poisoning episode found that delayed nerve damage (lesions) occurred.

The EPA has estimated that potential exposure to chlorpyrifos in foods substantially exceeds recommended limits, because of widespread agricultural use. Moreover, chlorpyrifos is one of the 10 most widely used household pesticides, and its use is expected to increase because chlorpyrifos is a major alternative to chlordane for termite control. Home-use exposures, combined with possibly excessive dietary exposure, may produce very high total exposure to this insecticide for some consumers.

Mutagenicity

There is some positive evidence of mutagenicity, but the studies need to be repeated.

Ecological Effects

Chlorpyrifos is extremely toxic to bees, birds, fish, and aquatic invertebrates. Great care should be taken when using this pesticide outdoors.

DEMETON

Demeton, which chemically is a mixture of O,O-diethyl O-[2-(ethylthio)ethyl]phosphorothioate and O,O-diethyl S-[2-(ethylthio)ethyl]phosphorothioate, is used to control a variety of sucking insects and mites in the garden and on ornamentals.

Mode of Action

Demeton is a systemic insecticide and acaricide that acts as a contact poison and stomach poison.

Toxicity—Acute Effects

Demeton is extremely toxic when ingested, and is a minor eye and skin irritant.

Toxicity—
Subacute and Chronic Effects

Data on subacute and chronic effects is largely lacking. There are no valid studies for carcinogenicity, reproductive effects, or tissue and organ damage. One study on mice found birth defects, but the study is flawed.

Mutagenicity

There are no valid studies.

Ecological Effects

Demeton is toxic to bees and fish, but has not been tested for other effects.

DIAZINON

Diazinon, known chemically as *O,O*-diethyl O-2-isopropyl-6-methylpyrimidin-4-yl phosphorothioate and commercially as Spectracide, is an insecticide and acaricide. It is used on a wide range of insect pests both indoors and outdoors, especially cockroaches, aphids, scale, mites, white grubs, Japanese beetles, fleas, and ticks. Diazinon is one of the ten most widely used household and garden pesticides and is one of the most popular lawn insecticides.

Mode of Action

Diazinon is both a contact poison and a stomach poison, and interferes with respiration.

Toxicity—Acute Effects

Considering its widespread use, the toxicity of diazinon has been very poorly studied. Diazinon is moderately toxic when ingested and is a minor eye irritant but not a skin irritant.

Toxicity—
Subacute and Chronic Effects

Data on subacute and chronic effects is sparse. Diazinon is known to cause birth defects in chick embryos. A few studies of marginal quality, involving mice, rats, and rabbits, found no birth defects or cancer, but these need to be repeated. There are no valid reproductive studies.

Mutagenicity

There are no valid studies available.

Ecological Effects

Diazinon is extremely toxic to birds, fish, and aquatic invertebrates even when used according to label directions. Birds are extremely sensitive, and numerous bird kills have occurred on lawns and golf courses; the EPA has canceled the use of diazinon on turf farms and golf courses for that reason. However, it is still sold in many products at hardware stores and garden shops. If you have birds in your yard, do not use diazinon on your lawn or garden. Finally, there have been problems with storing old containers of Spectracide; 126 pressurized container explosions have occurred in 27 states.

DICHLORVOS

Dichlorvos (or DDVP), known chemically as 2,2-dichlorovinyl dimethyl phosphate and commercially as Vapona, is used to control both household and garden pests. It is highly volatile (it readily evaporates) and is used in pest strips for fly control and flea collars that slowly release DDVP as a gas. It is also widely used in household ant and roach killers.

Mode of Action

Dichlorvos is a fast-acting insecticide and acaricide. It is both a contact poison and a stomach poison, and interferes with respiration.

Toxicity—Acute Effects

Dichlorvos has a moderate acute oral toxicity and is a minor eye irritant but not a skin irritant.

Toxicity—
Subacute and Chronic Effects

A methodologically sound study by the National Toxicology Program clearly demonstrated that DDVP causes tumors in mice and rats and causes significant damage to their livers as well. DDVP is also a potent acetylcholinesterase inhibitor, and is neurotoxic. There is some evidence of teratogenic effects in rabbits, but the studies are severely flawed. There are no good studies on reproductive effects, although DDVP is known to cross the placenta and lower acetylcholinesterase levels in the fetus.

Using existing residue data, the EPA has calculated that dietary intake of DDVP, assuming that all foods contain residues at tolerance levels, exceeds the Provisional Acceptable Dietary Intake (PADI). Thus, further exposure should be minimized.

On the basis of the NTP carcinogenicity studies and exposure studies, the EPA has estimated that home-use exposure, particularly via inhalation of fumes from pest strips, poses a very high cancer risk, roughly 1 cancer per 100 exposed people. The estimated risk of cancer was equally high for pets wearing flea collars containing DDVP. On the basis of this data,

Consumers Union recommends that consumers not use DDVP pest strips and flea collars; CU also recommends avoiding household insect sprays containing DDVP.

The EPA requires that products containing DDVP carry warnings that it causes cancer and liver effects in laboratory animals and is a potent acetylcholinesterase inhibitor. Labels must also state that protective clothing should be worn when using the product. However, household pesticides containing DDVP are exempt from the requirement to carry such warnings on the label.

Mutagenicity

Some studies have shown DDVP to be mutagenic in both microbial and mammalian cell culture systems, and it appears to be a direct-acting mutagen.

Ecological Effects

Dichlorvos is highly to very highly toxic to bees, natural enemies, birds, fish, aquatic invertebrates, and other wildlife.

DIMETHOATE

Dimethoate, known chemically as *O,O*-dimethyl *S*-methylcarbamoylmethyl phosphorodithioate, is used to control a variety of insects and mites in the garden.

Mode of Action

Dimethoate is a systemic insecticide and acaricide that kills via contact and as a stomach poison.

Toxicity—Acute Effects

Dimethoate is moderately toxic when ingested, and is a strong eye irritant but not a skin irritant.

Toxicity—
Subacute and Chronic Effects

Data on subacute and chronic effects is sparse. There are no valid studies for carcinogenicity, reproductive effects, or tissue and organ damage. One study in rats found evidence of blood vessel tumors, but more studies need to be done to confirm this effect. Valid rat studies and a rabbit study failed to find evidence of birth defects.

Mutagenicity

There is weak evidence from microbial systems that dimethoate may be a mutagen.

Ecological Effects

Dimethoate is toxic to bees, highly toxic to birds, and only moderately toxic to fish. Information on other ecological effects is not available.

DISULFOTON

Disulfoton, known chemically as *O,O*-diethyl *S*-2-ethylthioethyl phosphorodithioate, is used primarily against sucking insects and spider mites on outdoor plants and potted houseplants.

Mode of Action

Disulfoton is a systemic insecticide and acaricide, which is absorbed by the roots and spread throughout the plant to give long-lasting control.

Toxicity—Acute Effects

Disulfoton is extremely toxic when ingested and is a minor eye irritant but not a skin irritant. Because of its very high acute oral toxicity, products that contain over 2 percent disulfoton are restricted, which means they can be used only by certified pest control operators.

Toxicity—
Subacute and Chronic Effects

Data on subacute and chronic effects is sparse. A few studies have shown suggestive evidence of birth defects and negative reproductive effects, but the studies are flawed. Data on all other subacute and chronic effects is lacking.

Mutagenicity

There are no valid studies on file.

Ecological Effects

Disulfoton is very highly toxic to fish, aquatic invertebrates, mammals, and birds, and highly toxic to bees and beneficial insects. Given its long-lasting systemic effectiveness, consumers should not use disulfoton on vegetable gardens or other edible plants.

ISOFENPHOS

Isofenphos, known chemically as 1-methylethyl 2[[ethoxy[(1-methylethyl)amino]phosphinothioyl]oxy]benzoate, is applied to the soil to control insect pests in the lawn and vegetable gardens, such as carrot flies, corn rootworms, white grubs, Japanese beetles, wire worms, and termites. It is also widely used on turf farms and golf courses.

Mode of Action

Isofenphos is a systemic insecticide with both contact and stomach action.

Toxicity—Acute Effects

Isofenphos is extremely toxic when ingested, but is not an eye or skin irritant.

Toxicity—
Subacute and Chronic Effects

There is a dearth of adequate studies on subacute and chronic effects. Two long-term feeding studies—one involving dogs, the other rats—found evidence of decreased body weight. A recent study found evidence of delayed neurotoxicity in animal tests.

Mutagenicity

Only two studies on mutagenicity—one a microbial test, the other on mice—are on file with the EPA, and neither found evidence of mutagenicity. More tests involving different systems need to be done.

Ecological Effects

Isofenphos is quite toxic to fish and earthworms and toxic to birds. Bird kills have been noted, including a Chemlawn application in New York in 1985 linked with the deaths of more than 100 red-winged blackbirds. Due to such effects, Massachusetts has restricted use of isofenphos to commercial turf only. Isofenphos is not registered for general agricultural use in California. The finding of delayed neurotoxicity in experiments at the University of California at Davis helped persuade the state to halt its use as part of a Japanese beetle "eradication" program. Because of the problem of bird kills, the toxicity to earthworms, and potential health effects, consumers probably should avoid use of isofenphos. Information on other ecological effects is not available.

MALATHION

Malathion, known chemically as *S*-1,2-bis(ethoxycarbonyl)ethyl *O,O*-dimethyl phosphorodithioate, is used to control a wide range of sucking insect pests and mites outdoors. It is commonly used in home sprays for ornamentals and fruit trees and some pet products, and is one of the 10 most widely used household and garden pesticides.

Mode of Action

Malathion is an insecticide and acaricide that is both a contact poison and a stomach poison, and interferes with respiration.

Toxicity—Acute Effects

Malathion has a low acute oral toxicity, and is a minor eye irritant but not a skin irritant.

Toxicity—
Subacute and Chronic Effects

Good studies on subacute and chronic effects are sparse. At least three oncogenicity studies in rats and one in mice have been done under the auspices of the National Cancer Institute. Two studies found some evidence of tumors in male rats, and the mouse study found liver tumors at the highest dose. Although these studies are suggestive, more information is needed. One valid rabbit study turned up no evidence of birth defects, as did a poorly designed rat study. One study found reproductive toxicity in mice, but the study's design is poor.

Malathion may occur at excess levels in the diet. The EPA's Theoretical Maximum Residue Contribution, an estimate of worst-case human dietary exposure, is five times larger than the PADI.

Mutagenicity

Tests for mutagenicity have been negative, but methodological problems suggest that more data is needed.

Ecological Effects

Malathion is practically nontoxic to mammals, moderately toxic to birds, but is toxic to bees and natural enemies, and highly toxic to fish and aquatic invertebrates, with a number of fish kills having been reported. It is quite water-soluble and highly mobile in loam soils, and so may pose a threat to groundwater. Because of its effects on natural enemies, using it may exacerbate lawn and garden pest problems.

NALED

Naled, known chemically as 1,2-dibromo-2,2-dichloroethyl dimethyl phosphate, is used to control spider mites, mosquitoes, and a number of other insect pests both outdoors and indoors, and is a common ingredient of flea-and-tick collars for pets. The primary breakdown product of naled, both in the environment and in the body, is dichlorvos, which is very toxic in its own right (see *dichlorvos*, above).

Mode of Action

Naled is a fast-acting insecticide and acaricide that is both a contact poison and a stomach poison, and interferes somewhat with respiration.

Toxicity—Acute Effects

Naled has a moderate acute oral toxicity and is a severe eye irritant and a strong skin irritant.

Toxicity— Subacute and Chronic Effects

Good studies on all subacute and chronic effects except mutagenicity are available. The evidence on oncogenicity is contradictory. Naled does not appear to be teratogenic; no evidence of birth defects was found in a rat study and a rabbit study. A chronic toxicity study in dogs found evidence of mild testicular degeneration, some mineralization of the spinal cord, anemia, and mild iron deposits in the spleen.

All of the preceding studies involved exposure by ingestion, but flea collars work by giving off naled vapors. Pets and humans who live with them are more likely to be exposed to naled vapors by inhalation and skin contact than by ingestion. One inhalation study on rats found both corneal and nasal lesions even at the lowest dose tested. Unfortunately, there have been few other noningestion studies.

Mutagenicity

Some microbial studies indicate that naled is a mutagen, but more data is needed.

Ecological Effects

Naled is moderately to highly toxic to fish, very highly toxic to aquatic invertebrates, bees, and natural enemies, and slightly toxic to birds. Because of its effects on natural enemies, using it may exacerbate lawn and garden pest problems.

PHOSMET

Phosmet, known chemically as O,O-dimethyl S-phthalimidomethyl phosphorodithioate, is used against a wide range of insect and mite pests in the garden and yard, although we found it only for use against pests of fruit, shade, and ornamental trees.

Mode of Action

Phosmet is an insecticide and acaricide that kills primarily via contact action.

Toxicity—Acute Effects

Phosmet has a moderate acute oral toxicity, and is a mild eye irritant but not a skin irritant.

Toxicity—
Subacute and Chronic Effects

Good studies on subacute and chronic effects are sparse. A mouse study found evidence of liver tumors in male and female mice. A rat study found no evidence of tumors, but the study was too flawed to interpret. Thus, no definitive conclusions can be drawn about oncogenicity. A two-year study in dogs found no evidence of subacute effects. A three-generation reproduction study in rats found no adverse reproductive effects at the highest dose tested. A one-generation reproductive/teratology study in rabbits found no adverse reproductive or teratological effects at the highest dose tested. Four other studies—one using monkeys, two with rats, and one with rabbits—found no teratological effects at the highest dose tested. Phosmet has been shown to cause delayed neurotoxicity in hens.

Mutagenicity

Phosmet has tested positive for mutagenicity in a couple of microbial systems and in two different mammalian cell culture tests.

Ecological Effects

No data could be found.

TETRACHLORVINPHOS

Tetrachlorvinphos, known chemically as *(Z)*-2-chloro-1-(2,4,5-trichlorophenyl)-vinyl dimethyl phosphate, is used outdoors to control some caterpillar species and flies on garden crops and indoors to control flies, fleas, lice, and mites.

Mode of Action

Tetrachlorvinphos is an insecticide and acaricide that acts both as a contact poison and a stomach poison.

Toxicity—Acute Effects

Tetrachlorvinphos is practically nontoxic when ingested, but is a severe eye irritant and a skin irritant.

Toxicity—
Subacute and Chronic Effects

Good studies on subacute and chronic effects are sparse. Two National Cancer Institute–sponsored cancer studies, involving rats and mice, found tumors of various kinds. A different mouse study found tumors, but only at the two highest dosage levels. Neither mouse study is methodologically solid enough to resolve the issue convincingly. A three-generation reproduction study involving rats found evidence of increased liver weights in male pups. A long-term feeding experiment with dogs found negative effects on the kidneys of the females. An acceptable rabbit study found no evidence of birth defects.

Mutagenicity

Six tests—involving microbes, mammalian cell cultures, and mice—found no evidence of mutagenicity, but some of these tests were flawed.

Ecological Effects

Tetrachlorvinphos is toxic to bees and natural enemies, highly toxic to fish, and slightly toxic to practically nontoxic to birds and mammals. Information on other ecological effects is not available.

CARBAMATES

The carbamates, the most frequently used pesticides after the organophosphates, were developed in Switzerland by the Ciba-Geigy Corporation and began to appear commercially in the 1970s. Chemically, the carbamates are derivatives of carbamic acid, a nitrogen-containing acid, and include insecticides, herbicides, and fungicides.

Carbamates have biological and chemical properties similar to the organophosphates: they hydrolyze readily and thus degrade fairly rapidly; they have a wide range of toxic effects both in insects and in mammals; and they are quite toxic to beneficial insects. Carbamates are especially toxic to ants, wasps, bees, and other insects from the order Hymenoptera, which contains many natural enemies of insect pests, as well as pollinators.

Like organophosphates, carbamates block acetylcholinesterase. However, in this case the blocking is reversible, and acetylcholinesterase levels return to normal within a few hours after exposure. Because of this, carbamates lack the cumulative effect of organophosphates and so have a better poisoning safety record.

BENDIOCARB

Bendiocarb, known chemically as 2,3-isopropylidenedioxyphenyl methylcarbamate, is predominantly used to control pests such as ants, cockroaches, fleas, and crickets.

Mode of Action
Bendiocarb is a systemic insecticide that works as a stomach poison and a contact poison.

Toxicity—Acute Effects
Bendiocarb is extremely toxic when ingested, but is only a slight eye and skin irritant.

Toxicity— Subacute and Chronic Effects
There are still a number of data gaps on subacute and chronic effects. Two flawed oncogenicity studies—one in mice and one in rats—found no evidence of tumor formation at the dose levels tested. The rat oncogenicity study found chronic damage to the eyes (cloudiness) at the two highest dose levels. Three teratology studies in rats found no evidence of birth defects, but two of the studies suggested reproductive toxicity. A teratology study in rabbits also found no evidence of birth defects, but it did find a possible fetotoxic effect: an anomaly in the appearance of the eyes.

Mutagenicity
No acceptable data is available.

Ecological Effects
Bendiocarb is very highly toxic to fish, aquatic invertebrates, and mallard ducks, and highly toxic to bobwhite quail and honeybees. Bendiocarb use at permitted rates on turf poses significant risks to birds and mammals through ingestion, and to fish and aquatic invertebrates via runoff. The hazards to birds include reproductive toxicity and acute toxicity. Bendiocarb is a potential leacher, which may contaminate groundwater. Given these potential hazards, consumers should refrain from using bendiocarb on their lawns.

CARBARYL

Carbaryl, known chemically as 1-naphthyl methylcarbamate, is used to control insect pests on lawns, ornamental plants, and pets, and indoor pests, such as cockroaches, fleas, ticks, moths, various caterpillar species, and so forth. It is one of the ten most widely used household pesticides.

Mode of Action

Carbaryl acts as both a plant growth regulator and an insecticide with slight systemic properties. Its insecticidal action is both as a contact poison and as a stomach poison.

Toxicity—Acute Effects

Carbaryl is moderately toxic when ingested, and is a slight eye irritant but not a skin irritant.

Toxicity—
Subacute and Chronic Effects

There are numerous studies on subacute and chronic effects, though data gaps do exist. The EPA lists 10 oncogenicity studies, all of which were flawed but all of which failed to detect any evidence of oncogenic effects. Taken together, these studies suggest that carbaryl does not cause tumors. Two studies in dogs found birth defects at low dose levels; although the applicability of this data to humans is open to question, the EPA warns that carbaryl-containing products should not be used on pregnant dogs. In a study of manufacturing workers, different pathologists offered contradictory opinions about whether carbaryl caused adverse effects on sperm. Both a two-year rat study and a short-term human study indicated that carbaryl may cause kidney damage.

Mutagenicity

There is some evidence that carbaryl is a weak mutagen.

Ecological Effects

Carbaryl is moderately toxic to fish, highly toxic to aquatic invertebrates, bees, and natural enemies, and relatively nontoxic to birds. However, in California, more bee kills are associated with the use of carbaryl than with any other insecticide. Such bee-kill incidents have been reported from a number of states. Preliminary data also suggests that carbaryl and its residue(s) may bioaccumulate in catfish, crayfish, snails, duckweed, and algae. Furthermore, fish kills have been connected to carbaryl. Such data suggests that consumers should not use carbaryl near bodies of water.

METHOMYL

Methomyl, known chemically as S-methyl N-(methylcarbamoyloxy)thioacetimidate, is used against a wide variety of insects and spider mites in outdoor settings.

Mode of Action

Methomyl is a systemic insecticide and acaricide that acts as a contact poison and a stomach poison.

Toxicity—Acute Effects

Methomyl is extremely toxic when ingested, and is a severe eye irritant but not a skin irritant.

Toxicity—
Subacute and Chronic Effects

Acceptable rat and mouse oncogenicity studies found no evidence of oncogenic effects, but both studies found evidence of toxic effects on blood chemistry, and the rat study found adrenal lesions in females at the highest dose tested. Two rat reproduction studies have been done; one flawed study found no adverse effects at the highest dose tested, while the other found reduced litter size, pup survival, and pup growth as well as negative effects on blood chemistry. A chronic feeding study in rats found a dose-related decrease in hemoglobin and other adverse effects on the blood-forming system, as well as damage to the kidneys. A dog study found toxic effects on the kidneys, liver, spleen, and bone marrow, as well as enlargement of the prostate gland. Teratology studies done in rats and rabbits found no evidence of birth defects.

Mutagenicity

A battery of valid tests has failed to show that methomyl causes mutations.

Ecological Effects

Methomyl is moderately toxic to fish, highly toxic to birds, and toxic to bees (although it is not dangerous when the spray has dried) and to natural enemies.

PROPOXUR

Propoxur is known chemically as 2-isopropoxyphenyl methylcarbamate or O-isopropoxyphenyl methylcarbamate. It does not break down rapidly and thus has a long residual activity. Because of this persistence, it is used widely around the house, particularly for cockroach control.

Mode of Action

Propoxur works as both a contact poison and a stomach poison. It knocks insects down rapidly.

Toxicity—Acute Effects

Propoxur is only slightly toxic when ingested, and is not an eye or skin irritant.

Toxicity—
Subacute and Chronic Effects

A chronic feeding study in rats found bladder tumors in males and bladder and uterine tumors in females. No evidence of tumors was found in a mouse study and a hamster study. A rabbit and rat teratology study both found no evidence of birth defects, although both studies are methodologically flawed. A three-generation rat reproduction study found decreased pup numbers at the higher dose levels. Chronic dog feeding studies have found increased liver weights, hemolytic anemia, and adverse effects on blood parameters at higher doses.

Mutagenicity

A battery of tests has found no evidence that propoxur causes mutations or damages DNA.

Ecological Effects

Propoxur is moderately to highly toxic to fish, highly toxic to slightly toxic to birds, and highly toxic to bees and to natural enemies. The variability in toxicity to birds is a result of the differing susceptibilities of bird species to propoxur.

PYRETHROIDS

The botanical insecticide pyrethrum (a trade name for pyrethrins), which comes from the flowers of the painted daisy, *Chrysanthemum cinerariefolium*, a native of east Africa, has been used and traded for over 200 years. Pyrethrum is a mixture of substances, known chemically as cyclopropane carboxylic acids; each compound contains two of these, each combined with one of three naturally occurring alcohols.

Mode of Action

Pyrethrum is an ideal insecticide for household use, since it is practically nontoxic (in an acute sense) to humans, yet has spectacularly swift action against insects. It usually paralyzes insects within seconds after application as it is readily absorbed by the insect, which gives a homeowner harassed by insects great satisfaction. However, the paralysis often wears off as the chemicals are detoxified by enzymes called *mixed-function oxidases.* Mixing pyrethrum with a synergist such as piperonyl butoxide or MGK 264, which blocks the action of the mixed-function oxidases, increases the toxicity of pyrethrum by 1,000 to 2,000 percent. When mixed with a synergist, pyrethrum is effective at very low dosages; commercial preparations contain only 0.1 to 0.2 percent pyrethrum.

In recent years, chemists have produced a number of synthetic pyrethroids, some of which have properties similar to pyrethrum and are used in similar ways, while others have properties that make them more widely useful for agricultural applications. Examples of the former include allethrin, resmethrin, and tetra-

methrin. Like pyrethrum, these synthetic pyrethroids are fast-acting. They are generally more toxic to insects than is pyrethrum, because they are not detoxified as quickly. This makes them less susceptible to synergism, and addition of a synergist yields smaller increases in killing power than it does in the case of pyrethrum. Both pyrethrum and these synthetic pyrethroids degrade quickly in the presence of sunlight, which tends to restrict their use to indoor situations.

Other synthetic pyrethroids are photostable (i.e., are not readily degraded by sunlight). Examples include permethrin and phenothrin, as well as fenvalerate, a closely related synthetic that has similar biological properties. Like pyrethrum, these photostable pyrethroids are effective at much lower rates of application than most other insecticides.

A potential drawback of the pyrethroids is that they are highly toxic to fish and to most natural enemies.

ALLETHRIN

Allethrin is known chemically as 2-methyl-r-oxo-3-(2-propenyl)-2-cyclopent-1-yl 2,2-dimethyl-3-(2-methyl-1-propenyl)cyclopropanecarboxylate or (RS)-3-allyl-2-methyl-4-oxocyclopent-2-enyl (1RS) - *cis,trans* - chrysanthemate. Like pyrethrum, allethrin breaks down rapidly in sunlight and is used only to control indoor household pests.

Mode of Action

Allethrin is both a contact poison and a stomach poison, and interferes with respiration.

Toxicity—Acute Effects

Allethrin is practically nontoxic when ingested, and is a mild eye irritant and a slight skin irritant.

Toxicity— Subacute and Chronic Effects

Studies on subacute and chronic effects are quite sparse. A study in rats found no evidence of tumors but was considered flawed, and a test in another species is needed. No reproduction studies are available. One flawed rat teratology test found an increase in skeletal abnormalities.

Mutagenicity

A number of tests in both microbial and mammalian cell culture systems have found no evidence of mutagenicity, although these studies are not definitive.

Ecological Effects

Allethrin is toxic to bees, slightly toxic to practically nontoxic to birds, highly toxic to fish, and toxic to natural enemies.

FENVALERATE

Although fenvalerate, known chemically as (RS)-α-cyano-3-phenoxybenzyl (RS)-2-(4-chlorophenyl)-3-methylbutyrate, has very similar biological properties to the photostable synthetic pyrethroids, fenvalerate lacks the cyclopropane group, the defining characteristic of pyrethroids. It is used in pet flea-and-tick sprays and against termites.

Mode of Action

Fenvalerate is an insecticide and acaricide that works as both a contact poison and a stomach poison.

Toxicity—Acute Effects

Fenvalerate is only slightly toxic when ingested, and is a mild eye irritant and a slight skin irritant. However, a number of reports found that workers exposed to it can develop itchy, burning skin (usually without a rash), respiratory problems similar to allergies, and eye irritation. Some workers developed a peculiar tingling and burning sensation of the hands and face that usually subsided within a day or so. Such reports have caused Sweden to discontinue its use in forestry. Other work has found that application of vitamin E to exposed areas could block these effects.

Toxicity— Subacute and Chronic Effects

A number of oncogenicity studies of varying quality have been done in mice, rats, and hamsters, with mixed results. One study in hamsters and three in mice were negative for oncogenic effects, but one of the rat studies found good evidence of spindle-cell tumors and equivocal evidence of testicular tumors, which were also found in a second rat study. Both a rat study and a mouse study found evidence of chronic negative effects (infiltration and alteration of phagocytic cells) to the lymph system, liver, and spleen. A rat study also found some evidence of toxicity to the blood or blood-forming organs, and an apparent transitory neuropathological response at the highest dose tested. Two teratology studies, involving mice and rabbits, found no evidence of birth defects, although the mouse study found a developmental effect at a dosage that also caused maternal toxicity. In all, though, the EPA considers the data base to be essentially complete and has no chronic toxicity concerns.

Mutagenicity

A number of tests in microbial systems, mammalian cell cultures, and mice all failed to find evidence of mutagenicity, but the studies are all flawed.

Ecological Effects

Fenvalerate is extremely toxic to bees and fish, and slightly toxic to practically nontoxic to birds. Information on other ecological effects is not available.

PERMETHRIN

Permethrin, known chemically as 3-phenoxybenzyl (1RS) - cis,trans - 3 - (2, 2 - dichlorovinyl) - 2,2 - dimethylcyclopropanecarboxylate, is one of the newer synthetic pyrethroids, and is photostable. It is used primarily in the house for control of spiders, cockroaches, fleas, ticks, and mites.

Mode of Action

Permethrin is both a contact poison and a stomach poison and has a slight repellent effect.

Toxicity—Acute Effects

Permethrin is slightly toxic to practically nontoxic when ingested, and is a mild eye irritant and a slight skin irritant.

Toxicity— Subacute and Chronic Effects

Eight oncogenicity studies—three in rats, four in mice, and one in dogs—are on file with the EPA. None of the rat studies found evidence of oncogenicity, but all found negative liver effects, and one found disturbance in thyroid growth at the highest dose tested. Of the four mouse studies, two found lung tumors and one found liver tumors; one study

also found adverse effects on liver function. Because of the design of the mouse studies, however, these finding are not considered strong evidence of oncogenicity. The EPA's Scientific Advisory Panel concluded that the rat and mouse studies together suggest a very weak oncogenic potential. The dog study found no evidence of oncogenic effects, but found evidence of an adverse effect on the adrenal glands at medium and high dosages. One three-generation rat reproduction study found adverse effects on the eyes and liver at birth and on the kidneys at weaning. Six teratology studies—three in rats, two in rabbits, and one in mice—found no evidence of birth defects or fetal toxicity at dose levels that did not also cause maternal toxicity. Again, the studies are methodologically weak and not conclusive.

Mutagenicity

A range of mutagenicity tests involving microbial systems, mammalian cells, and mice all found no evidence of mutagenicity; taken together, they indicate that permethrin is not mutagenic.

Ecological Effects

Permethrin is very toxic to bees, natural enemies, and fish, and practically nontoxic to birds. It is also relatively resistant to degradation. Information on other ecological effects is not available.

PHENOTHRIN

Phenothrin, known chemically as 3-phenoxybenzyl 2, 2 - dimethyl - 3 - (2-methylprop-1-enyl)cyclopropanecarboxylate, or 3-phenoxybenzyl (±)-cis,trans-chrysanthemate is one of the newer synthetic pyrethroids, and is photostable.

It is used for a wide range of pests, including fleas, ticks, cockroaches, ants, hornets, flies, and aphids.

Mode of Action
Phenothrin is both a contact poison and a stomach poison, and gives rapid knockdown of pests.

Toxicity—Acute Effects
Phenothrin is practically nontoxic when ingested, and is a moderate to severe eye irritant and a slight skin irritant.

Toxicity—
Subacute and Chronic Effects
There are two oncogenicity studies—an invalid rat study and a mouse study—that found no evidence of oncogenic effects, but did find lung damage and liver weight changes at high doses. Three teratology studies—using rats, rabbits, and mice—found no evidence of birth defects, although there were fewer viable fetuses at the highest dose tested in the rabbit study, and increased skeletal ossification in the mouse study.

Mutagenicity
Phenothrin does not appear to be a mutagen, according to a battery of mutagenicity studies using different systems.

Ecological Effects
Phenothrin is toxic to bees and natural enemies, practically nontoxic to birds, and extremely toxic to fish.

PYRETHRINS

Pyrethrins are better known through their trade name, pyrethrum. The term is applied to the mixture of six substances extracted from the painted daisy flower. The six chemicals in the mixture are called pyrethrin I and II, jasmolin I and II, and cinerin I and II. These compounds break down rapidly in the presence of light, so they are generally used to control household insects, especially flies, mosquitoes, roaches, ants, and wasps.

Mode of Action
Pyrethrins kill insects by acting as a stomach poison. This mixture gives a rapid knockdown, paralyzing the insects immediately, while death occurs much later. Indeed, the knockdown effect is much stronger than the killing effect, and insects may initially be paralyzed but later recover unless a synergist is used to increase the killing action of this mixture.

Toxicity—Acute Effects
The pyrethrins are practically nontoxic when ingested. Data on skin and eye irritation is not available for pure pyrethrins. Formulated products with concentrations of pyrethrins ranging from 0.052 to 9.0 percent are slight to moderate skin irritants, while eye irritation of the products ranged from minimal to severe, with most in the moderate to severe range.

Toxicity—
Subacute and Chronic Effects
Four teratology tests—two in rats and two in rabbits—used a pyrethrin extract, and none found evidence of birth defects, although one rabbit study found reproductive toxicity at doses that also caused tremors and weight loss in the mothers. No other data is available on chronic and subchronic effects.

Mutagenicity

Information is not available.

Ecological Effects

There have been no studies on ecological effects, although pyrethrins are known to be highly toxic to fish and to aquatic invertebrates, and toxic to bees and natural enemies.

RESMETHRIN

Resmethrin is known chemically as 5-benzyl-3-furylmethyl (1*RS*)-*cis,trans*-chrysanthemate. Like pyrethrum, resmethrin is sensitive to light and is usually used only indoors to control household pests, especially flies, roaches, mealybugs, scale, moths, and wasps.

Mode of Action

Resmethrin is a contact poison and acts as a repellent for some insects.

Toxicity—Acute Effects

Resmethrin is practically nontoxic when ingested, and is a mild eye irritant and a slight skin irritant.

Toxicity—
Subacute and Chronic Effects

Of two available oncogenicity studies, a study on mice found no evidence of oncogenicity, but did find chronic effects on the adrenal glands, liver, kidneys, brain, and other tissues. A study on rats found liver lesions, liver weight changes, increases in thyroid weight, and thyroid cysts at high doses. Controversy exists over whether the thyroid effects are evidence of oncogenicity, and more studies are needed. A chronic feeding study in dogs found increased liver weights in fe-males and tremors in males. Two reproduction studies in rats found reduced litter size, litter weights, and other evidence of reproductive toxicity. Three teratology studies in rats found no evidence of birth defects, but a somewhat flawed study in rabbits found both birth defects and adverse reproductive effects.

Mutagenicity

Information is totally lacking.

Ecological Effects

Resmethrin is toxic to bees and natural enemies, practically nontoxic to birds, and extremely toxic to fish and aquatic invertebrates.

TETRAMETHRIN

Tetramethrin, known chemically as cyclohex-1-ene-1,2-dicarboximidomethyl (1*RS*)-*cis,trans*-2,2-dimethyl-3-(2-methylprop-1-enyl)cyclopropanecarboxylate, or 3,4,5,6-tetrahydrophthalimidomethyl (±)-*cis,trans*-chrysanthemate is sensitive to light and is usually used only indoors to control household pests.

Mode of Action

Tetramethrin is a contact poison that gives good knockdown.

Toxicity—Acute Effects

Tetramethrin is practically nontoxic when ingested, and is a mild eye irritant and a slight skin irritant.

Toxicity—
Subacute and Chronic Effects

Two oncogenicity studies involving different strains of rats found testicular tumors, and one of the studies found some evidence of kidney and thyroid tumors as well. Teratology studies involving rats and rabbits were both negative, but the rabbit study was flawed. A subchronic toxicity study in dogs found effects on various blood parameters, on the female estrus cycle, and on liver weights in males, at high dose levels. One reproduction study in rats found both reproductive toxicity and increased liver weights in males at all doses; another study found no reproductive effects, but found increased liver weights in females. Although most of the studies are flawed,

taken together they indicate that tetramethrin appears to cause liver problems.

Mutagenicity

There was no evidence of mutagenicity in three microbial studies and one mammalian cell culture study, but all were flawed, and data on this point is inconclusive.

Ecological Effects

Tetramethrin is toxic to bees, slightly toxic to birds, and highly toxic to fish and aquatic invertebrates. Other ecological data is not available.

 # BIOLOGICALS (OR BIORATIONALS)

As the ecological problems and health hazards associated with broad-spectrum synthetic organic pesticides became more widely known, interest increased in developing less harmful narrow-spectrum insecticides. What is needed are substances that affect only insects and—even better—only the targeted insect pest and not beneficial insects. Such pest-specific insecticides can be either synthetic or natural. The natural substances are either insect disease organisms, such as bacteria and viruses that kill the pest in natural ecosystems, or chemicals (called *pheromones*) that the insects use for a variety of purposes, including to communicate or locate mates. The synthetic substances are chemicals that disrupt biological processes specific to insects, or synthetic versions of important insect hormones.

BACILLUS THURINGIENSIS

Bacillus thuringiensis, also called *Bt*, is a widespread soil bacterium that is used primarily to control moth and butterfly larvae (caterpillars). Some strains of *Bt* also kill mosquitoes and some beetles.

Mode of Action

The bacteria produce a number of poisons, called *endotoxins*. One of these endotoxins is located inside a crystalline protein structure within the bacteria. A caterpillar's gut is acid enough to dissolve the protein crystal and liberate the endotoxin; the gut of most other insects is not, which explains why *Bt* is specific against caterpillars. *Bt* can therefore be sprayed on leaves where caterpillars will eat it, without fear that it will kill beneficial insects that feed on caterpillars.

Toxicity—Acute Effects

Despite *Bt*'s high toxicity to caterpillars, it is virtually nontoxic to mammals. It is a mild eye and skin irritant.

Toxicity—
Subacute and Chronic Effects

There is no evidence that *Bt* has subacute or chronic effects. *Bt* does not cause cancer, birth defects, reproductive problems, or damage to internal organs or tissues.

Mutagenicity

Bt does not cause mutations.

Ecological Effects

Most *Bt* strains do not harm nontarget organisms, except for other caterpillar species. A few strains will affect earthworms. While it may reduce numbers of esthetically pleasing butterflies or moths whose caterpillars feed on garden plants, it is very safe to use in the garden. If *Bt* were sprayed widely over large geographical areas, however, for instance to control gypsy moths or other forest moth pests, it would indiscriminately kill caterpillars, which could seriously affect birds that feed caterpillars to their young. For this reason, suggestions that *Bt* be used for large-scale pest control are not well founded.

Bt is very sensitive to ultraviolet light and quickly loses its efficacy, so it needs to be applied fairly frequently. Commercial formulas of *Bt* often include clays or other substances that help shield it from ultraviolet rays. Consumers can also lengthen its efficacy by applying it near dusk and spraying it on the undersides of leaves.

HELIOTHIS *NUCLEAR POLYHEDROSIS VIRUS* (HELIOTHIS *NPV*)

The NPVs attack primarily insects, and are known for their extreme specificity. The *Heliothis* NPV attack only moths in the genus *Heliothis*, and only the caterpillar, not the adult moth. This genus contains the fruitworms and bollworms, which are major agricultural and garden pests.

Mode of Action

Upon ingestion, the NPVs attack cells of the host's digestive system, where they multiply rapidly until they eventually kill the host.

Toxicity—Acute Effects

Although it is highly toxic to *Heliothis* caterpillars, the *Heliothis* NPV is not toxic to vertebrates.

Toxicity—
Subacute and Chronic Effects

There is no danger of any subacute or chronic effects.

Mutagenicity

Heliothis NPV does not cause mutations.

Ecological Effects

Because of their extreme specificity, the NPVs are free of potential ecological problems, except for reducing the caterpillar population that serves as food for natural enemies.

METHOPRENE

Methoprene is known chemically as isopropyl(*E,E*)-(*RS*)-11-methoxy-3,7,11-

trimethyldodeca-2,4-dienoate. Around the home and garden, methoprene is used against fleas, mosquitoes, pharaoh ants, cucumber beetles, leaf miners and hoppers, and pests of household plants.

Mode of Action

Methoprene is an insect growth regulator that mimics the action of a juvenile hormone. Juvenile hormones keep certain insects in the larval stage and prevent them from undergoing metamorphosis. Thus, methoprene keeps insects from reaching sexual maturity and effectively prevents reproduction. Given this mode of action, methoprene is most effective against insects in which the adult stage does the most damage. It can also be used to break the reproductive cycle and suppress population growth of insects that do their damage as larvae. It is usually applied at low dosages to either the larval or egg stages. It should not be used to control caterpillar pests in the garden, since it will extend the lifetime of the caterpillars and possibly increase the amount of damage they cause.

Toxicity—Acute Effects

Methoprene is practically nontoxic to mammals when ingested, and is not an eye or skin irritant.

Toxicity—
Subacute and Chronic Effects

Methoprene does not appear to cause any subacute or chronic toxic effects. There is sufficient evidence that methoprene does not cause cancer, birth defects, or negative reproductive effects. The EPA considers the data base essentially complete.

Mutagenicity

Available evidence suggests that methoprene does not cause mutations, but more definitive studies are needed.

Ecological Effects

Methoprene is practically nontoxic to bees and birds, but is moderately toxic to fish and amphibians, very highly toxic to some aquatic invertebrates, and can have the same effect on some beneficial insects that it has on insect pests. To conserve beneficial insects, it's best to avoid using methoprene in the garden. Most consumer uses of methoprene are for indoor pests, because it is not photostable. Indoors it is active against fleas for six months, but outdoors it degrades in a week. Thus, most environmental impacts are unlikely.

HYDROPRENE

Hydroprene, known chemically as ethyl (E,E)-3,7,11-trimethyldodeca-2,4-dienoate, is used to control various beetles, some plant-sucking insects (homopterans), and cockroaches.

Mode of Action

Hydroprene is an insect juvenile hormone mimic.

Toxicity—Acute Effects

Hydroprene is virtually nontoxic to mammals. It is a mild eye irritant but not a skin irritant.

Toxicity—
Subacute and Chronic Effects

Very little is known about subacute and chronic effects.

Mutagenicity

A few tests found no evidence of mutagenicity.

Ecological Effects

Hydroprene has a low toxicity to fish. Information on other ecological effects is not available.

FATTY ACIDS

Fatty acids, which consist of a mixture of the potassium salts of fatty acids in the C-8 to C-18 ranges (i.e., those that contain 8 to 18 carbon atoms), are used for insecticidal, herbicidal, and fungicidal purposes. The products are used against a number of houseplant and garden pests, including whiteflies, scale, mealybugs, aphids, thrips, and flea beetles.

Mode of Action

Fatty acids disrupt basic metabolic processes connected with the cell and organelle membranes, as well as respiration.

Toxicity—Acute Effects

Fatty acids are all practically nontoxic when ingested, and are a mild to moderate eye irritant, but not a skin irritant.

Toxicity— Subacute and Chronic Effects

All the fatty acids used in these products are on the FDA's GRAS (Generally Regarded as Safe) list, therefore much less data is needed to register them. For this reason, no studies on subacute and chronic effects are available.

Mutagenicity

No mutagenicity studies are required, owing to the GRAS status. Fatty acids do not cause mutations.

Ecological Effects

Fatty acids are practically nontoxic to birds, fish, and aquatic invertebrates, and have a very low toxicity to honeybees and most beneficial insects. Fatty acids are also nontoxic to earthworms. There should be no problem with residues, because fatty acids are naturally present in the soil and are biodegraded rapidly by soil microorganisms to yield water and carbon dioxide. Data on other ecological effects is not known. Consumer use of them should present no environmental problems, however.

 # REPELLENTS

This category contains chemicals that repel insects. Insects actively try to avoid such chemicals. The category includes chemicals that act as repellents (DEET, MGK 11, and MGK 874), serve as synergists by dramatically increasing the killing power of other chemicals (piperonyl butoxide and MGK 264), act as fumigants (paradichlorobenzene), or do not fit into any of the preceding categories (boric acid, petroleum distillates, and hydramethylnon).

BORIC ACID

As a pesticide, boric acid has uses as an insecticide, insect repellent, herbicide, and fungicide. Its uses against insects include control of cockroaches, ants, grain weevils, and several beetles, and as a repellent in insulation.

Mode of Action

Boric acid has a unique mode of action demonstrated by its use against cockroaches. Cockroaches are by nature fastidious insects that frequently clean their antennae and legs. Boric acid powder is spread where roaches hide; when the roaches walk through the boric acid, it sticks to their feet. When they clean themselves, the roaches ingest the crystals, which pierce the intestinal tract, eventually killing them.

Toxicity—Acute Effects

Boric acid has a low acute oral toxicity and is only mildly irritating to the skin and eyes.

Toxicity—
Subacute and Chronic Effects

Data is sparse. Studies in rats and dogs found decreased sperm count and/or sterility, while the rat studies also found smaller litter sizes and greater mortality.

Mutagenicity

There is some suggestive evidence that boric acid causes mutation, although more studies are needed.

Ecological Effects

There is a paucity of data on environmental effects. However, none is expected, since consumers do not usually use boric acid outdoors.

DEET

Known chemically as N,N-diethyl-m-toluamide, DEET is a widely used insect repellent. Products containing DEET are used on human skin, clothing, and pets.

Mode of Action

DEET penetrates the skin rapidly and is absorbed directly into the bloodstream. Mosquitoes coming close to or landing on the skin detect the chemical and try to avoid it.

Toxicity—Acute Effects

DEET has a low acute oral toxicity, but is a severe eye irritant and can be a severe skin irritant for people sensitive to it.

Toxicity—
Subacute and Chronic Effects

Although DEET has been used as an insect repellent for more than 30 years, there is still not enough information on its subacute and chronic effects. Products containing high concentrations (75 to 100 percent) have caused severe skin irritation and open sores in some people. At least a dozen cases of severe acute neurotoxicity, usually involving brain seizures, have been reported. All the victims were children; some involved heavy normal use of insect repellent, and some involved accidental ingestion. A few of the children died. Since the brain seizures all occurred in girls, such severe reactions may involve a genetic predisposition (deficiency of an enzyme that degrades DEET). Data on other subacute and chronic effects is generally inadequate or not available. Several studies in rats have found that DEET causes a variety of reproductive effects, but the studies are flawed. There is some slight evidence that

DEET may cause cancer, reproductive effects, and tissue damage. There are no adequate studies on birth defects.

Mutagenicity
There is some evidence that DEET is not a mutagen.

Ecological Effects
There is little information.

PETROLEUM DISTILLATES

Petroleum distillates are a mixture of chemicals that are by-products of the oil industry. Except for some acute toxicity data, there is no readily available information on subacute, chronic, or environmental effects.

PIPERONYL BUTOXIDE

Piperonyl butoxide, known chemically as 5 - [2 - (2-butoxyethoxy)ethoxymethyl] - 6 - propyl - 1,3 - benzodioxole, is found in virtually all products containing a pyrethroid insecticide; an EPA study estimated that it was used in some 4,200 products.

Mode of Action
Piperonyl butoxide is a synergist used to increase the killing power of pyrethrins and the pyrethroid insecticides by blocking the enzymes that detoxify these insecticides.

Toxicity—Acute Effects
Piperonyl butoxide has a low acute oral toxicity and causes only minor skin and eye irritation.

Toxicity—
Subacute and Chronic Effects
Little is known. One study found some positive evidence of lymphomas in female rats, but the study was flawed and additional oncogenicity studies are needed. Piperonyl butoxide's potential to cause mutations, birth defects, reproductive effects, or damage to tissues or organs is basically unknown.

Mutagenicity
There is no real data available.

Ecological Effects
It is not toxic to fish or bees. Information on other environmental effects is not available.

MGK 264

MGK 264, known chemically as *N*-octyl bicycloheptene dicarboximide, is used predominantly with the pyrethroids and related insecticides.

Mode of Action
MGK 264 is an insecticide synergist.

Toxicity—Acute Effects
MGK 264 has a low acute oral toxicity and is a minor skin irritant, but is considered a severe eye irritant.

Toxicity—
Subacute and Chronic Effects
There is virtually no data on the potential of MGK 264 to cause tumors, birth defects, reproductive effects, or damage to tissues or organs.

Mutagenicity
No information is available.

Ecological Effects
No information is available.

MGK 11

MGK 11 is known chemically as 2,3:4,5-bis (2-butylene) tetrahydro-2-furaldehyde.

Mode of Action
MGK 11 is an insect repellent.

Toxicity—Acute Effects
MGK 11 is practically nontoxic when ingested, and is a minor skin and eye irritant.

Toxicity—
Subacute and Chronic Effects
There is virtually no information.

Mutagenicity
No information is available.

Ecological Effects
No information is available.

MGK 874

MGK 874, known chemically as 2-hydroxyethyl-n-octylsulfide, is used to repel flies, mosquitoes, moths, and other flying insects, and is used in some outdoor foggers.

Mode of Action
MGK 874 is an insect repellent.

Toxicity—Acute Effects
MGK 874 is practically nontoxic orally, but is a minor skin and eye irritant.

Toxicity—
Subacute and Chronic Effects
There is virtually no valid data.

Mutagenicity
No information is available.

Ecological Effects
No information is available.

HYDRAMETHYLNON

Hydramethylnon, known chemically as 5,5-dimethylperhydropyrimidin-2-one 4-trifluoromethyl-α-(4-trifluoromethylstyryl)-cinnamylidenehydrazone, is used primarily against certain ant species and cockroaches, although its use against termites, wasps, and houseflies is being studied. The ants feed on a bait containing hydramethylnon and, after returning to the nest, feed the bait to their nestmates. After a few days, the poison takes effect and kills all the exposed ants.

Mode of Action
Hydramethylnon is a very selective stomach poison with delayed action.

Toxicity—Acute Effects
Hydramethylnon has a low acute oral toxicity. It is not a skin irritant, but is a mild eye irritant.

Toxicity—
Subacute and Chronic Effects
There is some equivocal evidence for oncogenicity, reproductive toxicity, and damage to internal organs and tissues. Rat studies suggest that hydramethylnon does not cause birth defects, but studies in other species need to be done.

Mutagenicity

Sufficient tests have been done to demonstrate that hydramethylnon is not a mutagen.

Ecological Effects

Hydramethylnon is nontoxic to honeybees, and only slightly toxic to birds. It is quite toxic to both fish and aquatic invertebrates under controlled laboratory conditions, but probably poses little risk to aquatic life since it has a low solubility in water and breaks down rapidly in sunshine.

PARADICHLOROBENZENE

Paradichlorobenzene (or PDB), both the chemical and common name, is the fumigant frequently used in mothballs to kill various insect pests, especially clothes moths, and to deodorize toilets and urinals.

Mode of Action

PDB enters through an insect's respiratory system and has a toxic effect on the nerves.

Toxicity—Acute Effects

PDB is moderately toxic when ingested, and is an eye and skin irritant.

Toxicity—
Subacute and Chronic Effects

A recent oncogenicity study by the National Toxicology Program using mice and rats found evidence of liver tumors in both sexes of mice, adrenal and thyroid tumors in female mice, and a dose-dependent increase in kidney tumors and increased cell leukemia in male rats. The import of these findings is unclear, because the PDB was force-fed to the animals, not delivered by inhalation, the normal route of human exposure. A previous inhalation study, of questionable value, found no tumors, and mutagenicity tests were negative. However, the EPA classifies PDB as a possible (Class C) human carcinogen. And exposure to it is quite high, owing to its volatile nature. A draft cancer risk estimate circulated by the EPA showed cancer risks of 5 in 10,000 for exposure to toilet deodorants, up to 1 in 1,000 for space deodorants, and 6 in 100 for workers handling PDB. There are no valid studies on birth defects or reproductive effects. Several studies involving rats and mice found toxic effects on the liver, kidneys, and possibly the brain. Reports of chronic exposure in workers have been linked with adverse liver and kidney effects, respiratory problems, cataracts, neurological effects, and blood effects.

Mutagenicity

A good number of studies have not found PDB to cause mutations, but the studies are flawed.

Ecological Effects

Information is not available.

APPENDIX B

HERBICIDES

Herbicides are chemicals used to fight plant pests or weeds. Interestingly, in terms of poundage, herbicides are the most frequently used pesticides in the United States, accounting for almost 69 percent of the total poundage of pesticides used. But most of that usage occurs in the areas that the EPA refers to as agriculture and industry, commerce, and government.

In the home and garden, herbicide use falls behind insecticide use. This probably means that homeowners view weeds as less of a problem than insects.

There are 16 different herbicide active ingredients covered in this section, compared to 48 different insecticide active ingredients. But remember that most insecticides consumers purchase are used inside the home, rather than in the garden. In addition, the home and garden figures do not include pesticides applied by commercial pest control operators or lawn-care companies. Finally, many gardeners deal with weeds by physically removing them, either by hand or using some kind of tool.

In general, there are two classes of herbicides—*inorganic* and *organic*—and two broad types of action—*contact* and *systemic.* Inorganic herbicides came into being around the turn of the century. The organic herbicides have now largely replaced them, because they tend to be cheaper and more effective, and cause fewer environmental problems. Indeed, virtually all the herbicides available for use around the home and garden are organic. All the inorganic herbicides work primarily via contact action, as do some of the organic ones. Contact herbicides have their killing or inhibition action upon coming into contact with a plant. Systemic herbicides are absorbed by the plant and moved to the particular site of activity in the plant, where they do their killing and/or inhibition. The systemic herbicides tend to have more complex modes of action and/or greater specificity in terms of the types of plants affected.

Both the physical form and the specific formulation of an herbicide product can influence both its physiological/biochemical activity and its toxicity. This section, however, covers only the active ingredient itself and does not deal with the various formulations.

 # INORGANIC HERBICIDES

Only one herbicide covered on our list is inorganic: diquat dibromide. Diquat dibromide belongs to the chemical family of quaternary ammonium salts, which also includes the herbicide paraquat.

Mode of Action

All inorganic herbicides have the same basic mode of action: during photosynthesis in the plant, they generate toxic peroxides, which are particularly damaging to cell membranes and cytoplasm, through the reoxidation of free radicals created by the ammonium salts themselves.

DIQUAT DIBROMIDE

Diquat dibromide, known chemically as 1,1'-ethylene-2,2'-dipyridylium dibromide, is a nonselective contact herbicide, applied to leaves or water, and absorbed by leaves. It works by disrupting photosynthesis and damaging cell membranes and cytoplasm. It is used to control annual broad-leaf weeds in gardens, ornamental plants, and shrubs, as well as emergent and submerged aquatic weeds.

Toxicity—Acute Effects

Diquat dibromide has a moderate acute oral toxicity and is a mild skin and eye irritant.

Toxicity— Subacute and Chronic Effects

The evidence is suggestive of negative subacute effects and most chronic effects. Because of some severe birth defects in animal studies, the CDFA considers it a possible adverse effect.

Mutagenicity

There are 18 studies involving microbial, mammalian cell cultures, and whole animals and testing for either gene mutations, chromosome mutations, or DNA damage. The results of these studies are somewhat conflicting. The CDFA concluded that diquat dibromide appears to be a mutagen in cellular systems (*in vitro* studies), but not when tested in whole animals (*in vivo* studies).

Environmental Effects

Diquat dibromide is not considered toxic to fish and bees, but is moderately toxic to chickens. It is fixed to the soil-exchange complex and is degraded by microbes and light. Other ecological data is not known or is unavailable.

 # ORGANIC HERBICIDES

Organic herbicides are molecules that are based primarily on carbon, hydrogen, oxygen, and nitrogen atoms. This chemical characteristic refers to the fact that such compounds have been ultimately derived, in some fashion, from living organisms. Organic herbicides vary widely in chemical structure and mode of action. There are numerous different chemical classes of organic herbicides.

The group of herbicides known as *s-triazines* share a similar chemical structure and tend to work, in part, by inhibiting photosynthesis. In our list there are three chemicals from this class: *atrazine*, *prometon*, and *simazine*.

ATRAZINE

Atrazine, known chemically as 2-chloro-4-ethylamino-6-isopropylamino-s-triazine, is a selective systemic herbicide, applied to both the soil and leaves, that is absorbed primarily through roots but also through leaves. It is used for pre- and postemergence control of annual grass and broad-leaf weeds around fruit trees, roses, and turf, among others. Care must be taken with its use as it is phytotoxic to most vegetables.

Mode of Action

Atrazine is transferred to the active growing tips and to leaves, and acts by interfering in enzymatic processes, including the synthesis of an energy-storing molecule (ATP) and by inhibiting photosynthesis.

Toxicity—Acute Effects

Atrazine has a slight acute oral toxicity and is a moderate eye irritant but not a skin irritant.

Toxicity— Subacute and Chronic Effects

Some evidence exists for oncogenic effects. An acceptable rat study found evidence of a statistically significant, dose-related increase in female mammary tumors (both fibrodenomas and adeno-carcinomas) and some evidence of testicular cell tumors, but only at the highest dose. Another rat study, done for the International Agency for Research on Cancer, found a dose-related increase in combined leukemia/lymphoma incidence in both sexes, although the results were statistically significant only for females. Statistically significant increases were also seen in uterine tumors and in benign mammary tumors in males. Data on reproductive effects and birth defects is flawed.

Mutagenicity

A number of mutagenicity tests have been done, but most are highly flawed. None of these studies, a couple of which are of acceptable quality, has found evidence of mutagenicity, but more studies are needed to confirm this result.

Ecological Effects

Atrazine is considered slightly toxic to mammals, moderately toxic to fish, and practically nontoxic to birds and bees. Atrazine persists in the soil, and may have residual activity for one year or more. Sensitive plants, which includes most vegetables, may be harmed or killed if planted within one year of use of atrazine. Atrazine is also a significant contaminant of groundwater in a number of agricultural states. This is not surprising, given that a greater quantity of atrazine was used in the United States in 1987 than any other single pesticide. Indeed, atrazine residues have turned up in such a large number of wells in Iowa that the state has proposed a tax on atrazine sales, the revenues of which will be used to research alternatives to atrazine.

PROMETON

Prometon, known chemically as 2,4-bis(isopropylamino) - 6 - methoxy-*s*-triazine or 2,4-bis(isopropylamino)-6-methoxy-1,3,5-triazine, is a nonselective systemic herbicide, applied to both soil and foliage, and absorbed by both roots and leaves. It works against most weed species and is used primarily on land not used for growing crops, as it persists in the soil for several years.

Mode of Action

Prometon moves throughout the plant and works by inhibiting photosynthesis.

Toxicity—Acute Effects

Prometon has a slight acute oral toxicity and is a slight eye and skin irritant.

Toxicity— Subacute and Chronic Effects

There have been two teratogenicity studies, both of marginal quality, involving rats and rabbits. Neither found evidence of birth defects, but the results are not considered conclusive. There are no useful studies on file for all the other subacute and chronic effects.

Ecological Effects

Prometon is considered slightly toxic to mammals, moderately to slightly toxic to fish, slightly toxic to birds, and practically nontoxic to bees. Like the other *s*-triazines, prometon may persist in the soil for some time and can have residual activity for up to several years before it is eventually degraded by microbes. Information on other ecological effects is not available.

SIMAZINE

Simazine, known chemically as 2-chloro - 4, 6 - bis(ethylamino)-1, 3, 5 - triazine, or 2-chloro-4,6-bis(ethylamino)-*s*-triazine, is a selective systemic herbicide, applied primarily to the soil and absorbed by the roots. Its main uses are to control annual grasses and broad-leaf weeds in vegetables and ornamentals, and to control algae and submerged weeds in ponds. It also is used as an algicide in swimming pools.

Mode of Action

It moves throughout the plant, but concentrates in the leaves and the main growing tip, where it works by inhibiting photosynthesis.

Toxicity—Acute Effects

Simazine is practically nontoxic to mammals when ingested, but is a slight eye and skin irritant.

Toxicity—
Subacute and Chronic Effects

An interim report of a rat feeding study found evidence for increases in mammary gland hyperplasia in males, mammary gland tumors in females, and pituitary gland tumors in females, all at the highest dosage level. These effects were not seen in another rat study, which used a 50 percent formulation of simazine and whose highest dosage level was 10 times lower than that of the study in progress. Data gaps exist for birth defects, and an additional reproduction study is needed.

Mutagenicity

None of the various studies, the majority of which are significantly flawed, has found evidence of mutagenicity, but more studies are needed to confirm these findings.

Ecological Effects

Simazine is considered practically nontoxic to mammals, moderately to slightly toxic to fish and aquatic invertebrates, slightly toxic to practically nontoxic to birds, and practically nontoxic to bees. Like the other s-triazines, simazine may persist in the soil for more than a year. Information on other ecological effects is not available.

The class of herbicides known as *dinitroanilines* are structurally related chemically and tend to work in similar fashion. The bulk of them are used for preemergence weed control and are applied to the soil, where they affect germinating seeds and seedlings. Our study includes three herbicides from this group: *benfluralin* (or *benefin*), *pendimethalin*, and *trifluralin*.

BENFLURALIN

Benfluralin (or benefin), known chemically as N-butyl-N-ethyl-α,α,α-trifluoro-2,6-dinitro-p-toluidine, is a selective herbicide, applied to the soil and absorbed by the roots as well as stems of germinating seedlings. Its major use is to control annual grasses and some annual broadleaf species in established lawns and turf.

Mode of Action

Benefin works by inhibiting seed germination, cell division, and elongation of roots.

Toxicity—Acute Effects

Benefin is practically nontoxic to mammals when ingested, and is a moderate eye irritant but not a skin irritant.

Toxicity—
Subacute and Chronic Effects

The EPA appears to have no studies for oncogenicity, mutagenicity, or reproductive effects on file. There is a rabbit teratology study, but it is of such low quality that the EPA is requiring additional studies before ruling on developmental toxicity.

Ecological Effects

Benefin is considered practically non-toxic to mammals, slightly toxic to birds, and extremely toxic to fish and aquatic invertebrates. It may remain active in the soil, under the right conditions, for four to eight months. Information on other ecological effects is not available.

PENDIMETHALIN

Pendimethalin, known chemically as N-(1-ethylpropyl)-3,4-dimethyl-2,6-dinitrobenzenamine or N-(1-ethylpropyl)-2,6-dinitro-3,4-xylidine, is a selective herbicide, applied to the soil and absorbed by both the roots and shoots of seedlings. It is used primarily for pre-emergence, pretransplanting, or early postemergence control of annual grasses and annual broad-leaf weeds in gardens and established turf.

Mode of Action

Pendimethalin works primarily by inhibiting cell division and cell elongation, causing plants to die shortly after germination or emergence from the soil.

Toxicity—Acute Effects

Pendimethalin is slightly toxic to mammals when ingested, and is a minor eye and skin irritant.

Toxicity—
Subacute and Chronic Effects

Some data gaps exist. Although all the studies—two oncogenicity studies using mice and rats, a reproduction study using rats, two teratology studies using rats and rabbits, and a chronic feeding study using dogs—found evidence of adverse effects, they were considered of unacceptable quality to either the EPA or the CDFA. The results are mildly suggestive.

Mutagenicity

Three acceptable studies, with both microbial and mammalian cells, have found no evidence of mutagenicity.

Ecological Effects

Pendimethalin is considered slightly toxic to mammals, slightly toxic to practically nontoxic to birds, highly toxic to fish and aquatic invertebrates, and practically nontoxic to bees. It may remain active in the soil, under the right conditions, for just under a year, with a half-life of four to five months. Information on other ecological effects is not available.

TRIFLURALIN

Trifluralin, known chemically as α,α,α-trifluoro-2,6-dinitro-N,N-dipropyl-p-toluidine, is an herbicide, applied to the soil and absorbed by the roots and shoots of germinating seedlings. It is normally incorporated into the soil prior to planting for preemergent control of many annual grasses and broad-leaf weeds in a variety of vegetables and ornamentals.

Mode of Action

Trifluralin works by inhibiting both root and shoot development and indirectly inhibiting seed germination.

Toxicity—Acute Effects

Trifluralin is slightly toxic to mammals when ingested, and is a severe eye irritant but a minor skin irritant.

Toxicity—
Subacute and Chronic Effects

There are a few data gaps. A mouse study and a rat study found evidence of tumors in a number of organs, but both studies suffered from contamination with a carcinogen (an *N*-nitrosamine), and so are considered invalid. One good oncogenicity study, using rats, found evidence of malignant tumors of the kidneys and thyroid gland of males and urinary bladder tumors in females, with all effects, except for the kidney tumors, being statistically significant at the highest dose. Structural analogs of trifluralin have been reported to cause similar effects in rodents. The EPA has classified trifluralin as a possible human carcinogen, although more data is needed to determine trifluralin's oncogenic potential. A number of rat and rabbit studies found no evidence that trifluralin is teratogenic. An acceptable rat reproduction study found no evidence of impairment of reproductive ability. Finally, a couple of rat feeding studies found evidence of negative effects on the kidneys and thyroid, although the thyroid effects may be due to the negative kidney effect.

Mutagenicity

The available data suggests that trifluralin is not a mutagen, but the EPA thinks that, owing to effects on *spindle fibers* (components of a cell crucial for orderly cell division), there is a theoretical potential for genetic effects.

Ecological Effects

Trifluralin is considered slightly toxic to mammals, practically nontoxic to birds, extremely toxic to fish and aquatic invertebrates, and practically nontoxic to bees. It may remain active in the soil for 16 to 24 weeks. Data on other ecological effects is not available.

Herbicides in the class known as *chlorophenoxy alkanes*, also as *chlorophenoxy herbicides*, have a similar chemical structure and tend to have a similar mode of action. The vast bulk of them have a hormone-type action in that they inhibit formation of *nucleic acids*, the building blocks of genetic material, as well as interfering with the process of energy creation and storage known as *oxidative phosphorylation*. There is also some epidemiological evidence (which is somewhat controversial) from both the United States and Sweden that this class of herbicides in general, and 2,4-D in particular, may be associated with cancers in humans, particularly soft-tissue sarcomas (from the Swedish studies) and non-Hodgkin's lymphoma (from the U.S. studies). Our study contains two herbicides from this group: *2,4-D* and *mecoprop*.

2,4-D

Known chemically as (2,4-dichlorophenoxy)acetic acid, 2,4-D is a selective systemic herbicide that comes in numerous forms—an acid, various salts, or an ester. It is applied to both soil and foliage, with the plant roots absorbing the salt forms while the leaves absorb the ester. It moves throughout the plant and accumulates at the growing tips in the shoots and roots. It is used as a postemergence control of annual and perennial broadleaf weeds in lawns and established turf.

Mode of Action

Acting like a hormone, 2,4-D inhibits formation of nucleic acids and interferes with oxidative phosphorylation, thereby inhibiting growth.

Toxicity—Acute Effects

Depending on the form used, 2,4-D is slightly to moderately toxic to mammals when ingested, and is a slight eye and skin irritant.

Toxicity—
Subacute and Chronic Effects

There are a few data gaps. A rat feeding study found some evidence of brain tumors in high-dose males and negative kidney effects in both sexes. Human epidemiological studies in the United States and Sweden purport to show evidence of cancer. Swedish studies on forestry workers found increased incidences of soft-tissue sarcomas and non-Hodgkin's lymphoma. A recent study with Kansas farmers, done by a researcher from the National Cancer Institute (NCI), found chlorophenoxy herbicide use to be linked with a significantly increased risk of non-Hodgkin's lymphoma, but not soft-tissue sarcoma. Since the vast bulk of chlorophenoxy herbicides used by farmers in Kansas is 2,4-D, it seems reasonable to assume that 2,4-D is the main culprit. Both industry and the EPA's Scientific Advisory Panel have heavily criticized this study. A follow-up study of Nebraska farmers, which tried to answer some of the objections about the design of the Kansas study, also found an increase, although not quite statistically significant, of non-Hodgkin's lymphoma among farmers using 2,4-D. Furthermore, a jury in Texas decided that the defendant's case of non-Hodgkin's lymphoma was caused by exposure to 2,4-D in his work environment and awarded him over $1 million. Given the notorious weakness of human epidemiological studies, we consider 2,4-D very likely to be a human carcinogen.

A recent NCI study found increased risk of lymphoma in dogs whose owners applied 2,4-D to their lawns. A rat reproduction study found evidence of a dramatic reduction in fetal survival at dose levels causing only marginal effects in the parents. A teratogenicity study using rats found no evidence of birth defects, but did note dramatic increases in early embryonic death.

Mutagenicity

A study involving male mice found an inhibition of testicular DNA synthesis, while a few bacterial and mammalian cell cultures found no evidence of mutagenicity.

Ecological Effects

2,4-D is considered slightly toxic to birds and practically nontoxic to bees. Toxicity to fish depends on the formulation, ranging from highly toxic to slightly toxic. It may remain active in the soil for up to four to six weeks. It may also affect the physiology of crops such as corn, rice, wheat, and others, making them significantly more susceptible to insect pests and plant pathogens. Information on other ecological effects is not available.

MECOPROP

Mecoprop, known chemically as (RS)-2-(4-chloro-o-tolyloxy)propionic acid, is a selective systemic herbicide, applied to foliage and absorbed by the leaves. It also readily moves to the roots. It is used to control broad-leaf weeds in lawns and turf.

Mode of Action

Mecoprop has a hormone-type action in that it inhibits both formation of nucleic acids and oxidative phosphorylation, and thus inhibits growth.

Toxicity—Acute Effects

Mecoprop is slightly toxic to mammals when ingested, and is a slight eye and skin irritant.

Toxicity—
Subacute and Chronic Effects

Large data gaps exist. There are no oncogenicity or reproduction studies on file. A feeding study involving dogs is considered invalid. A teratology study of minimum quality, using rabbits, found evidence of birth defects.

Mutagenicity

The three existing mutagenicity studies, all involving bacteria, are of unacceptable quality to the EPA.

Ecological Effects

Mecoprop is considered slightly toxic to mammals, slightly toxic to fish, and practically nontoxic to bees. It can remain active in the soil for up to two months. Information on other ecological effects is not available.

The herbicides known as *benzoics* are all chemically related to benzoic acid, although their mode of action may differ among the herbicides in this class. We list only one herbicide, *dicamba*, from this class.

DICAMBA

Dicamba, known chemically as 3,6-dichloro-*o*-anisic acid or 3,6-dichloro-2-methoxybenzoic acid, is a selective systemic herbicide, applied to the leaves but absorbed by both leaves and roots. It is used primarily to control annual and perennial broad-leaf species and brush in lawns and turf.

Mode of Action

Dicamba acts like the natural plant growth regulator *auxin*, preventing growth of the primary growing tip and promoting tissue proliferation.

Toxicity—Acute Effects

Dicamba is slightly toxic to mammals when ingested, and is a severe eye irritant and a minor skin irritant.

Toxicity—
Subacute and Chronic Effects

There are numerous data gaps and few studies. A rat study found suggestive evidence of decreased time for tumor formation in the thyroid and pituitary. Further information is needed to resolve the potential for oncogenicity in rodents. One rat reproduction study, of questionable quality, found no evidence of negative reproductive effects. Three flawed teratology studies, involving rabbits, found no evidence of birth defects, while a rat teratology study found a marginal effect. More teratology studies need to be done.

Mutagenicity

A number of studies of varying quality, involving bacterial or mammalian cells, have been done. Only one study, involving bacteria, found evidence of damage to the DNA. It is difficult to interpret the importance of this finding, since there was no evidence of DNA damage in a study involving mammalian cells, or evidence of other forms of genotoxicity in the other studies.

Ecological Effects

Dicamba is considered slightly toxic to mammals, practically nontoxic to fish and aquatic invertebrates, slightly toxic to birds, and practically nontoxic to bees. Under the right conditions, the half-life is less than two weeks, although activity can persist for up to six months. Information on other ecological effects is not available.

Herbicides in the group known as *nitriles* have a similar chemical structure, but their modes of action can differ. Only one compound in our study, *dichlobenil*, belongs to this group.

DICHLOBENIL

Dichlobenil, known chemically as 2,6-dicholorobenzonitrile, is a selective systemic herbicide applied to the soil, although it is absorbed by both leaves and roots. It is used for pre- and postemergence control of annual and many perennial weeds in fruit trees, roses, and ornamental shrubs and trees.

Mode of Action

Dichlobenil works by inhibiting active growing tips, inhibiting seed germination, damaging roots, and inhibiting cellulose synthesis.

Toxicity—Acute Effects

Dichlobenil is slightly toxic to mammals when ingested, and is a slight eye irritant but not a skin irritant.

Toxicity—
Subacute and Chronic Effects

There are numerous data gaps. An acceptable rat study found some evidence of liver tumors in high-dose females,

renal tumors in males, and liver tumors, malignant lymphoma, and leukemia in females. The results are suggestive, but more studies are needed to confirm the results. A rat reproduction study found some evidence of increased infertility in females, but the study is considered flawed and needs to be repeated. A chronic feeding study in dogs found some evidence of liver damage, while a rat feeding study found evidence of increased liver and kidney weights.

Mutagenicity

Several mutagenicity tests, most of low quality, have not found evidence of mutagenicity, but more tests are needed.

Ecological Effects

Dichlobenil is considered slightly toxic to mammals, moderately to slightly toxic to fish, slightly toxic to practically nontoxic to birds, and practically nontoxic to bees. It can remain active in the soil from one month to one year, depending on soil conditions. Information on other ecological effects is not available.

The group called *glycines* contains only one compound, *glyphosate*, which is a modified form of the amino acid glycine. It has a number of modes of action, but the main one is inhibition of synthesis of certain amino acids.

GLYPHOSATE

Glyphosate, known chemically as *N*-(phosphonomethyl)glycine, is a very non-selective systemic herbicide, applied to and absorbed by the foliage. It can be used on a wide array of weeds, including most annual, biennial, and perennial grasses, sedges, broad-leaf weeds, and woody shrubs.

Mode of Action

Glyphosate moves quickly throughout the plant and works by interfering with various enzymes, leading to inhibition of the synthesis of amino acids and other important compounds.

Toxicity—Acute Effects

Glyphosate is practically nontoxic to mammals when ingested, and is a slight eye irritant but not a skin irritant.

Toxicity—
Subacute and Chronic Effects

Glyphosate appears relatively benign. There is some question whether it can cause cancer, and the EPA is requiring more animal studies. Acceptable rat and rabbit teratogenicity studies failed to find evidence of birth defects. A rat reproduction study, of questionable quality, failed to find evidence of negative reproductive effects, but another study is needed to clarify this result.

Mutagenicity

A number of acceptable studies show that glyphosate is not a mutagen.

Ecological Effects

Glyphosate is considered relatively benign. It is considered practically nontoxic to mammals, slightly to practically nontoxic to birds, only slightly toxic to fish and aquatic invertebrates, and practically nontoxic to bees. It strongly adsorbs to soil and is usually rapidly broken down by microbes. Data on other ecological effects is not available.

The class of herbicides known as *phenylureas*, although structurally related, can have different modes of action. Only one herbicide in our study, *siduron*, comes from this class.

SIDURON

Siduron, known chemically as 1-(2-methylcyclohexyl)-3-phenylurea, or N-(2-methylcyclohexyl)-N'-phenylurea, is a selective herbicide applied to the soil, absorbed by the roots, and moved throughout the plant. Siduron is used for preemergent control of annual grass weeds in newly seeded grass lawns or established turf.

Mode of Action

Siduron works primarily by inhibiting root growth and photosynthesis.

Toxicity—Acute Effects

Siduron is practically nontoxic to mammals when ingested, but is a slight eye and skin irritant.

Toxicity—
Subacute and Chronic Effects

Significant data gaps exist. A flawed rat reproduction study found no evidence of negative reproductive effects, but it needs to be repeated to verify the results. There are no acceptable studies on file for any of the other subacute or chronic effects, including oncogenicity, mutagenicity, and teratogenicity.

Ecological Effects

Siduron is considered practically nontoxic to mammals and birds, and moderately to slightly toxic to fish. Under the right conditions, it can remain active in the soil for up to a year. Information on other ecological effects is not available.

The class of herbicides known as *pyridines* all contain pyridine groups as a major part of their structure, but can differ in terms of their modes of action. Only one compound from this class, *triclopyr*, is found on our list.

TRICLOPYR

Triclopyr, known chemically as 3,5,6-trichloro-2-pyridyloxyacetic acid, is a selective systemic herbicide, applied to the soil and foliage and absorbed by both roots and leaves.

Mode of Action

Triclopyr moves easily throughout the plant, accumulating in growing tips, where it acts like a growth regulator, preventing growth of the primary growing tip and promoting tissue proliferation.

Toxicity—Acute Effects

Triclopyr is slightly toxic to mammals when ingested, and is a slight eye and skin irritant.

Toxicity—
Subacute and Chronic Effects

Some data gaps exist. As for oncogenicity, a two-year rat feeding study found evidence of potential negative impacts on the kidneys in males, but no evidence of oncogenicity. A two-year mouse feeding study found evidence of lung tumors at all doses tested, but there was also a question about relatively high incidence of lung tumors in the control animals. EPA considers the study acceptable, but further studies are needed. One acceptable rat reproduction study found no evidence of reproductive problems, but a mutagenicity study using rats found a trend toward increased resorption of fetuses in females. Rat and rabbit teratology tests found no evidence of birth defects, at least at levels that did not cause maternal toxicity.

Mutagenicity

A study using rats found evidence of a weak mutagenic effect (dominant lethal mutation). No evidence of mutagenicity was found in a number of tests involving bacterial and mammalian cells.

Ecological Effects

Triclopyr is slightly toxic to mammals, birds, and fish, and practically nontoxic to bees. The half-life in the soil is about 46 days. Information on other ecological effects is not available.

The two following herbicides, *chlorthal dimethyl* and *fluazifop-butyl*, are in a miscellaneous class not chemically similar to any other class of herbicides, or similar to more than one class.

CHLORTHAL DIMETHYL

Chlorthal dimethyl (or DCPA), known chemically as dimethyl tetrachloroterephthalate, is a selective nonsystemic herbicide applied to the soil and absorbed by germinating seeds of grasses and some annual broad-leaf weeds. It is used for preemergence control of annual grasses and some annual broad-leaf weeds in vegetables, ornamentals, and established turf.

Mode of Action

Chlorthal dimethyl interrupts cell division, thereby killing the germinating seed.

Toxicity—Acute Effects

Chlorthal dimethyl is practically nontoxic to mammals when ingested, and is neither an eye irritant nor a skin irritant.

Toxicity—
Subacute and Chronic Effects

Numerous data gaps exist. Two flawed oncogenicity studies, both negative, exist, but they need to be repeated. A flawed rat reproduction study found evidence of a reduced fertility index and lower pup weights at the highest dose tested and in the absence of overt parental toxicity. Limited evidence—an acceptable rat study and a very flawed rabbit study—suggests that chlorthal dimethyl is not a teratogen, but more studies are needed to show this conclusively.

Mutagenicity

One study, considered acceptable by the EPA (but not by the CDFA), involving pregnant rats and looking at chromosome damage, found evidence of a possible weak adverse effect. More studies are needed.

Ecological Effects

Chlorthal dimethyl is considered practically nontoxic to mammals, practically nontoxic to fish, and slightly toxic to bees. Its half-life in the soil is about 100 days. Data on other ecological effects is not available.

FLUAZIFOP-BUTYL

Fluazifop-butyl, known chemically as butyl (RS)-2-[4-(5-trifluoromethyl-2-pyridyloxy)phenoxy]propionate, is a selective systemic herbicide applied to foliage and absorbed by the leaves. It is used for postemergence control of annual and perennial grasses in broad-leaf crops such as vegetables and ornamentals.

Mode of Action

Upon absorption by the leaf, fluazifop-butyl is converted to fluazifop, which is transported throughout the plant, accumulating in growing tips, where it acts by interfering with the synthesis of an energy-storing molecule (ATP).

Toxicity—Acute Effects

Fluazifop-butyl is slightly toxic to mammals when ingested, and is a slight eye and skin irritant.

Toxicity—Subacute Effects

Both positive and negative evidence exists. The available evidence suggests that fluazifop-butyl is not oncogenic, although more information is needed to confirm these findings. Acceptable rabbit studies show fluazifop-butyl to cause birth defects, while acceptable rat and mouse studies show that it causes negative reproductive effects. Rat feeding studies also show evidence of liver damage.

Mutagenicity

Available evidence suggests that fluazifop-butyl is not a mutagen.

Ecological Effects

Fluazifop-butyl is considered slightly toxic to mammals, moderately to highly toxic to fish, slightly toxic to aquatic invertebrates, and practically nontoxic to both birds and bees. Its half-life in the soil is less than a week. Information on other ecological effects is not available.

APPENDIX C

FUNGICIDES

A number of organisms—fungi, bacteria, and viruses—cause plant diseases, with fungi causing the majority. Fungal diseases also wreak havoc on lawns, turf, flowers, and vegetables in the home garden. The chemicals used to combat fungal diseases are referred to as *fungicides.*

In spite of the frequency and severity of many fungal diseases, fungicides account for only about 16 percent of the volume of pesticide active ingredients used in the home and garden. Unlike insect, plant, and mammal pests, little is known about the life cycle and ecology of many fungal pests, making it harder to come up with alternatives to synthetic pesticides. Ignorance of fungal pest biol-

ogy, coupled with the fact that a broad class of fungicides can be used only to prevent infection, means that there is inordinate prophylactic use of them, which is often justifiable, compared to other pesticides. In spite of this, fungicides, like all other pesticides, tend to be used more often than they are needed.

In some ways, fungicide active ingredients present more of a danger to human health than do insecticide and herbicide active ingredients. Unlike insecticides and herbicides, virtually no fungicides have high acute mammalian toxicity. Indeed, the vast bulk of fungicides (and all the ones covered here) have very low acute toxicities and so do not represent a threat of acute poisoning in humans. This may lull people into thinking that they are safe. The real risk, however, may be the chronic effects, particularly from the most widely used ones. According to the National Academy of Sciences, about 90 percent of the fungicides used in the United States (in terms of volume) are thought to be carcinogenic. This figure is higher than for any other class of pesticides.

Modes of Action

In general, there are two classes of fungicides—*inorganic* and *organic*—and two modes of action—*surface protectants* and *systemics.* All the inorganic fungicides are surface protectants, as are the majority of organic fungicides. Surface protectants work by inhibiting or killing fungal spore germination and growth, but, as the name implies, only on a plant's surface. Thus, a surface protectant cannot control an established fungal infection that has spread into a plant's tissues; it can be used only before an infection happens. Such prophylactic use often leads to overuse, especially when one does not know when, or if, a particular disease will attack. Systemic fungicides penetrate plant tissue and so can kill or control an established infection as well as act as a surface protectant. Chemically more complex, they tend to be more species-specific and less subject to weathering than surface protectants. However, owing to their more specific mode of action, systemics also tend to be more susceptible to fungal resistance.

 # INORGANIC FUNGICIDES

Only two fungicides in our survey, copper sulfate and Bordeaux mixture, are considered inorganic—i.e., they contain no carbon atoms. Furthermore, all inorganic fungicides are surface protectants.

BORDEAUX MIXTURE

Bordeaux mixture, a mixture of copper sulfate and calcium hydroxide, acts as a surface protectant and is a general-purpose fungicide; it also has a repellent effect against many insects. The addition of lime in the form of $Ca(OH)_2$ to copper sulfate causes the precipitation of the vast bulk of the copper ions in the form of a copper hydroxide. This mixture, known as Bordeaux mixture, does not have copper sulfate's phytotoxicity and is effective against a wide array of fungal diseases. While Bordeaux mixture has been used on many crops for over 100 years, and is considered a very useful fungicide, it does have a couple of minor drawbacks. Especially during hot weather, Bordeaux mixture has been reported to cause damage to some crops, particularly tomatoes and cantaloupes. At low temperatures, Bordeaux mixture can also damage peach and plum foliage. In addition, the fungicidal efficiency of Bordeaux mixture declines with time after preparation (it comes as a powder and needs to be mixed with water before application), so it should be applied to the plant soon after mixing. Runoff into streams and lakes may be hazardous.

Toxicity—Acute Effects

Bordeaux mixture has low acute oral toxicity to mammals, but is a severe eye irritant and a strong skin irritant.

Toxicity— Subacute and Chronic Effects

There is a lack of methodologically solid studies.

Mutagenicity

Studies are lacking on the mutagenicity of Bordeaux mixture.

Ecological Effects

Copper sulfate is toxic to fish and aquatic invertebrates. There is a lack of data on other ecological impacts.

COPPER SULFATE

Copper sulfate is both an algicide and a fungicide that works as a surface protectant. It is used against most species of algae in ponds, lakes, and swimming pools. It is also injected into elm trees to control Dutch elm disease. Since the copper ion itself is toxic to leaves, copper sulfate cannot be used effectively as a foliar spray.

Toxicity—Acute Effects

Copper sulfate has moderate acute oral toxicity to mammals, and is a severe eye irritant and a strong skin irritant. It causes nausea at levels significantly lower than the lethal dose.

Toxicity— Subacute and Chronic Effects

Copper sulfate can cause damage to the kidneys, liver, and other organs of experimental rats. Other than such damage, there is a lack of methodologically solid studies on subacute and chronic effects.

Mutagenicity

Studies on the mutagenicity of copper sulfate are lacking.

Ecological Effects

Copper sulfate is toxic to fish, aquatic invertebrates, and some natural enemies. There is no information on other ecological impacts.

 # ORGANIC FUNGICIDES

Ethylenebisdithiocarbamate (EBDC)

The fungicides in this group are derivatives of ethylenediamine and act as surface protectants. They are all polymers with a virtually identical structure—that of an ethylenebisdithiocarbamate—and all contain a heavy metal atom at a given place in the polymer. Only the identity of the metal atom differs among the various EBDCs. It appears that the EBDCs work by inhibiting a number of vital sites, which makes it more difficult for fungi to evolve resistance to them.

The EBDCs have a serious drawback: an impurity, breakdown product, and metabolite known as *ethylene thiourea* (ETU).

Recent studies have shown ETU to be a very strong carcinogen in both rats and mice, causing malignant tumors at a number of different sites and showing dosage dependency. (Older studies had found evidence of thyroid tumors in rats and liver tumors in mice.) Also, some epidemiological work has associated ETU exposure in workers with increased levels of thyroid cancers. ETU has been shown to cause birth defects in rats and hamsters, and thyroid damage in rats, mice, and monkeys. Since ETU is associated with the manufacture of the EBDCs as well as with their metabolism, and since the data on carcinogenicity, teratogenicity, and thyroid damage of ETU is clear-cut, we recommend that the EBDC fungicides not be used.

There are three EBDCs in our list: *mancozeb* (containing both manganese and zinc atoms), *maneb* (containing manganese atoms), and *zineb* (containing zinc atoms).

MANCOZEB

Mancozeb, known chemically as manganese ethylenebis(dithiocarbamate), a polymeric complex with zinc salt, is a foliar fungicide that works as a surface protectant. It is used to control many fungal diseases of fruit trees, garden fruits, vegetables, and ornamentals. Its major agricultural uses are on potatoes, tomatoes, apples, and onions.

Toxicity—Acute Effects
Mancozeb has very low acute oral toxicity to mammals, and is a mild eye and skin irritant.

Toxicity— Subacute and Chronic Effects
There is a lack of valid oncogenicity studies on mancozeb per se. Evidence of birth defects was found in a rat study, but the effect might have been a result of ETU contamination. A single three-generation rat reproduction study found evidence of decreased fertility, but there was debate whether this effect was caused by a husbandry problem. There is no evidence of damage to internal organs, although there is little data.

Mutagenicity
The bulk of the mutation studies, although usually not up to guideline standards, suggest that mancozeb is not a mutagen. One study found evidence of sister chromatid exchanges related to mutagenicity. More studies need to be done.

Ecological Effects
Mancozeb is practically nontoxic to mammals and birds, moderately to highly toxic to fish, and nontoxic to bees.

MANEB

Maneb, known chemically as manganese ethylenebis(dithiocarbamate) in polymeric form, differs from mancozeb in that it lacks the zinc salt. It is a foliar fungicide that works as a surface protectant. It is used to control many fungal diseases of fruit trees, garden vegetables, turf, and ornamentals.

Toxicity—Acute Effects
Maneb has very low acute oral toxicity to mammals, and is a mild eye and skin irritant.

Toxicity—
Subacute and Chronic Effects

There is a lack of valid oncogenicity studies on maneb per se. Evidence of birth defects was found in a rat study, but the effect might have been a result of ETU contamination. Valid reproductive studies are lacking. There is evidence of damage to the thyroid in both rats and monkeys, as well as muscle damage in rats. These studies are not considered definitive.

Mutagenicity

There is some evidence from both bacterial and mammalian cell culture systems that maneb is a mutagen, but this data is considered weak.

Ecological Effects

Maneb is practically nontoxic to mammals and birds, highly toxic to fish, and nontoxic to bees.

ZINEB

Zineb, known chemically as zinc ethylenebis(dithiocarbamate) in polymeric form, is a foliar fungicide that works as a surface protectant. It is used to control many fungal diseases of fruit trees, garden vegetables, and ornamentals.

Toxicity—Acute Effects

Zineb has very low acute oral toxicity to mammals, and is a mild eye and skin irritant. There is a lack of valid oncogenicity studies on zineb per se.

Toxicity—
Subacute and Chronic Effects

Evidence of birth defects was found in a rat study, but the effect might have been a result of ETU contamination. There is a lack of valid reproductive studies. There is evidence of damage to the thyroid in both rats and monkeys as well as muscle damage in rats. These studies are not considered definitive.

Mutagenicity

There is evidence that zineb is a mutagen, but this data is considered weak.

Ecological Effects

Zineb is practically nontoxic to mammals and birds, moderately to highly toxic to fish, and nontoxic to bees.

Phthalimides

These compounds have roughly similar structures and are all surface protectants. One group—captan and its relatives captafol and folpet—are structurally very similar to one another and are quite reactive. Our survey contains only three phthalimides: captan, folpet, and chlorothalonil.

Mode of Action

Captan and its relatives all cause inhibition of biochemical processes at numerous sites, which slows fungal resistance to them. Chlorothalonil appears to inhibit one or more biochemical processes, which should make it more susceptible to the evolution of fungal resistance than captan and its relatives.

CAPTAN

Captan, known chemically as N-trichloromethylthio-4-cyclohexene-1,2-dicarboximide, or 1,2,3,6-tetrahydro-N-

(trichloromethylthio)phthalimide, is a fungicide that has both protective and curative action. It is a broad-spectrum fungicide used to control many diseases of fruit trees, garden vegetables, turf, and ornamentals. Captan is also frequently used as a seed treatment on corn and other crops and on ornamentals. It is not considered phytotoxic when used as directed, although there is evidence of injury to Red Delicious and Winesap apples and D'Anjou and Bosc pears.

Toxicity—Acute Effects

Captan has very low acute oral toxicity to mammals, and is a severe eye irritant but not a skin irritant.

Toxicity— Subacute and Chronic Effects

Captan has been linked to a modest dose-related increase in renal tumors in male rats, a small dose-related increase in kidney tumors in rats, and a dose-related increase in duodenal tumors in mice. Evidence of birth defects (fused ribs) was found in a hamster study, although at a dosage level significantly higher than that which caused significant maternal effects. Both a rat study and a monkey study found evidence of negative fetal effects. Although suggestive, these studies need to be repeated. There is weak evidence of negative effects on the stomach of female rats.

Mutagenicity

It is considered mutagenic in bacterial systems and in mammalian cell culture systems under some conditions. Although the bulk of these studies are flawed in one way or another, it is generally accepted that captan is mutagenic in microbial systems.

Ecological Effects

Captan is considered slightly toxic to practically nontoxic to birds and mammals, very highly toxic to fish, moderately toxic to aquatic invertebrates, and nontoxic to bees when used as directed.

FOLPET

Folpet, known chemically as N-(trichloromethylthio)phthalimide, is a foliar fungicide that works via a protective action. Its chemical structure and activity are similar to captan's. Folpet is used to control a number of fungal diseases in fruit trees, some vegetables (potatoes, lettuce, squashes, onions, leeks, celery, tomatoes), and ornamentals.

Toxicity—Acute Effects

Folpet has very low acute oral toxicity to mammals, and is a moderate eye irritant but not a skin irritant.

Toxicity— Subacute and Chronic Effects

There is equivocal evidence for an oncogenic effect in the thyroid of rats, while two mice studies found evidence of a dose-related increase in intestinal tumors in both sexes. One of the mouse studies also found evidence of stomach tumors in both sexes and malignant lymphomas in females, but only at the highest dosage. There is some suggestive evidence for birth defects from one rat study and three rabbit studies, but more data is needed. A few rat and rabbit studies found some evidence of negative reproductive effects (incomplete ossification in the fetus), but the effect usually occurred at levels that also caused toxicity to the mother. Evidence of effects on the stomach, spleen, testes, and thyroid were seen in rat studies, but the evidence is equivocal and more data is needed before a definitive conclusion can be reached.

Mutagenicity

Folpet is clearly a mutagen, at least in bacterial systems.

Ecological Effects

Folpet is considered slightly toxic to practically nontoxic to birds and mammals, very highly toxic to fish and aquatic invertebrates, and nontoxic to bees.

CHLOROTHALONIL

Chlorothalonil, known chemically as 2,4,5,6-tetrachloroisophthalonitrile, is a foliar fungicide that has strong protective action and some curative action. It is a broad-spectrum protectant that is used to control many fungal diseases of fruit trees, some garden vegetables, turf, and ornamentals.

Toxicity—Acute Effects

Chlorothalonil has very low acute oral toxicity to mammals, and is a severe eye irritant but not a skin irritant.

Toxicity—
Subacute and Chronic Effects

Two rat studies found evidence of renal tumors, while one also found evidence of forestomach tumors. A mouse study found evidence of renal tumors in males and evidence of tumors in the esophagus and stomach of both sexes. Another mouse study found evidence of possible squamous papillomas (potential evidence of oncogenicity) of the forestomach in males at the highest dose. A

number of rat studies and a mouse study consistently found evidence of damage to the kidneys and stomach. Some of the rat studies also found evidence of damage to the liver and thyroid. A couple of dog studies found evidence of, among other things, kidney and liver damage, but these studies are considered flawed, although the data is suggestive and consistent with the rat and mouse data. A few rat reproduction studies found evidence of decreased pup weights. One of the reproduction studies also found evidence of esophageal and stomach damage in pups, but this occurred at levels that also caused significant maternal toxicity. As for birth defects, no positive evidence was found in an acceptable rat study or in two unacceptable rabbit studies, which were considered substantially flawed.

Mutagenicity

Chlorothalonil consistently tests negative in a range of bacterial tests of mutagenicity. Two studies using hamsters found evidence of chromosomal aberrations, although the effect occurred only at high doses and the evidence is considered equivocal.

Ecological Effects

Chlorothalonil is considered slightly toxic to birds and mammals, and very highly toxic to fish and aquatic invertebrates. Information on other ecological impacts is not available.

DINOCAP

Dinocap, known chemically as 2(or 4)-(1-methylheptyl)-4,6(or 2,6)-dinitrophenyl crotonate, is a complex mixture of six different crotonate phenols. Dinocap has uses as both an acaricide and a surface-protectant fungicide with both protective and curative action. It is used primarily for the control of powdery mildews in fruits, some vegetables, and ornamentals. Until recently, baking soda, dinocap, and sulfur were about the only fungicides that could be used against powdery mildews.

Toxicity—Acute Effects

Dinocap has relatively low acute oral toxicity to mammals, and is a slight eye and skin irritant.

Toxicity—
Subacute and Chronic Effects

Two rat studies found no evidence of oncogenicity, while a mouse study found evidence of reticulum cell carcinoma. All three studies are considered very flawed, although the mouse data is suggestive since reticulum cell carcinoma is not that common. There is evidence of birth defects in both rabbits and mice at dosages that did not cause toxic maternal effects, but no positive evidence from rats or hamsters. There is evidence of adverse reproductive effects—decreased pregnancy rates in rats and lower birth weight and higher neonatal pup mortality in mice—but the studies contain flaws of one sort or another. A couple of studies involving dogs found evidence of damage to the eyes, heart, and liver.

Mutagenicity

On the basis of eleven studies, which gave conflicting results, the CDFA concluded that dinocap is a weak mutagen in *Salmonella* bacteria. Studies involving mammalian cell systems have not found evidence of mutagenicity, however.

Ecological Effects

Dinocap is considered slightly toxic to birds and mammals, toxic to fish, and nontoxic to bees. Other ecological information is not available.

Benzimidazoles

These fungicides, currently the most important group of systemic fungicides in use, are effective against all the major classes of fungi. Only the best-known member of this group, *benomyl*, is treated here.

Mode of Action

Although benzimidazoles are systemics, the bulk of their action arises from surface protection. They work by preventing cell division (by disrupting division of the nucleus) and disrupting growth of the vegetative stage of fungi.

BENOMYL

Benomyl, known chemically as methyl 1 - (butylcarbamoyl)benzimidazol - 2 - ylcarbamate, is a systemic fungicide, absorbed through the roots and leaves, which has both protective and curative action. It is a broad-spectrum protectant that is used to control many fungal diseases of fruit trees, vegetables, turf, trees, and ornamentals. A primary metabolite of benomyl is MBC (methyl-2-benzimidazole carbamate), another benzimidazole fungicide commonly known as carbendazim.

Toxicity—Acute Effects

Benomyl has extremely low acute oral toxicity to mammals, and is a moderate eye irritant but not a skin irritant.

Toxicity—
Subacute and Chronic Effects

Studies in rats have failed to show evidence of oncogenicity from benomyl or one of its metabolites, MBC, while three mouse studies—one using benomyl and two using MBC—found liver tumors in both sexes. Since there is some question about the three positive mouse studies involving either benomyl or MBC, the EPA has classified both benomyl and MBC as "possible" human oncogens. Birth defects have been found in two rat studies and a mouse study, all at levels of benomyl lower than that which caused maternal toxicity. Furthermore, MBC has been found to cause birth defects in a rabbit study. A few rat studies have found evidence of negative reproductive effects—primarily increased resorption of fetuses, decreased fetal and weaning weights, and decreased neonatal weight gain. One rat study found evidence of decreased sperm production. As for subacute effects, benomyl has been shown to cause liver damage in both rats and dogs. One dog study found evidence of testicular damage, but a later, better defined study, found no such evidence. Because a couple of the birth defects in rats were associated with the eyes, there is concern that benomyl may cause eye damage. Eye damage was not looked for, however, in any of the subacute studies, which are considered deficient in this regard.

Mutagenicity

Both benomyl and MBC have tested positive in a number of mutagenicity tests. Numerous bacterial mutagenicity tests have found contradictory results. Three of four studies designed to detect chromosomal aberrations found positive evidence for them. One of three designed to detect evidence of DNA damage obtained positive results.

Ecological Effects

Benomyl is considered slightly toxic to birds and mammals, highly toxic to fish and aquatic invertebrates, and nontoxic to bees. MBC is highly toxic to fish and aquatic invertebrates. Other ecological data is not available.

Inhibitors of Sterol Biosynthesis

These systemic fungicides, rather than being structurally related, all appear to have the same primary mode of action: they inhibit sterol biosynthesis. Most of the compounds, including the two covered here—*triademefon* and *triforine*—act by inhibiting ergosterol synthesis. Ergosterol is an important compound that helps regulate synthesis as well as breakdown of molecules (primarily lipids) associated with the cell's membrane. Although used to control a number of diseases, these fungicides are particularly useful in controlling powdery mildews.

TRIADIMEFON

Triademefon, known chemically as 1-(4-chlorophenoxy)-3,3-dimethyl-1-(1H-1,2,4-triazol-1-yl)butanone, is a systemic fungicide that has both curative and protective effects. It is readily absorbed by the roots and leaves and then moves to young growing tissue. It is used to control a number of diseases, particularly powdery mildews attacking fruits, vegetables, turf, ornamentals, flowers, shrubs, and trees, and rusts attacking turf, ornamentals, flowers, shrubs, and trees. If used excessively, damage to ornamentals may occur.

Toxicity—Acute Effects

Triademefon has moderate acute oral toxicity to mammals, and is a slight eye and skin irritant.

Toxicity—
Subacute and Chronic Effects

Neither a rat study nor a mouse study found evidence of oncogenicity. Both were considered flawed; a new mouse oncogenicity study is being done. Although negative for oncogenicity, both studies found evidence of negative blood effects, while the mouse study found a negative effect on the liver. A dog study also found evidence of negative liver effects, although interpretation of this result is hard, because the data is equivocal. Two rat reproduction studies—one two-generation, the other three-generation—found evidence of negative reproductive effects, including decreases in pup body weights, postnatal survival, litter size, and male fertility, but only the decrease in pup body weights occurred at a dosage less than that which caused some maternal toxicity. The results of three rat studies, while somewhat conflicting, show triademefon to be a weak teratogen. No evidence of birth defects was found in two rabbit studies, both of which were considered marginal.

Mutagenicity

Available data, most of poor quality, points to triademefon not being mutagenic.

Ecological Effects

Triademefon is considered slightly toxic to birds and mammals, moderately toxic to fish, and nontoxic to bees. Other ecological data is not available.

TRIFORINE

Triforine, known chemically as 1,1'-piperazine-1,4-diyldi-[N-(2,2,2-trichloroethyl)formamide], is a systemic fungicide that has both a curative and a protective effect. It is readily absorbed by the roots and leaves and then moves to the growing tips. In addition to being used against a wide array of diseases in both the lawn and the garden (primarily powdery mildews, rusts, and fairy rings), it suppresses spider mite activity.

Toxicity—Acute Effects

Triforine has slight acute oral toxicity to mammals, and is a severe eye irritant but not a skin irritant.

Toxicity—
Subacute and Chronic Effects

In one study each on rats and mice, neither found evidence of oncogenicity, but the results are not conclusive as the rat study is considered of only minimal quality. The rat study found evidence of anemia, while a dog study, also of minimal quality, found evidence of negative effects on bone marrow and Kupfer cells. Evidence of negative reproductive effects, resorption of fetuses, and an increase in fetal variations (ossification of sternebrae), was found in a rat teratogenicity study. This rat study, along with a rabbit study, failed to find evidence of birth defects, although the studies were not considered definitive.

Mutagenicity

Only one of eight studies for mutagenicity, the one involving mouse bone marrow, found any evidence of mutagenicity, but this data could not be repeated.

Ecological Effects

Triforine is considered slightly toxic to birds and mammals, practically nontoxic to fish, and nontoxic to bees. Other ecological data is not available.

APPENDIX D

RODENTICIDES and MOLLUSCICIDES

 ## RODENTICIDES

Mice and rats are by far the most serious rodent pests, and they transmit a number of diseases to people. Chemical substances or mixtures of substances that kill, prevent, repel, or mitigate rodents are known as *rodenticides.* Although the rodenticides are represented by a wide variety of chemical classes, some scientists have divided them into five functional categories, based on their primary effects: *acute rodenticides, chronic rodenticides, fumigants, repellents,* and *chemosterilants.* The rodenticides discussed below fall into the first two categories.

Acute Rodenticides

Acute rodenticides are those that act quickly as single-dose toxicants. Although useful for rapid population reduction, they have the potential for development of "bait shyness," particularly if the poison is used more than twice per year. *Bait shyness* refers to the rodents, particularly rats, not accepting the bait or poison. On encountering a new food, rats eat only a small portion of it. If they become ill within the next 18 to 24 hours, they will not eat that food again. In addition, female rats may bring their young to the food so that they can sample it and thereby learn to avoid it.

Acute rodenticides belong to three broad chemical categories—*inorganics*, *botanicals*, and *synthetic organics*; our list includes an example of each one: zinc phosphide, strychnine, and cholecalciferol.

ZINC PHOSPHIDE

Zinc phosphide is an inorganic rodenticide that delivers a toxic dose in a single feeding. It is used as a bait rodenticide for control of rats, mice, voles, ground squirrels, and gophers.

Mode of Action

Zinc phosphide works by reacting with water and stomach acids to form the highly toxic and flammable gas phosphine, which enters the bloodstream, causing damage to the liver, kidneys, and heart.

Toxicity—Acute Effects

Zinc phosphide is extremely toxic to mammals when taken orally, and is a minor eye irritant but not a skin irritant. Secondary poisoning is usually not a problem with zinc phosphide, although it has been reported in cats that had ingested rats that had eaten a bait containing 5 percent zinc phosphide. Because it so readily reacts with acids and water to form a toxic, flammable gas, it is dangerous to use.

Toxicity— Subacute and Chronic Effects

There are large data gaps. The EPA has only a single mouse teratology study on file. This study, of minimal quality,

found no evidence of birth defects. Phosphine gas causes characteristic poisoning symptoms in the liver and lungs of animals that ingest it.

Mutagenicity

The EPA has no studies on file.

Ecological Effects

Zinc phosphide is extremely toxic to birds, mammals, and fish. Information on other ecological effects is not available.

STRYCHNINE

Strychnine, known chemically as strychnidin-10-one, is an alkaloid that comes from seeds of plants in the genus *Strychnos* (particularly *S. nux-vomica*), which are found in Southeast Asia. Although used as the alkaloid, strychnine is also used in the form of one of its salts, such as strychnine sulfate. It is usually mixed with cereals as pellets, or applied as a bright green or red dye to the bait grain itself.

Mode of Action

Strychnine is a nonspecific poison that acts directly on the central nervous system by interfering with an inhibitory neurotransmitter in the spinal cord and medulla. Signs of poisoning are similar in all species and consist of uncontrolled and relatively diffuse, unchecked reflex activity, eventually leading to convulsions and death. Soon after ingestion the animal appears apprehensive, nervous, tense, and stiff. It then becomes hypersensitive to external stimuli and may undergo violent seizures either spontaneously or triggered by touch, sound, or sudden bright light. Death occurs from anoxia during seizures.

Toxicity—Acute Effects

Strychnine sulfate is moderately toxic to mammals when taken orally, and is a minor eye and skin irritant.

Toxicity— Subacute and Chronic Effects

The EPA has no studies on file.

Mutagenicity

The EPA has no studies on file.

Ecological Effects

Strychnine sulfate is highly toxic to birds. Information on other ecological effects is not available.

CHOLECALCIFEROL

Cholecalciferol, known chemically as activated 7-dehydrocholesterol (vitamin D_3), is very closely related to calciferol (vitamin D_2). Both are nonspecific poisons developed for use against anticoagulant-resistant mice and rats.

Mode of Action

Cholecalciferol and calciferol work by disrupting calcium metabolism, causing increased calcium resorption from bones, increased intestinal calcium absorption, and a consequent deposition of calcium salts in soft tissues such as the kidneys, blood vessels, heart, and lungs. Poisoning symptoms include anorexia, diarrhea, and thirst.

Toxicity—Acute Effects

Cholecalciferol is moderately toxic to mammals when taken orally, and is a minor eye and skin irritant.

Toxicity— Subacute and Chronic Effects

The EPA has no studies on file, although cholecalciferol does cause internal bleeding problems in mammals.

Mutagenicity

The EPA has no studies on file.

Ecological Effects

The EPA has no readily accessible information.

Chronic Rodenticides

Unlike acute rodenticides, chronic rodenticides come from only two chemical classes—*coumarins* and *indandiones*—both of which are oral anticoagulants. Oral anticoagulants have three characteristics that differentiate them from acute rodenticides. First, poisoning symptoms are delayed for several days, which means that the rodents continue to feed on the baits and consume a lethal dose, so that bait shyness ceases to be a problem. Second, concentration of the anticoagulant in the bait is very low, usually necessitating multiple feedings before a toxic dose is ingested. Finally, vitamin K_1 is an effective antidote.

Mode of Action

The effects of coumarins and indandiones are qualitatively similar, as are their mechanisms of action. Both groups of chemicals work by disrupting formation of one or more blood clotting factors. They all interfere with vitamin K, which plays a key role in the synthesis of various blood clotting factors, but particularly prothrombin, in the liver. Toxic symptoms do not appear for a few days, as clotting factors circulate in the blood and gradually disappear when not steadily replaced by the liver. Death results from hemorrhaging.

COUMARINS

The chemicals in this class are derivatives of coumarin. Their development goes back to the 1920s, when the sweet clover disease that causes bleeding in cattle that eat sweet clover was described. The chemical causing the bleeding was identified as *dicoumarol*, which is used in human medicine. Field trials were done with the chemically similar warfarin, which proved more effective at controlling rodents than dicoumarol. Warfarin and three other chemicals—coumafuryl, coumachlor, and coumatetralyl—are referred to as first-generation coumarins. After resistance to warfarin appeared in mice and rats, further research led to the second-generation coumarins—difenacoum, brodifacoum, and bromadiolone. Unlike the first-generation coumarins, these three compounds have extremely low LD_{50}s (less than 5 mg/kg in wild Norway rats), are highly effective after a single feeding, and are effective against warfarin-resistant rats.

WARFARIN

Warfarin, known chemically as 4-hydroxy-3-(3-oxo-1-phenylbutyl)-2H-1-benzopyran-2-one, or 3-(α-acetonylbenzyl)-4-hydroxycoumarin, was the first anticoagulant developed as a rodenticide.

Mode of Action

Warfarin functions as an antivitamin, competing with vitamin K, which plays an essential role in the production of various blood clotting factors in the liver. Repeated ingestion is necessary to produce toxicity.

Toxicity—Acute Effects

Warfarin is moderately to highly toxic to mammals when taken orally, and is a minor eye and skin irritant.

Toxicity— Subacute and Chronic Effects

The EPA has no animal studies on file. However, there are a number of published reports of clinical sodium warfarin causing fetal abnormalities in humans. Because it interferes with vitamin K in the liver, warfarin can cause severe hemorrhaging in mammals.

Mutagenicity

The EPA has no studies on file.

Ecological Effects

The EPA has no readily accessible data. Although most birds are susceptible to warfarin, poultry are relatively resistant to it.

BRODIFACOUM

Brodifacoum, known chemically as 3-[3-4'-bromo-[1,1-biphenyl]-4-yl)-1,2,3, 4 - tetrahydro - 1 - naphthalenyl] - 4 - hydroxy-2H-1-benzopyran-2-one, or 3-[3-(4'-bromobiphenyl-4-yl)-1,2,3,4-tetrahydro-1-naphthyl]-4-hydroxycoumarin, is one of the second-generation coumarins and is a brominated version of difenacoum.

Mode of Action

Like warfarin, brodifacoum works by inhibiting vitamin K. It is usually formulated as bait, either in block form or as grain baits. Only a single feeding is needed for rodent death to occur.

Toxicity—Acute Effects

Brodifacoum is extremely toxic to mammals when taken orally, and is a minor eye and skin irritant. This extreme toxicity has been associated with a number of accidental poisonings of animals as well as people. According to a telephone hot line set up by a manufacturer of brodifacoum, ICI Americas, Inc., 64 poisonings of domestic animals occurred from 1982 to 1985. Three-quarters of the incidents involved dogs either ingesting baits or end-use products containing .005 percent brodifacoum or carcasses of poisoned rodents. At least 71 people in Indonesia ate a poisoned bait consisting of rice and .005 percent brodifacoum between December 1982 and January 1983, and at least 20 of them died.

Toxicity— Subacute and Chronic Effects

The EPA has virtually no animal studies on file. However, there are two teratology studies—one with rats, the other with rabbits. Neither study found evidence of birth defects, and both were of minimal quality.

Mutagenicity

There is limited evidence from bacterial studies that brodifacoum is not mutagenic.

Ecological Effects

Brodifacoum is extremely toxic to birds, mammals, and fish. Information on other ecological impacts is not available.

BROMADIOLONE

Bromadiolone, known chemically as 3-[3,(4′-bromo[1,1′-biphenyl]-4-yl)-3-hydroxy-1-phenylpropyl]-4-hydroxy-2H-1-benzopyran-2-one, or 3-[3-(4′-bromobiphenyl-4-yl)-3-hydroxy-1-phenylpropyl]-4-hydroxycoumarin, is one of the second-generation coumarins.

Mode of Action

Bromadiolone works by inhibiting vitamin K. It is usually formulated as bait, either in block form or as soaked seed.

Toxicity—Acute Effects

Bromadiolone is extremely toxic to mammals when taken orally, and is a minor eye and skin irritant.

Toxicity— Subacute and Chronic Effects

The EPA has no animal studies on file, except for two teratology studies—one with rats, the other with rabbits. Neither study found evidence of birth defects, but both were of poor quality.

Mutagenicity

The EPA has no studies on file.

Ecological Effects

Bromadiolone is highly toxic to fish, extremely toxic to mammals, slightly toxic to quail, and nontoxic to bees when used as directed. Information on other ecological impacts is not available.

Indandiones

These chemicals are anticoagulants that usually require multiple feedings in small doses to produce the toxic effect.

CHLOROPHACINONE

Chlorophacinone, known chemically as 2-[(4-chlorophenyl)phenylacetyl]-1*H*-indene-1,3(2*H*)-dione, or 2-[(*p*-chlorophenyl)phenylacetyl]-1,3-indandione is used to control all types of rodents, including those resistant to warfarin. It is formulated as a bait concentrate, bait, grain bait, block bait, or tracking powder.

Mode of Action

Chlorophacinone acts both as an anticoagulant and to block the process of energy production and storage by uncoupling oxidative phosphorylation, the process whereby energy is stored in the form of the molecule ATP (adenosine triphosphate).

Toxicity—Acute Effects

Chlorophacinone is extremely toxic to mammals when taken orally, and is a minor eye irritant but not a skin irritant.

Toxicity—
Subacute and Chronic Effects

The EPA has no animal studies on file.

Mutagenicity

The EPA has no animal studies on file.

Ecological Effects

Chlorophacinone is nontoxic to bees when used as directed. Information on other ecological impacts is not available.

DIPHACINONE

Diphacinone, known chemically as 2-(diphenylacetyl)-1*H*-indene-1,3(2*H*)-dione, is an oral anticoagulant that requires multiple feedings to produce a lethal effect. It is used to control a wide variety of rodents and comes in a bait or bait concentrate form.

Toxicity—Acute Effects

Diphacinone is extremely toxic to mammals when taken orally, and is a minor eye and skin irritant.

Toxicity—
Subacute and Chronic Effects

The EPA has no animal studies on file.

Mutagenicity

Diphacinone is considered nonmutagenic in bacterial tests, but the evidence is not decisive.

Ecological Effects

Diphacinone is toxic to fish, extremely toxic to mammals, slightly toxic to mallard ducks, and nontoxic to bees when used as directed. Information on other ecological impacts is not available.

PINDONE

Pindone, known chemically as 2-pivalyl-1,3-indandione, is an oral anticoagulant.

Toxicity—Acute Effects

2-pivalyl-1,3-indandione is extremely toxic to mammals when taken orally.

Toxicity—
Subacute and Chronic Effects

The EPA has no studies on file.

Mutagenicity

The EPA has no studies on file.

Ecological Effects

The EPA has no readily accessible data.

MOLLUSCICIDES

Both snails and slugs are members of the gastropod family, belonging to the same order as mollusks; hence the use of the term *molluscicide* to describe chemicals used to control or kill them.

At present, there are basically two chemicals used as molluscicides: *methiocarb* and *metaldehyde*. Both cause the mollusk to stop feeding, but a debate rages over which is the better slug and snail killer. Both are most frequently formulated as bran pellets containing the molluscicide. Studies have shown that snails and slugs stop feeding sooner when the bait contains metaldehyde compared to methiocarb, thereby reducing the mollusk's chance of ingesting a lethal dose. On the basis of this work, some scientists claim that methiocarb is superior to metaldehyde because the slugs have a greater chance of ingesting a lethal dose. However, field studies have shown the two chemicals equally effective as baits. Why? The frequency of the dose is also important. By leaving pellets out in the field or garden, the farmer or gardener repeatedly challenges the snail/slug population, until a lethal dose is ultimately ingested.

METHIOCARB

Methiocarb, known chemically as 4-methylthio-3,5-xylyl methylcarbamate, is a multipurpose pesticide used as a molluscicide, insecticide, acaricide, and bird repellent.

Mode of Action

Chemically, methiocarb is a carbamate, a neurotoxin that acts primarily by inhibiting the neurotransmitter acetylcholine, which normally conveys nerve impulses across the junction between neurons. Methiocarb works in slugs and snails by disrupting nerves in the *central pattern generator*, the part of the brain charged with making the mouth move in a coordinated fashion as the animal feeds. After exposure to methiocarb, these nerves fire irregularly and then stop, causing the animals to stop feeding and, if a high enough dose has been ingested, to die.

Toxicity—Acute Effects

Methiocarb is very highly toxic to mammals when taken orally, and is a minor eye and skin irritant. A primary metabolite, methiocarb sulfoxide, is even more acutely toxic to mammals than is methiocarb itself.

Toxicity— Subacute and Chronic Effects

Data on subacute and chronic effects of methiocarb is sparse. In an oncogenicity study on rats, the result was negative. A second oncogenicity study, in mice, is required to evaluate methiocarb's oncogenic potential fully. The EPA has no valid studies for reproductive toxicity. On the basis of two minimum-quality studies utilizing rats and rabbits, methiocarb does not appear to be teratogenic. Finally, animal studies have shown that methiocarb inhibits plasma, red blood cell, and brain cholinesterase activity.

Mutagenicity

The EPA has no valid studies on file.

Ecological Effects

Methiocarb appears to be very highly acutely toxic to fish, aquatic invertebrates, and birds, and toxic to bees. Less than one granule from the registered granular or pelletized bait products contains enough methiocarb to kill a blackbird or house finch. Methiocarb is slightly to practically nontoxic to birds on a subacute dietary basis. Methiocarb is also highly toxic to worms and carabid beetles. Since these beetles are predators of slugs, methiocarb may actually lead to increases in slug populations in areas where their numbers are controlled by carabid beetles.

METALDEHYDE

Metaldehyde, known chemically as r-2,c-4,c-6,c-8-tetramethyl-1,3,5,7-tetra-oxocane, is used in a bran pellet form to control slugs and snails in the garden.

Mode of Action

Metaldehyde has both stomach and contact action. When it comes into contact with the mollusk's foot, the snail or slug becomes torpid and stops feeding. In addition, metaldehyde induces increased mucus secretion, which leads to dehydration.

Toxicity—Acute Effects

Metaldehyde is moderately toxic to mammals when taken orally, and is a minor eye irritant but not a skin irritant.

Toxicity— Subacute and Chronic Effects

There are no available studies for carcinogenicity or teratogenicity, and it appears that only two long-term studies have been done. Both involved rats—one was a feeding study and the other a three-generation reproduction study. Both studies found evidence of posterior paralysis and transverse spinal cord lesions. The reproduction study also found decreases in male and female fertility, lactation, and viability.

Mutagenicity

No evidence of mutagenicity has been found in the few available studies, but more studies are needed to determine that metaldehyde is not a mutagen.

Ecological Effects

Metaldehyde is not toxic to fish and is moderately toxic to mammals. For some reason, dogs seem attracted to metaldehyde baits and can be accidentally poisoned. Information on other ecological effects is not available.

APPENDIX E

INERT INGREDIENTS

The pesticide products that you use around the home and garden are actually a mixture of active ingredients and inert ingredients. This appendix deals with the so-called inert ingredients.

According to the EPA, an inert ingredient is added to a pesticide to give it a desired attribute, such as to dissolve, dilute, propel, stabilize, or enhance the action of the active ingredient(s). At present there are more than 1,200 inert ingredients used in pesticides.

Just because ingredients are called "inert" does not mean that they are non-toxic, however. Indeed, they may be quite hazardous. In spite of this, inert ingredients did not have to be listed on the labels of pesticides. Yet, particularly in household insecticides, inert ingredients often constitute the bulk of the actual product. With increased concern over avoiding toxic chemicals, more and more consumers are demanding to know everything that is in the products they use.

LAWS GOVERNING PESTICIDES

Our national pesticide law, called the Federal Insecticide, Fungicide, and Rodenticide Act (FIFRA) and enacted in 1911, regulates all pesticide products and requires that they be registered before they can be sold or distributed commercially. Now they must be registered by the EPA, but before the creation of the EPA in 1970, the Department of Agriculture registered them.

Most of the data requirements and regulatory activities mandated by FIFRA have focused on the active ingredients and neglected the inerts. Prior to the EPA, there was little testing for subacute and chronic effects of pesticides, in part because there was little realization that problems existed. The EPA, recognizing this, pushed for reform of FIFRA, and in 1972 FIFRA was amended so that all pesticides had to be reregistered and undergo testing for a range of subacute, chronic, environmental, and ecological effects as part of the reregistration process. The task was daunting. At the time there were about 45,000 pesticides on the market. The EPA decided to focus on the active ingredients, since there were only about 600 of them. The inert ingredients were basically ignored.

As concern increased over the environmental and health impacts of pesticides, and as more pesticides were revealed to cause chronic health effects, people began to realize that the inert ingredients that were not being scrutinized could pose potential health and environmental problems. The EPA developed a strategy for regulating inerts, making a draft available to the public in the spring of 1986. In the spring of 1987 the agency officially announced this strategy in the *Federal Register*.

First, the EPA divided the approximately 1,200 intentionally added inert ingredients into four toxicity categories:

- inerts of toxicological concern (List 1)
- potentially toxic inerts/high priority for testing (List 2)
- inerts of unknown toxicity (List 3)
- inerts of minimal concern (List 4)

Inerts on List 1 are chemicals that are known to be toxic and/or have significant ecological or environmental effects. The criteria included demonstrated evidence of one or more of the following: carcinogenicity, adverse reproductive effects, birth defects, neurotoxicity or other chronic effects, adverse ecological effects, and the potential for bioaccumulation.

Inerts on List 2 are those thought to be potentially toxic, either because some toxicity data already exists, or because they are structurally similar to known toxic chemicals.

Compounds that the EPA traditionally regarded as innocuous—cookie crumbs, corn cobs, substances "generally regarded as safe (GRAS)"—are on List 4. List 3 is a "wastebin" category, composed of inerts that do not fit one of the other lists. List 3 is larger than the other three lists combined.

The EPA announced that it would take action. For inerts on List 1, companies were encouraged to substitute in their products inerts from Lists 3 and 4 voluntarily. Further, all pesticides that contain an inert from List 1 must contain the following statement on the label: "This product contains the toxic inert ingredient [name of inert]." The EPA also demanded an array of data about health, environmental, and ecological effects. Finally, since a number of compounds on List 1 are actually active ingredients— such as dichlorvos, dinitrophenol, formaldehyde, pyrethrins, and pyrethroids—the EPA plans to reclassify them. Finally, the EPA plans to revoke all exemptions from tolerances for List 1 inerts. This means that if a List 1 inert remains in a product, the manufacturer will have to submit toxicity data for that chemical to gain a *tolerance* (amount of chemical permitted legally) for that chemical in a given food.

For List 2 inerts, the EPA "is monitoring ongoing testing and gathering existing information on the potential adverse effects of these substances and will require additional testing from industry if it is needed." However, the EPA does "not plan to issue any specific requirements in the near future for inert ingredients on List 2."

As for Lists 3 and 4, the EPA is "taking no particular regulatory actions with respect to these inert ingredients at this time."

In other words, the EPA is taking no real concrete actions to deal with inerts on Lists 1 and 2. Indeed, they have suggested pesticide manufacturers should consider petroleum distillates as inert ingredients and, therefore, *not* list them as active ingredients.

Initially, only Lists 1 and 2 were publicly published in 1987; List 4 became available in 1988. Through a Freedom of Information Act (FOIA) request, the Northwest Coalition of Alternatives to

Pesticides (NCAP) obtained a general list of *all* the inert ingredients, plus updated versions of Lists 1, 2, and 4. By a process of elimination, they determined which compounds were on List 3.

It is chilling to look at the compounds on List 1. A number are highly toxic, including dichlorvos, asbestos fibers, benzene, various cadmium compounds, carbon tetrachloride, ethylene thiourea, formaldehyde, five different lead compounds, and pentachlorophenol. One wonders why so many of these compounds were ever permitted to be used.

Pentachlorophenol, an insecticide/fungicide/herbicide that was banned for most uses in the United States a number of years ago, was permitted to continue being used as an inert ingredient. Indeed, one wonders why a number of known pesticide active ingredients were permitted to be called "inert" ingredients.

List 2 also contains a number of very suspect compounds.

List 3 is also quite disturbing. These are purported to be inerts of unknown toxicity, yet the list contains at least six different known pesticide active ingredients—including copper sulfate, malathion, naphthalene, piperonyl butoxide, and two pyrethrins—as well as compounds known to have some toxic effect, including coal tar, saccharin, and sulfuric acid. How does a known pesticide active ingredient get on a list of inerts of unknown toxicity? Furthermore, what about compounds such as Chicago sludge, chlorinated rubber, and chlorinated wax? What is in Chicago sludge? Since it is "inert," the label doesn't have to say.

List 4 is made up of what appear to be mostly innocuous compounds, but we notice, for example, polyvinyl chloride resin (PVC). There are no plans to test the compounds on Lists 3 or 4 routinely.

Finally, it should be noted that NCAP's FOIA request also netted a list of some 296 "deleted" inert ingredients—which the EPA doesn't make public because they are considered "confidential business information." If we do not even know the names of these compounds, how can we be sure that they are safe? Judging from some of the compounds on the other lists, we are reluctant to trust the EPA on this issue.

The whole question of inert ingredients is a pressing one that casts grave doubts on the utility of ratings schemes for pesticides. We can't rate products if their contents are kept secret. Consumers should demand that the identities of *all* ingredients, as well as concentrations, be put on the label so they can make truly a informed choice.

For a complete list of inert ingredients, contact NCAP at P.O. Box 1393, Eugene, OR 97440.

REFERENCES
and RESOURCES

BOOKS

Flint, M. L. *Pests of the Garden and Small Farm: A Grower's Guide to Using Less Pesticide.* Davis, Calif.: University of California Press, 1990.

Franklin, S. *Building a Healthy Lawn.* Pownal, Vt.: Storey Communications, Inc., 1988.

Kourik, R. *Designing and Maintaining Your Edible Landscape Naturally.* Santa Rosa, Calif.: Metamorphic Press, 1986.

Levy, M. P. *A Guide to the Inspection of Existing Homes for Wood-Inhabiting Fungi and Insects.* Washington, D.C.: U.S. Department of Housing and Urban Development, 1975.

MacCaskey, M. *Lawns and Ground Covers: How to Select, Grow and Enjoy.* Tucson, Ariz.: HP Books, 1982.

Mallis, A. *Handbook of Pest Control,* 6th ed. Cleveland: Franzak and Foster, 1982.

Olkowski, William, Sheila Daar, and Helga Olkowski. *Common-Sense Pest Control.* Newtown, Conn.: The Taunton Press, 1991.

Schultz, W. *The Chemical-Free Lawn.* Emmaus, Pa.: Rodale Press, 1989.

Smith, M., and A. Carr. *Rodale's Garden Insect, Disease, and Weed Identification Guide.* Emmaus, Pa.: Rodale Press, 1988.

Tashiro, H. *Turfgrass Insects of the United States and Canada.* Ithaca, N.Y.: Cornell University Press, 1951.

Tuttle, M. D. *America's Neighborhood Bats: Understanding and Learning to Live in Harmony with Them.* Austin: University of Texas Press, 1988.

Yepson, Roger B., Jr., ed. *Encyclopedia of Natural Insect and Disease Control.* Emmaus, Pa.: Rodale Press, 1984.

NONCHEMICAL PEST CONTROL PRODUCTS

LAWN
Information on Turfgrass Varieties

The Lawn Institute
County Line Road
P.O. Box 108
Pleasant Hill, TN 38578-0108

Core Aerator

Alsto's Handy Helpers
P.O. Box 1267
Galesburg, IL 61401
800-447-0048

Gardener's Supply Co.
128 Intervale Road
Burlington, VT 05401
802-863-1700
800-548-4784

Ringer Corp.
9959 Valley View Road
Eden Prairie, MN 55344-3585
800-654-1047

Microbial Amendments

Agro-Chem Corp.
(Soil Aid, Green Magic,
Strengthen & Renew)
1150 Addison Avenue
Franklin Park, IL 60131
708-455-6900

Gardens Alive! (Alive! products)
Highway 48
P.O. Box 149
Sunman, IN 47041
812-623-3800

Interstate Brokers, Inc.
Box 254
Westhampton, NY 11977
516-288-1598

Medina Agricultural Products
P.O. Box 309
Highway 90 West
Hondo, TX 78861
512-426-3011

Necessary Trading Co.
(BioActivator)
P.O. Box 305
New Castle, VA 24127
703-864-5103
800-447-5354 (for orders only)

Ringer Corp. (Restore products)
(address on this page)

Nematodes

BioLogic
(Scanmask)
(*Steinernema carpocapsae*)
P.O. Box 1777
Willow Hill, PA 17201
717-349-2789

Biosis
(*Steinernema* spp.)
1057 E. Meadow Circle
Palo Alto, CA 94303
714-783-2148
714-685-7681

Gardens Alive!
(*S. carpocapsae* and *Heterorhabditis heliothidis*)
(address on page 345)

Gardener's Supply Co. (Biosafe)
(address on page 345)

Hydro-Gardens, Inc.
(Lawn Patrol [*H. heliothidis*],
Guardian [*S. carpocapsae*])
P.O. Box 9707
Colorado Springs, CO 80932
800-634-6362

M&R Durango, Inc.
(*S. carpocapsae*)
P.O. Box 886
Bayfield, CO 81122
800-526-4075

Nematec
P.O. Box 93
Lafayette, CA 94549
415-735-8800

Safer, Inc.
(BioSafe, [*S. carpocapsae*])
9959 Valley View Road
Eden Prairie, MN 55344-3585
800-423-7544

Spiked Shoes

Alsto's Handy Helpers
(lawn sandals, 1½-inch spikes)
(address on page 345)

Gardener's Supply Co.
(lawn aerator sandals 1½-inch spikes)
(address on page 345)

Mellinger Nursery
(lawn aerator sandals 1½-inch spikes)
2310 W. South Range Road
North Lima, OH 44452-9731
216-549-9861

Bug Vacuum Cleaner

BioQuip Products
17803 La Salle Avenue
Gardena, CA 90248
213-324-0620

Ringer Corp.
(address on page 345)

Flamers

Ben Meadows Co.
P.O. Box 80549
Atlanta, GA 30366
800-241-6401

Flame Engineering, Inc.
P.O. Box 577
LaCrosse, KS 67548
800-255-2469

Forestry Suppliers, Inc.
205 W. Rankin Street
P.O. Box 8397
Jackson, MS 39284-8397
800-752-8460

Plow & Hearth
Weed Destroyer
301 Madison Road
P.O. Box 830
Orange, VA 22960-0492
800-866-6072

GARDENS
Weed Mat

Gardener's Supply Co.
(address on page 345)

Gardens Alive!
(address on page 345)

Necessary Trading Co.
(address on page 345)

Diatomaceous Earth

Beneficial BioSystems
P.O. Box 8461
Emeryville, CA 94662
415-655-3928

W. Atlee Burpee & Co.
300 Park Avenue
Warminster, PA 18991-0001
215-674-9633
800-888-1447

Eco-Safe Laboratories
P.O. Box 1177
St. Augustine, FL 32085
800-274-7387

Gardener's Supply Co.
(address on page 345)

Gardens Alive!
(address on page 345)

Necessary Trading Co.
(address on page 345)

Pristine Products Inc.
2311 E. Indian School Road
Phoenix, AZ 85016
602-955-7031
800-266-4968

Universal Diatoms, Inc.
410 12th Street, NW
Albuquerque, NM 87102
505-247-3271

Floating Row Covers

Gardener's Supply Co.
(address on page 345)

Gardens Alive!
(address on page 345)

Necessary Trading Co.
(address on page 345)

Sticky Traps

W. Atlee Burpee & Co.
(whitefly)
(address on this page)

Gardens Alive!
(whitefly)
(address on page 345)

Great Lakes IPM
(whitefly, thrips)
10220 Church Road, NE
Vestaburg, MI 48891
209-268-3417

Growing Naturally
(whitefly, thrips)
149 Pine Lane
P.O. Box 54
Pinesville, PA 18946
215-598-7025

GTF Laboratories
(whitefly)
P.O. Box 2646
Fresno, CA 93745-2646
517-268-5693

Hydro-Gardens, Inc.
(whitefly thrips)
(address on page 346)

IPM Laboratories, Inc.
(whitefly, thrips)
Main Street
Locke, NY 13092-0099
315-497-3129

Necessary Trading Co.
(whitefly, thrips)
(address on page 345)

Viruses

Sandoz Crop Protection
(Elcar, for *Heliothis* spp. caterpillars)
1300 E. Touhy Avenue
Des Plaines, IL 60018
708-699-1616

Natural Enemies

Arizona Biological Control
(aphid midge [*Aphidoletes
    aphidimyza*]; lacewings;
    Trichogramma pretiosum; whitefly
    parasitoid [*Encarsia formosa*])
P.O. Box 4247 CRB
Tucson, AZ 85738
602-825-9785

Beneficial Insectary
(lacewings; predatory mites;
    Trichogramma pretiosum; whitefly
    parasitoid [*Encarsia formosa*])
14751 Oak Run Road
Oak Run, CA 96069
916-472-3715

Beneficial Insects, Inc.
(lacewings; *Trichogramma
    pretiosum*; whitefly parasitoid
    [*Encarsia formosa*])

P.O. Box 40634
Memphis, TN 38174-0634
901-276-6879

Foothill Agricultural Research, Inc.
(predatory mites; *Trichogramma
    pretiosum*; whitefly parasitoid
    [*Encarsia formosa*])
510½ West Chase Drive
Corona, CA 91720
714-371-0120

Gardener's Supply Co.
(lacewings)
(address on page 345)

Gardens Alive!
(lacewings; spined soldier bugs)
(address on page 345)

Great Lakes IPM
(lacewings)
10220 Church Road, NE
Vestaburg, MI 48891
517-268-5693

Hydro-Gardens, Inc.
(aphid midge [*Aphidoletes
    aphidimyza*]; whitefly parasitoid
    [*Encarsia formosa*]; lacewings;
    predatory mites)
(address on page 346)

IPM Laboratories, Inc.
(aphid midge [*Aphidoletes
    aphidimyza*]; lacewings; predatory
    mites; whitefly parasitoid [*Encarsia
    formosa*])
(address on this page)

Rincon-Vitova Insectaries, Inc.
(lacewings; *Trichogramma
    pretiosum*; whitefly parasitoid
    [*Encarsia formosa*])
P.O. Box 95

Oakview, CA 93022
805-643-5407

Artificial Diet for Natural Enemies

Beneficial Insectary
(address on page 348)

Gardens Alive!
(Bug Pro)
(address on page 345)

Ringer Corp.
(Pred Feed)
(address on page 345)

Pheromones

Gardens Alive!
(for stored product pests)
(address on page 345)

Insects Limited
(for stored product pests)
10540 Jessup Boulevard
Indianapolis, IN 46280-1438
800-992-1991

Santa Cruz Horticultural Supply
(for stored product pests)
P.O. Box 1534
Morro Bay, CA 93442
805-772-8262

Trece, Inc.
(for stored product pests)
Salinas, CA 93907-1817
408-758-0205

Antitranspirant

Aquatrols Corp. of America, Inc.
1432 Union Avenue
Pennsauken, NJ 08110
609-665-1130

Gardener's Supply Co.
(Wilt Pruf)
(address on page 345)

Green Pro Services
380 S. Franklin Street
Hempstead, NY 11550
516-538-6444

Nature's Touch
11150 W. Addison Street
Franklin Park, IL 60131
708-455-8600

PBI-Gordon Corp.
P.O. Box 4090
Kansas City, MO 64101
800-821-7925

Precision Laboratories, Inc.
P.O. Box 12
Northbrook, IL 60065
708-498-0800

Safer, Inc.
(address on page 346)

Slug and Snail traps

Cedar Pete, Inc.
(Snailproof)
P.O. Box 969
Mt. Shasta, CA 96067

Gardener's Supply Co.
(The Garden Sentry Trap,
  Snailer)
(address on page 345)

W. Atlee Burpee & Co.
(Slug Bar)
(address on page 347)

TERMITES
Inspection Dogs

Beacon Dogs, Inc.
Andrew K. Solarz
1409 Bayhead Road
Annapolis, MD 21401
301-757-44999

Industrial Discovery Systems
P.O. Box 130
Kenner, LA 70063
504-466-9964

TADD Services
1617 Old Country Road
Route 4
Belmont, CA 94002
415-595-5171
800-345-TADD

Flea Traps

Gardener's Supply Co.
(Happy Jack Flea Trap)
(address on page 345)

Necessary Trading Co.
(Happy Jack Flea Trap)
(address on page 345)

Plow & Hearth
(Ultimate Flea Trap)
(address on page 346)

Gopher Traps

Gardens Alive!
(address on page 345)

Guardian Trap Co.
P.O. Box 1935
San Leandro, CA 94577
415-357-0900

Z.A. Macabee Gopher Trap Co.
110 Loma Alta Avenue
Los Gatos, CA 95030
408-354-4158

Woodstream Corp.
Lititz, PA 17543
717-626-2125

Mole Traps

Nash Products
(Nash Mole Trap)
5716 East S Avenue
Vicksburg, MI 49097-9990
616-323-2980

Woodstream Corp.
(address on this page)

Alsto's Handy Helpers
(The Mole Eliminator)
(address on page 345)

STATE COOPERATIVE EXTENSION SERVICES

Local Cooperative Extension agents are located in nearly every county. Look up Cooperative Extension Service under County Government in the white pages of the local telephone directory. The state headquarters for the Cooperative Extension Service are located at the state land-grant universities listed below.

College of Agriculture
Auburn University
Auburn, AL 36830

School of Agriculture and
Land Resources Management
University of Alaska
Fairbanks, AK 99701

College of Agriculture
University of Arizona
Tucson, AZ 85721

College of Agriculture
University of Arkansas
Fayatteville, AR 72701

College of Agricultural Sciences
University of California
Berkeley, CA 94720

College of Agricultural Sciences
Colorado State University
Fort Collins, CO 80523

College of Agriculture
and Natural Resources
University of Connecticut
Storrs, CT 06368

College of Agricultural Sciences
University of Delaware
Newark, DE 18711

Institute of Food and
Agricultural Sciences
University of Florida
Gainesville, FL 32611

College of Agriculture
University of Georgia
Athens, GA 30602

College of Agriculture and
Life Sciences
University of Guam
Agana, Guam 96910

College of Tropical Agriculture
University of Hawaii
Honolulu, HI 96822

College of Agriculture
University of Idaho
Moscow, ID 83843

College of Agriculture
University of Illinois
Urbana, IL 61801

School of Agriculture
Purdue University
West Lafayette, IN 47907

College of Agriculture
Iowa State University
Ames, IA 50011

College of Agriculture
Kansas State University
Manhattan, KS 66505

College of Agriculture
University of Kentucky
Lexington, KY 40506

College of Agriculture
Louisiana State University
Baton Rouge, LA 70893

College of Life Sciences
and Agriculture
University of Maine
Orono, ME 04473

College of Agriculture
University of Maryland
College Park, MD 20742

College of Food and
Natural Resources
University of Massachusetts
Amherst, MA 01002

College of Agriculture and
Natural Resources
Michigan State University
East Lansing, MI 48824

College of Agriculture
University of Minnesota
St. Paul, MN 55108

College of Agriculture
Mississippi State University
Mississippi State, MS 39762

College of Agriculture
University of Missouri
Columbia, MO 65201

College of Agriculture
Montana State University
Bozeman, MT 59717

Institute of Agriculture
and Natural Resources
University of Nebraska
Lincoln, NB 68503

College of Agriculture
University of Nevada
Reno, NV 89507

College of Life Sciences
and Agriculture
University of New Hampshire
Durham, NH 03824

Cook College
Rutgers State University
New Brunswick, NJ 08903

College of Agriculture and
Home Economics
New Mexico State University
Las Cruces, NM 88003

College of Agriculture and
Life Sciences
Cornell University
Ithaca, NY 14853

School of Agriculture
and Life Sciences
North Carolina State University
Raleigh, NC 27607

College of Agriculture
North Dakota State University
Fargo, ND 58102

College of Agriculture
Ohio State University
Columbus, OH 43210

College of Agriculture
Oklahoma State University
Stillwater, OK 74074

School of Agriculture
Oregon State University
Corvallis, OR 97331

College of Agriculture
Pennsylvania State University
University Park, PA 16802

College of Agricultural Sciences
University of Puerto Rico
Mayaguez, Puerto Rico 00708

College of Resource Development
University of Rhode Island
Kingston, RI 02881

College of Agricultural Sciences
Clemson University
Clemson, SC 29631

College of Agriculture and
Biological Sciences

South Dakota State University
Brookings, SD 57007

College of Agriculture
University of Tennessee
P.O. Box 1071
Knoxville, TN 37901

College of Agriculture
Texas A&M University
College Station, TX 77843

College of Agriculture
Utah State University
Logan, UT 84322

College of Agriculture
University of Vermont
Burlington, VT 05401

College of the Virgin Islands
P.O. Box L
Kingshill, St. Croix
Virgin Islands 00850

College of Agriculture and
Life Sciences
Virginia Polytechnic Institute
Blacksburg, VA 24061

College of Agriculture and
Home Economics
Washington State University
Pullman, WA 99164

College of Agriculture
and Forestry
West Virginia University
Morgantown, WV 26506

College of Agriculture
and Life Sciences
University of Wisconsin
Madison, WI 53706

College of Agriculture
University of Wyoming
Laramie, WY 82071

INDEX